Perfect Symmetry

Perfect Symmetry

The Accidental Discovery of Buckminsterfullerene

JIM BAGGOTT

OXFORD UNIVERSITY PRESS
OXFORD NEW YORK TOKYO
1994

Valton Street, Oxford OX2 6DP
New York
d Bangkok Bombay
n Dar es Salaam Delhi
ng Istanbul Karachi
ras Madrid Melbourne
Mexico City Nairobi Paris Singapore
Taipei Tokyo Toronto
and associated companies in
Berlin Ibadan

Oxford is a trade mark of Oxford University Press

Published in the United States
by Oxford University Press Inc., New York

A catalogue record for this book is available from the British Library

Library of Congress Cataloging in Publication Data
Baggott, J. E.
Perfect symmetry: the accidental discovery of buckminsterfullerene/
Jim Baggott.
Includes bibliographical references and indexes.
1. Buckminsterfullerene. I. Title.
QD181.C1B28 1994 546'.68142—dc20 94–14068
ISBN 0 19 855790 6

Typeset by Cotswold Typesetting Ltd, Gloucester

Printed in Great Britain on acid-free paper by
The Bath Press, Avon

To Emma

'Are there, in nature, behaviours of whole systems unpredicted by the parts? This is exactly what the chemist has discovered to be true.'

Richard Buckminster Fuller

'If God would give me the grace to make one molecule, what would that molecule be?'

Orville Chapman

Preface

It is said that chemistry is a mature science.

Now by its very nature, a mature science is going to be rather conservative, perhaps somewhat staid, stuffy, and predictable. It will not hold much by way of genuine surprises. To those not confronted every day by the mysteries of chemistry at the frontiers, it may even appear a little dull.

Setting aside the contributions of the alchemists, chemistry as an investigative science has an illustrious history stretching back some 300 years or so. Even for those scientists working away at the edges, the sheer volume of accumulated knowledge, wisdom, and understanding can weigh heavily. The scientists expect to find some intriguing new facts about the chemical world; they expect to confirm what they think they already know or to have their prejudices reinforced. But looking back over a 300-year legacy does little to raise their expectations of discovering fundamental discrepancies between the world as they understand it and the world as it is. They do not greatly expect to discover something that does not fit.

When that something extraordinary eventually happens, it shocks the scientists from their complacency and they begin to realize that they do not fully understand everything. In some very fundamental respects, they do not understand the world as it is. Instead of answers that fit the pattern or confirm the theories, they are faced only with questions. Science comes alive once again because, of course, answers are of only fleeting interest. It is the questions that fascinate.

In 1966 the notion of hollow graphite 'balloons' was no more than an amusing idea—a distraction for the entertainment of readers of the weekly popular science magazine *New Scientist.* Now this is not the kind of periodical that the experts hold much store by. If asked, they would have probably dismissed these structures as interesting, even theoretically possible, but obviously wholly impractical as candidates for synthesis in the laboratory. 'We should know,' they might have said, 'We are the experts.'

Then, nearly twenty years later, a team of American and British chemists made a rather startling, if accidental, discovery. Out of the chaos of a plasma of carbon atoms and ions produced in an intense pulse of laser light there emerged spontaneously what they believed was a perfectly spherical molecule made up of exactly sixty carbon atoms. A molecule shaped like a soccer ball. There was more. Buckminsterfullerene (as it was to be called) was but one of a whole series of new, hollow-cage molecules. It seemed only slightly less likely a possibility than a bunch of monkeys toying randomly with words of the English language, and coming up with a Shakespeare play.

Some scientists were sceptical, but when in 1990 a team of physicists—physicists!—announced that they had accidentally discovered a way to make buckminsterfullerene and a few of its relatives in amounts that you could see and touch, all scepticism vanished. Literally overnight countless encyclopaedias and all the textbooks on the chemistry, physics, and materials science of carbon were rendered instantly out of date. There were not two, but *three* basic forms of carbon: diamond, graphite, and now fullerite.

To put all this in context, try to imagine what it might be like if diamond itself was to be discovered tomorrow. In our hypothetical world of today there are no diamond cutting tools and no diamond engagement rings. Some of the less pragmatic scientists dream of diamond's elegant structure—imagine! a *tetrahedral* arrangement of carbon atoms!—and even calculate the kinds of properties it might have if it existed. Experts in carbon chemistry and materials science with their feet planted more firmly on the ground dismiss these notions as imaginative but irrelevant to the real world. By morning, their world has been turned upside-down.

Of course, this is fantasy. Diamond has been known to mankind for thousands of years and it is very difficult to imagine a world in which twentieth century science had somehow managed to overlook its existence. And yet, in the case of buckminsterfullerene, this is exactly what modern science has done for most of the twentieth century.

There were yet further surprises in store. By 1994 a thriving 'three-dimensional' fullerene chemistry (and biochemistry) had developed. A new range of high-temperature organic superconductors had been found, many with properties that set new records. New large-scale carbon architectures had been discovered—including the hollow graphite balloons of 1966—at once both marvellous and elegant. A great many questions remain unanswered and nobody knows if there are more surprises still to come.

The story of the discovery of buckminsterfullerene and its aftermath serves as an important case study, with object lessons for the organization and funding of basic science, for the sponsors of commercial R&D, for the scientists' own perceptions of themselves and their science, and for the perceptions of ordinary people about how scientific progress is made. It is a good story, and in *Perfect symmetry* I have tried my best to do it justice.

I could not have even begun to tell this story without the support and encouragement of many busy scientists who gave of their time to talk to me or write to me about their experiences. I therefore acknowledge a debt of gratitude to: Mike Alford, Don Bethune, Bob Curl, François Diederich, Kosta Fostiropoulos, Patrick Fowler, Robert Haddon, Jonathan Hare, Jim Heath, Don Huffman, Sumio Iijima, David Jones, Wolfgang Krätschmer, Harry Kroto, Lowell Lamb, David Manolopoulos, Sean O'Brien, Eric Rohlfing, Rick Smalley, Daniel Ugarte, David Walton, Robert Whetten, and Fred Wudl.

I have dedicated this book to my daughter, who arrived on the scene just as I was about to get stuck into the first draft. Without her, my life would certainly have been emptier, but the book would have been finished a darn sight more quickly.

Reading J. E. B.
May 1994

Contents

Prologue

There is a curious discontinuity between the densities of gases around 0.001 relative to water, and liquids and solids from 0.5 to 25 or so; this week Daedalus has been contemplating ways of bridging this gap and has conceived the hollow molecule, a closed spherical shell of a sheet-polymer like graphite, whose molecules are flat sheets of benzene-hexagons. He proposes to modify the high-temperature graphite process by introducing suitable impurities into the sheets to warp them (rather like 'doping' semiconductor crystals to cause discontinuities), reasoning that the curvature thus produced will be transmitted throughout the sheet to its growing edges so that it will ultimately close on itself. The radii of the molecules thus produced would be controlled by the level of impurities included in them, and Daedalus calculates that a hollow molecule 0.1 micrometres across would give rise to a bulk density of about 0.01. These low-density substances would constitute a vague fifth state of matter, for such big molecules (molecular weight about 100 million) could hardly evaporate, but would interact so weakly at their few points of contact as not to be solid or even liquid! They would behave as tenuous fluids, retainable in open vessels though without any definite surface, and if heated would expand steadily without boiling into a gas-like state. Such fascinating new materials would have a host of uses, in novel shock-absorbers, thermometers, barometers, and so on, and possibly in gas-bearings where the rolling contact of the spherical molecules would lower friction even further. Daedalus was worried that they might deform under pressure until he realized that if synthesized in a normal atmosphere they would be full of gas and resilient like little footballs: he is now seeking ways of incorporating 'windows' into their structure so that they can absorb or exchange internal molecules, so as to produce a range of super molecular-sieves capable of entrapping hundreds of times their own weight of such small molecules as can enter the windows.

David Jones, in 'Ariadne', *New Scientist*, November 3, 1966

PART I

From space to symmetry

nature

INTERNATIONAL WEEKLY JOURNAL OF SCIENCE

Volume 318 No 6042 14-20 November 1985 £1.90

SIXTY-CARBON CLUSTER

AUTUMN BOOKS

Harvey Brooks
(transformation of MIT)

P. N. Johnson-Laird
(brain and mind)

Anthony W. Clare
(psychoanalysis as religion)

A. O. Lucas
(war on disease)

Hendrik B. G. Casimir
(physics and physicists)

Gordon Thompson
(dimensions of nuclear proliferation)

Jacques Ninio
(origins of life)

Edward Harrison
(steps through the cosmos)

1

The last great problem in astronomy

By the late 1960s, the sciences of astronomy, physics, chemistry, and biology had combined in the attempt to tell the truly remarkable story of the origin and evolution of ourselves and our universe, in all its mind-boggling complexity. Sure, there were still many uncertainties, many gaps, and much speculation. Perhaps it wasn't quite the complete answer to the eternal question of life, the universe, and everything, but it was still potentially the greatest story ever told.

The universe was the domain of the astronomer and the physicist. Their part of the story concerned the physics of the 'big bang' and the expanding universe, the mechanics of stellar formation and evolution, the formation and clustering of galaxies, the properties of interstellar clouds of gas and dust, and the dynamics of planetary formation.

The astrophysicists couldn't be exactly sure how the large-scale structure of the universe had come about, but they had made a lot of progress with theories of the birth and evolution of stars. Their story went like this. The origin of the universe in a big bang some 15 billion years before had given us hydrogen and helium. Over time, these gases had accumulated into clouds, and the self-gravity of the clouds had caused them to condense to form stars. Elements heavier than hydrogen and helium were synthesized in the interiors of the stars through thermonuclear fusion reactions.

The stars lived and died. A star with a mass close to that of our own sun would slowly burn its way through its core reserves of hydrogen, making more helium in the process. With its hydrogen exhausted, the star would expand and cool to become a red giant, before triggering helium fusion reactions to make beryllium and carbon. Further fusion reactions would produce heavier elements up to iron. Under the conditions likely to be found in the interior of a star, nuclear reactions involving iron lead to fission, not fusion—the physics said that iron was as far as you could go. Finally, with an iron core and no fuel remaining, a star with a mass greater than about eight times that of the sun would suffer gravitational collapse and spectacular death in a supernova explosion. The star would be torn apart in a cataclysmic release of energy, making still heavier elements and scattering debris through space, sowing the seeds of the next generation of stars.

Now the clouds from which new stars could form contained heavy elements. A second generation of stars would condense from interstellar clouds of gas and dust complete with their own planetary systems. Planets like our own earth were formed from the accretion of heavy silicon-containing compounds, iron and magnesium oxides, and other simple inorganic chemicals. The pattern of planet size and composition in our solar system resulted from the dynamics of the rotating gas cloud that gave birth to the sun.

Once the scientists had a mechanism connecting the big bang with the formation of a planet like the earth, the rest was supposed to be reasonably straightforward. The theories of the origin of life prevalent in the late 1960s could tell us how it might be possible to get from a lifeless lump of rock to a fertile world teeming with life forms. Here the physicists handed the job of story-teller over to the chemists and biologists, and concentrated instead on the many exotica that modern astronomy and theoretical physics were revealing to them: pulsars, quasars, and black holes.

The chemists and biologists addressed themselves to the question of the origin of life on earth. What they believed was needed were plenty of complex organic molecules floating around in the warm seas of the young earth, or what J. B. S. Haldane had referred to as the earth's 'primeval soup'. In 1923, the Russian scientist A. I. Oparin suggested that the accumulation of biologically important molecules in microscopic colloidal droplets had led, two thousand million years before, to the spontaneous generation of the first primitive single-celled organisms. But where had these biologically important molecules come from? In the 1950s, chemists Harold Urey and Stanley Miller had an answer. Begin with a primordial atmosphere of methane, ammonia, hydrogen, and a little water vapour, add ultraviolet light (from the young sun) and an electrical discharge (lightning), and lo—you got amino acids.

Take some single-celled organisms and a few thousand million years of evolution through natural selection and, it could be argued, we would find ourselves: intelligent, carbon-based life forms able to theorize on our universe and our origins. The fact that every carbon atom in our bodies had been forged in the interior of a star—that we are made of stardust—added a poignant twist to the scientists' tale.

If the astrophysicists weren't going to have all the fun, then they tried hard to make sure they had most of it. According to them, chemistry and biology were things that supposedly couldn't happen until planets had formed and settled down to relatively mundane existence. Space was believed to be simply too hostile a place for fragile organic molecules to survive for very long. There could be no complex chemistry in space. Chemists should rather stick to their test tubes or whatever, and let the physicists get on with the serious business of stargazing.

At least, this was the view until Charles Townes and his colleagues at the University of California in Berkeley decided to look more intently at interstellar space using a radio telescope. Townes was a microwave spectroscopist.* He had shared the Nobel Prize in 1964 for his invention of the maser (a forerunner to the laser), which is based on the stimulated emission of microwave radiation from ammonia molecules. When he and his colleagues from Berkeley's Department of Physics and the Radio Astronomy Laboratory looked at the heavens through the eyes of the 6.3-metre diameter radio telescope at Hat Creek, California, they were looking for microwave signals.

The wavelengths of radio waves are measured in metres (short wave) or kilometres (long wave). Townes and his colleagues wanted to push radio astronomy to shorter wavelengths of the order of a centimetre or so (microwaves). In order to detect faint signals in this part of the spectrum, the telescope must have a dish which is both large (to collect as much of the signal as possible) and smooth to within a wavelength of the radiation being detected (in this case a centimetre). Fortunately, by the late 1960s the technology of microwave amplification and detection had benefited enormously from the development of communications satellites.

In the autumn of 1968, Townes and his colleagues used the newly installed Hat Creek radio telescope to look for microwave radiation from space. They found signals with a wavelength of 1.25 centimetres, characteristic of ammonia molecules, coming from a dense cloud lying in the direction of the centre of the galaxy. Ammonia thus became the first polyatomic molecule (a molecule with more than two atoms) to be identified in the interstellar medium, and was but the beginning. Within the next three years, another twenty molecules had been identified. These included water, formaldehyde, hydrogen cyanide, and acetylene. Other microwave signals were found which could not be assigned to any molecule that had been studied in the laboratory.

The list grew longer as ever more complex molecules were found. Though composed of about 70 per cent hydrogen and 28 per cent helium, the remaining two per cent of matter in the interstellar medium was beginning to look very interesting indeed. The new subject of astrochemistry was born, and one of the best hunting grounds for astromolecules was a dark interstellar cloud in Sagittarius that had been named Sagittarius B2.

It was here that Barry Turner, an astronomer at the National Radio Astronomy Observatory at Green Bank, West Virginia, detected microwave signals due to cyanoacetylene, H—C≡C—C≡N (or HC_3N). By the middle of the decade, a fundamental change had taken place. Far from being devoid of chemical interest, the interstellar medium was turning up a large number of molecules, many of them more complex than anyone had previously thought possible. The fact that the dense clouds were also the birthplaces of stars meant

*A brief introduction to molecular spectroscopy and quantum mechanics can be found in the Appendix.

that these molecules were present in the swirling mixtures of gas and dust from which the stars condensed. It seemed that we may not after all have to call on the chemistry of the earth's primitive atmosphere to explain the presence of complex organic molecules in the primeval soup. The maverick astronomer Fred Hoyle and his colleague Chandra Wickramasinghe argued that life on earth may have been seeded from space. In their book *Lifecloud*, first published in 1978, they pointed out that two molecules known to be present in the interstellar medium—formaldehyde and methylenimine (H_2CNH)—could react to give glycine, the simplest of the amino acids.

But the task of the radio astronomer was now getting increasingly tricky. More complicated molecules could be looked for and positively identified only if the astronomers knew what their microwave spectra looked like and could calculate the frequencies of microwave signals likely to be coming from space. This presented no problem for stable molecules like water, ammonia, and cyanoacetylene, since their microwave spectra were already well documented or easily measured. But the interstellar medium held the promise of more exotic molecules which did not occur naturally on earth and had never been made in terrestrial laboratories.

To continue with the search meant that more and more of these complex molecules had to be synthesized in the laboratory and their microwave spectra measured. Chemists, who had been busily occupying themselves with just this task, began to wake up to the fact that astronomers were picking up their results and searching for these molecules in space. Before too long, chemists were beginning to join in the fun that had thus far been the preserve of the astrophysicists: they began making molecules to order and helping to search for them in space. Their efforts became focused on what was then a relatively mature and thriving area of scientific activity: microwave spectroscopy.

In 1975, a young chemistry lecturer at the University of Sussex began to take more than a passing interest in interstellar molecules. Harry Kroto was also a microwave spectroscopist, and he followed closely the reports of Townes and others who were finding large, polyatomic molecules in interstellar space using microwave techniques.

Kroto also had a long-standing fascination with chain molecules containing multiply bonded carbon atoms. This was an interest that had begun with some (largely unsuccessful) studies on carbon suboxide, O═C═C═C═O, that he had carried out in the early 1960s as part of his Ph.D. research under the supervision of Richard Dixon at Sheffield University, and which he had nurtured during his years as a postdoctoral scientist.

David Walton, one of Kroto's Sussex colleagues and a specialist in chemical synthesis, had devised some elegant methods for making very long chain molecules. Walton had successfully produced dilute solutions of minute quantities of molecules with 24 and 32 carbon atoms strung in a chain like so many beads on a thread. These molecules conjured up an image in Kroto's mind of a rather flexible baton hurled in the air by a cheer leader. Normally, such a baton is a rigid rod—its rotations are relatively easy to predict and, with a little practice, the baton can be readily caught as it descends. However, Walton's chain molecules wouldn't behave like that. Although the multiple bonds between the carbon atoms in the chain would provide a strong 'backbone', Kroto expected the molecules to bend as they rotated. A microscopic cheer leader would need more than a little practice to catch a baton that was bending along its length as it rotated.

In quantum mechanical terms, the bending of a long-chain molecule along its length is manifested as a low-frequency (low-energy) vibrational state. Such a vibration is of sufficiently low energy that it will become excited at room temperature through collisions with other molecules or with surfaces. Whereas the microwave spectrum of a small, linear polyatomic molecule shows a series of 'lines' corresponding to transitions between rotational states, the spectrum of a long-chain molecule shows a series of extra lines known as vibrational 'satellites'. These extra lines correspond to rotational transitions in molecules that are already bending. Because the bending of the molecule changes its shape (on average), a vibrationally excited molecule has a slightly different moment of inertia, a measure of its resistance to rotational acceleration. The same rotational transition in the vibrationally excited molecule therefore requires a slightly different amount of energy compared to the unexcited molecule. The more rungs on the ladder of bending states that are excited, the more vibrational satellite lines that appear in the microwave spectrum.

Kroto was particularly keen to measure this kind of spectrum experimentally in order to unravel the intriguing vibrational and rotational motions of long-chain molecules. The extent and nature of the mixing of the different types of motions could in principle be predicted using quantum mechanics, and Kroto saw that these molecules would provide an interesting experimental testing ground for the theory's predictions.

Kroto and Walton agreed to collaborate. Walton devised a synthesis for cyanobutadiyne, H—C≡C—C≡C—C≡N (or HC_5N for short), which Kroto could then study by microwave spectroscopy. They decided that this research project would provide ideal material for a student taking the new Chemistry by Thesis degree that had recently been introduced in the School of Chemical and Molecular Sciences at the University of Sussex. The basic idea of this new degree was to build an element of real research activity into an undergraduate chemistry curriculum, and allow students the opportunity to cross the boundaries between traditional scientific disciplines.

Details of the project were posted, and an undergraduate student by the name of Anthony Alexander applied. Kroto was initially apprehensive. Alexander did not have a strong background in mathematics, yet the project demanded a good understanding of some mathematically challenging aspects of molecular quantum theory. But, as is sometimes found in experimental science, an ability to spew up a year's accumulated knowledge in written examinations can be a very poor guide

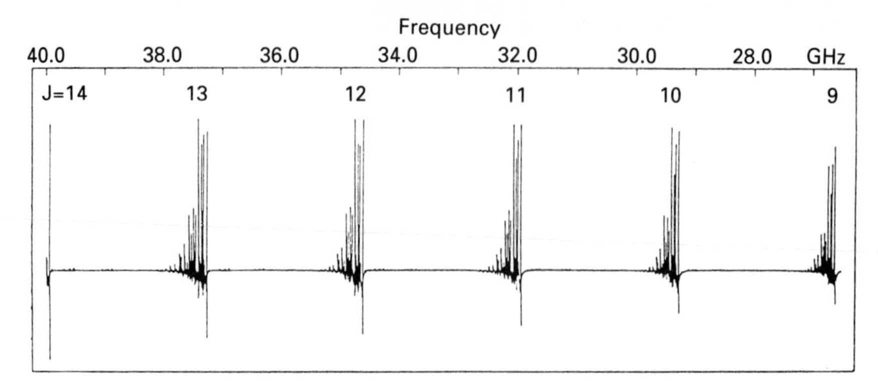

Part of the microwave spectrum of HC_5N. Each group of lines represents a transition between adjacent rotational states, characterized by the quantum number J. The sequences of lines within each group are the vibrational satellites—transitions between rotational states of molecules that are also bending. Adapted, with permission, from Alexander, A. J., Kroto, H. W., and Walton, D. R. M. (1976). *Journal of Molecular Spectroscopy*, **62**, 175.

to a student's academic ability or suitability for research. Kroto's fears turned out to be groundless: Alexander proved himself to be an outstanding research student. The project was a complete success.

Under Kroto and Walton's supervision, Alexander synthesized a sample of HC_5N and measured its microwave spectrum between 26.5 and 40.0 gigahertz using Kroto's newly acquired Hewlett Packard 8460A microwave spectrometer. The spectrum showed the expected satellite lines. It was, in all, a nice piece of work.

Kroto then started to make connections between this work on HC_5N and the observation five years earlier by Barry Turner that the molecule cyanoacetylene, HC_3N, is present in the interstellar medium. Could it be that HC_5N is also present? Kroto now had the microwave spectrum of HC_5N, so all that was needed was to choose a strong feature in the spectrum and look for the corresponding microwave signals from space using a radio telescope. As a postdoctoral scientist, Kroto had worked for two years at the National Research Council (NRC) in Ottawa, Canada, before moving on to Bell Telephone Laboratories in 1966. At the NRC, he had met a bewildering number of top-class spectroscopists and theoreticians and established some life-long friendships. Some of these spectroscopists also collaborated with radio astronomers at the NRC's Herzberg Institute for Astrophysics. Kroto decided to write to one of his former NRC colleagues, Takeshi Oka, to ask if he was interested in looking for HC_5N in interstellar space. Oka wrote back to say that he was 'very, very, very, very, very much interested'.

Armed with the information derived from Kroto's analysis of the microwave spectrum of HC_5N, Oka and his colleagues Lorne Avery, Norman Broten and

John McLeod searched for its tell-tale microwave signals in Sagittarius B2 using the 43-metre diameter radio telescope at Algonquin Park in Ontario. It was quite a long shot. Space was still supposed to be too hostile an environment for complex chemistry. If they were present at all, polyatomic molecules with three or four heavy atoms like carbon, oxygen or nitrogen were presumed to be much too rare to be detected. Nevertheless, and much to Kroto's delight, the Canadian astronomers were successful in their search. In November 1975, HC_5N became the largest molecule then known to be present in the interstellar medium.

What hopes for cyanohexatriyne, H—C≡C—C≡C—C≡C—C≡N (HC_7N)? Walton devised a synthesis and a new graduate student, Colin Kirby, was assigned the difficult task of making it. Kroto's ambitions had changed. He had originally been interested in these long-chain molecules (collectively called cyanopolyynes) because they posed some fundamental questions for the quantum theory of molecular vibrations and rotations. That interest was not forgotten, but it had become secondary to the more immediate desire to discover ever larger and more complex molecules in space. To do this, the radio astronomers needed to know what characteristic microwave frequencies they should be searching for.

This time Kroto was determined to get in on the fun. With the aid of a grant from the Scientific Affairs Division of NATO, Kroto joined Oka and Lorne Avery at the observatory at Algonquin Park in March 1977 to look for microwave emission from HC_7N. There was only one problem. Kirby and Walton were having problems with the synthesis. Like other long-chain polyynes, HC_7N tended to polymerize. They succeeded in making small samples only to watch them turn to a sticky goo as they tried to transfer them from solution into the gaseous state required for measuring the microwave spectrum. The time slot allocated at the observatory was fixed months in advance and could not be changed. Kroto had been forced to leave Sussex without the crucial spectroscopic information he needed for the search.

The session on the telescope at Algonquin Park had already started when Kirby, back in Sussex, finally succeeded in measuring the microwave spectrum of a sample of HC_7N. A quick calculation gave him the parameters that Kroto, Oka, and Avery were waiting for impatiently. Kirby telephoned Kroto's wife, Margaret, who passed on the numbers to Fokke Creutzberg, a family friend in Ottawa. Creutzberg telephoned the numbers through to Kroto, who was by this time with Oka and Avery at the telescope. A quick back-calculation gave them the microwave frequencies they needed to look for. They now knew the frequency and the receiver was made ready. The constellation Taurus rose in the early evening sky. Within Taurus lay their target—the dark Heiles's Cloud 2.

They tracked the dark cloud in Taurus until it set around 1 a.m. Although the telescope and receiver electronics were controlled by computer, it was not possible for the scientists to see if the signal was accumulating as expected. The system did display integrations of the total signal on an oscilloscope at ten-minute intervals, but they could not be sure from these all-too-brief glimpses that

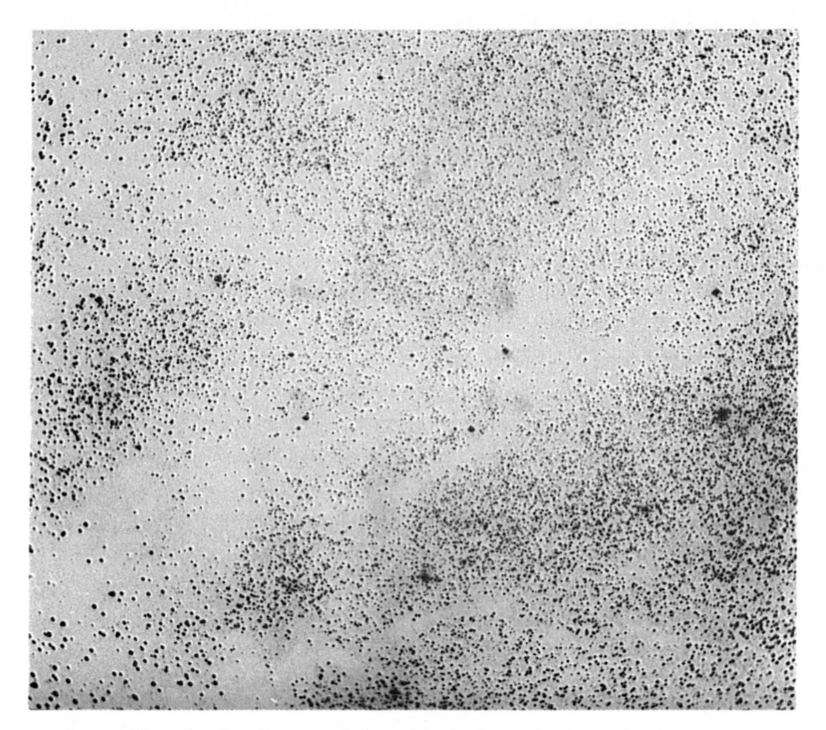

Interstellar clouds of gas and dust block the light from distant stars, as shown in this photograph of Taurus. From Frost, E. B. and Calvert, R. (eds.), *Atlas of Selected Regions of the Milky Way*, Carnegie Institute, Washington D.C., 1927. Heiles's Cloud 2 is located in the bottom left-hand corner.

the signal they were seeking was above the background noise. Finally, as Taurus began to set, Avery stopped the run and instructed the computer to do a final processing of the data. In the end there could be no mistaking the signal: HC_7N had now taken over as the largest molecule known to be present in the interstellar medium. Kroto has made a habit of watching oscilloscope traces ever since.

How far could they go? The problems that Kirby and Walton had experienced with the synthesis of HC_7N suggested that HC_9N would be well-nigh impossible to make. But then Oka hit on the idea of extrapolating the data already obtained for HCN, HC_3N, HC_5N and HC_7N and predicting the likely parameters for the next member of the series, HC_9N. It worked. Oka's predicted microwave frequencies matched observation: HC_9N joined the list of interstellar molecules.

This all seemed to be getting a bit out of hand. The list of astromolecules drawn up in the early 1970s consisted of a handful of molecules with no more than three, four, or at most five atoms. By the end of the decade, radio astronomers had added to this list a variety of complex organic molecules containing up to 11 atoms. How were these molecules being formed?

There was a number of theories kicking around. In 1973, Eric Herbst and Bill Klemperer at Harvard University had put forward a mechanism to explain the formation of some of the small molecules then known to be present in dense interstellar clouds. They argued that the role of dust grains in these clouds was to shield atoms and molecules from the disruptive effects of high energy ultra-violet radiation, and to provide surfaces on which hydrogen atoms could combine to produce hydrogen molecules, H_2. Their task was to start with H_2, helium, and carbon monoxide, known to be abundant, and devise ways in which larger molecules might be formed.

The key was to explain how reactions between atoms and small molecules could take place at anything like reasonable rates in the environment of an interstellar cloud. Reactions between electrically neutral atoms (atoms with an equal number of protons and electrons) required too much energy and were simply too slow.* However, there was in principle still enough high-energy cosmic radiation around to remove one or more of the outer electrons from an atom or molecule, converting it to a positively charged ion. Reactions involving ions proceed much more quickly.

Apart from the process in which H_2 is formed, Herbst and Klemperer's ion–molecule mechanism was a homogeneous one: reactions take place between ions, atoms and molecules exclusively in the gaseous state. However, one well-known way of increasing the speed of a chemical reaction is to use a catalyst, and some catalysts work by providing a surface on which adsorbed atoms or molecules can react. Reactions involving this kind of catalyst are called heterogeneous reactions since they involve more than one sort of physical state (gas and solid, say). It was accepted that molecular hydrogen was being formed from its constituent atoms on the surfaces of dust grains. Could it be that larger interstellar molecules were being formed in the same way?

Kroto was convinced that this kind of chemistry, taking place in cold interstellar clouds of gas and dust, could not provide an easy explanation for the presence of molecules like HC_5N, HC_7N and HC_9N. The proposed mechanisms offered no clue as to why long, straight-chain molecules should build up in preference to branched chains. No large branched-chain molecules had been found: if they were there in the dark clouds in the same kind of abundance, then there were more molecules in interstellar space than had ever been thought possible. The possibility of forming the long-chain molecules on the surfaces of dust grains also left some unanswered questions. One problem was that at the very low temperatures of some interstellar clouds, it would be extremely difficult for these molecules to unstick from the surface.

Kroto gradually came to the conclusion that some answers were to be found in the properties of red giant stars. Having exhausted their core reserves of hydrogen, these huge stars undergo a period of dramatic expansion and

*Recent studies suggest that these reactions are, in fact, much faster than expected.

cooling. Depending on their history, they may be rich in carbon or oxygen. Carbon formed in the interior of a carbon-rich red giant is churned up to the surface and, with modest surface temperatures of only 3000 kelvin, the carbon-rich gases condense to form grains. Like huge cosmic snuffed candles, these stars pour smoke and soot into the interstellar medium. The outer atmospheres of these stars would be ideal places for forming long-chain carbon molecules.

There was one spectacular infra-red object, catalogued as IRC + 10°216, that seemed particularly promising. This cool star is extremely bright in the infra-red region of the spectrum and appears to be pumping out vast quantities of carbon-containing grains. Kroto presented a lecture on 'semistable' molecules in space at a meeting of the Faraday Division of the Royal Society of Chemistry held in London on 29 October, 1981. In the written version of this lecture, he speculated that the long chains might be formed in the expanding outer shell of gas and dust surrounding such a star. He went on to ponder on the role of the long-chain molecules as possible intermediates between atoms and small molecules such as C, C_2 and C_3 and soot particles themselves.

In February 1982, radio astronomers M. B. Bell, P. A. Feldman, Sun Kwok, and H. E. Matthews published an article in the British journal *Nature* giving details of their study of microwave signals from IRC + 10°216. They assigned some weak features, lying close to the (by now well-known) emission signals from ammonia, as belonging to the molecule $HC_{11}N$. Of the 17 molecules known to be present in the atmosphere surrounding this red giant star, seven had been found to possess carbon–carbon bonds and 10 had —C≡N substituents. It seemed as though the red giants did indeed hold the key. The trick now was to find more conclusive evidence linking long-chain carbon molecules like the cyanopolyynes to the physical conditions—temperatures and pressures—likely to be found in the shell of gas and dust surrounding a red giant star.

Kroto's attention had also been drawn to a series of papers, published in the early 1960s by a team from the Max Planck Institute for Chemistry in Mainz. In one article which appeared in the German journal *Zeitschrift für Naturforschung* in 1963, H. von Hintenberger, J. Franzen and K. D. Schuy described how they had passed a high voltage through two electrodes made of graphite. Analysis of the material produced in the resulting carbon arc discharge revealed a series of carbon molecules increasing in size up to C_{33}. The conditions in an arc discharge were not too different in many ways from those in the outer atmosphere of a red giant. If Kroto could find a controlled way to make cyanopolyynes under these kinds of conditions, perhaps he could measure the spectra of chain molecules much longer than $HC_{11}N$. He dreamt of finding $HC_{33}N$ in space.

There was another reason to be interested in the chemistry and properties of long-chain carbon molecules. In 1977, the NRC physicist Alec Douglas suggested that these molecules might provide an explanation for the so-called diffuse interstellar bands. This was not a trivial suggestion. The problem of identifying the origin of the diffuse interstellar bands had plagued astronomers

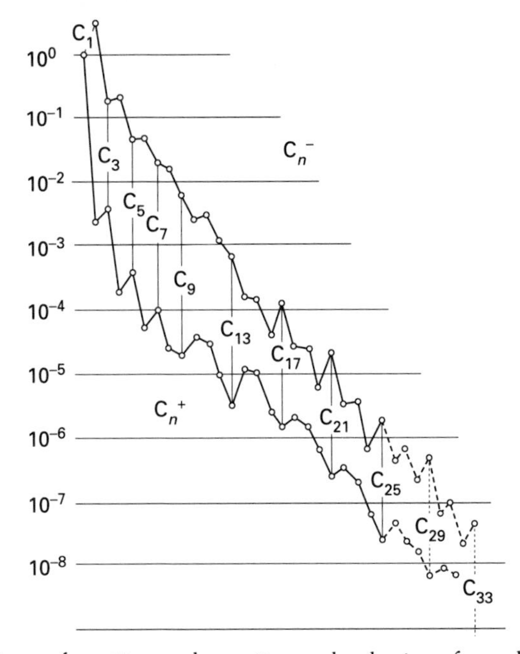

Distributions of positive and negative molecular ions formed in an arc discharge between two graphite electrodes. Adapted, with permission, from Hintenberger, H. von, Franzen, J., and Schuy, K. D. (1963). *Zeitschrift für Naturforschung A*, **18**, 1236.

for 50 years. From the perspective of the chemical spectroscopist, this was (as Kroto put it) 'the last great problem in astronomy'.

When we disperse sunlight through a prism, we obtain the familiar rainbow pattern of colours, from red to violet. We also repeat an experiment that Isaac Newton carried out in his darkened Cambridge laboratory in 1666. What Newton didn't notice is that this spectrum of colours is crossed by a series of narrow, dark lines. These lines appear because atoms present in the outer layers of the sun absorb visible light at certain characteristic frequencies, thereby preventing these particular 'colours' from reaching us on earth. Discovering this explanation led to the development of the science of spectroscopy and, ultimately, to quantum mechanics.

When we do the same with visible light from distant stars, we find a similar phenomenon. The spectrum is again crossed with dark lines, of which some are the familiar atomic lines seen in the sun's spectrum, but there are others that are much broader (or more 'diffuse'). For this reason they are sometimes referred to as 'bands' rather than lines. One of the strongest of these diffuse bands appears at 443 nanometres (in the blue region of the spectrum), and was identified to be of interstellar origin in the 1930s. By the mid-1970s, the diffuse interstellar bands numbered around 40. They were a complete mystery.

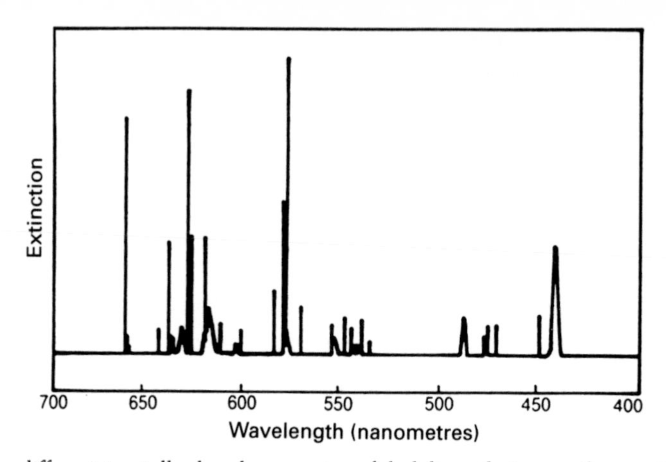

The diffuse interstellar bands—a series of dark lines that cross the spectra of light dispersed from distant stars. In 1977, their origin was a complete mystery.

It was clear that the bands were somehow associated with interstellar dust grains, since there was a strong correlation between their intensities and regions in space where the grains were known to be present. Just as dust particles in the earth's atmosphere produce a red sunset, so starlight is reddened as it passes through clouds of interstellar dust. The diffuse bands are much stronger in the light from reddened stars.

There was good evidence to suggest that the diffuse bands result from a single absorbing material or series of related materials. It was known that whatever this material was, it must be composed of very small particles, smaller than the grains themselves and possibly even of molecular dimensions. Many explanations had been put forward: excited molecules in the process of falling apart, molecules stuck to the surfaces of the grains, a variety of possibilities in which light interacted directly with solid particles. All were unsatisfactory.

Alec Douglas had followed the work that his NRC colleagues had done in collaboration with Kroto. He realized that the cyanopolyynes had become the focus of attention simply *because* they could be detected so readily. These molecules possess a permanent dipole moment: the —C≡N group draws the molecule's cloud of negative charge provided by its outer bonding electrons towards itself, making one end of the molecule slightly negatively charged and the other end slightly positively charged. Molecules with permanent dipole moments are able to emit microwave radiation as they lose energy, much like the antenna of a radio transmitter. Douglas reasoned that there must be many more long-chain carbon molecules present in the interstellar medium which cannot be detected in this way because they have no dipole moment. This restriction does not necessarily apply to higher energy transitions between states which involve a change in the distribution of the electron cloud in the molecule. A molecule without a dipole moment may still absorb visible or ultraviolet light if a dipole is created in the excited state. In other words, interstellar molecules which do not possess permanent dipole moments may still absorb starlight. Whatever was responsible for the diffuse bands might be invisible to the radio astronomer.

Building on some theoretical calculations, Douglas argued that a mixture of C_n molecules, with n in the range 5–15, might collectively give an absorption spectrum to match the diffuse interstellar bands. These molecules have no permanent dipole moment and so would not be detectable by radio astronomy. That these carbon chains would be closely connected with dust grains appeared entirely logical. Perhaps they were being formed together in the smoke and soot in the expanding outer shells of carbon-rich red giant stars.

Kroto found this explanation difficult to accept. However, such was the nature of the problem that almost any attempt at an explanation was well worth looking into. In the early 1980s, the long-chain carbon molecules were very much in Kroto's thoughts as a possible solution to the last great problem in astronomy.

2

Some kind of junk

It was an offer he couldn't refuse. In 1968, Wolfgang Krätschmer had moved from the Technical University in Berlin to the Max Planck Institute for Nuclear Physics in Heidelberg, where he studied for a Ph.D. on radiation damage in solids caused by high energy ions. He had acquired his doctorate three years later, and had spent the following six years as a postdoctoral researcher working on various problems related to cosmic rays. When it was suggested by the Institute's directors in 1976 that he might put his experimental talents to use in the increasingly fashionable field of interstellar dust—and thereby secure a permanent contract from the Institute—Krätschmer didn't need much time to think it over.

The field may have been fashionable, but in physics it was dominated largely by theoreticians. Of course, experimentalists were needed to make the necessary spectroscopic measurements on light coming to earth from distant stars. They were also needed for making painstaking measurements of the properties of the assorted elements and compounds that were thought to constitute interstellar dust. But attempts to prepare samples of dust particles in the laboratory and measure their properties for comparison were difficult and often inconclusive. And theoreticians were quick to use their computers to fill any vacuum in experimental knowledge and understanding.

Krätschmer entered the field as an experimental physicist, and so joined the ranks of the few. The leading exponent of the experimental approach to interpreting the spectroscopic role played by interstellar dust particles was an American based in the Department of Physics at the University of Arizona in Tucson. His name was Donald Huffman.

In 1976, Huffman was on sabbatical at the Max Planck Institute for Solid State Physics in Stuttgart and was entertaining audiences in various physics institutes in West Germany with his accessible lecturing style. Now the quickest way to become familiar with an unfamiliar field of scientific activity is to spend some time talking to the experts. When Huffman delivered a lecture on interstellar dust in Heidelberg, Krätschmer listened carefully, and decided that his education would benefit enormously by establishing direct contact.

Shortly afterwards, Krätschmer joined Hugo Fechtig, his director in the Institute's Cosmophysics Department, in a visit to Huffman in Stuttgart. They explained that they wanted to enter the field of interstellar dust and asked who would be the best people to talk to or work with. Huffman, who by then had worked in the field for eight years, set aside his modesty and gave them the names of three leading scientists, including his own.

This was to be the beginning of a long and eventful collaboration, and it developed very quickly. On returning to Arizona, Huffman received a letter from Krätschmer asking if he could spend some time working with him in his lab, at the Institute's expense. Huffman, who had no funding to support his studies on interstellar dust and no students to carry them out, was delighted to accept.

Huffman had committed his life's work in physics to the study of the optical properties of solids and small particles. Over a period of ten years of activity, he had brought his expertise to bear in many subjects, such as atmospheric physics and biophysics, where small particles play key roles. His interest in interstellar dust had developed in the late 1960s when, as a newly appointed assistant professor, he had begun to talk to the astronomers at the University of Arizona. They had asked him all sorts of questions about the optical properties of the kinds of very small particles they were encountering in interstellar space, and he hadn't known the answers.

Solid state physics taught at undergraduate and postgraduate level was based on the assumption of solids effectively infinite in extent. Conditioned to think of solids in this way, Huffman couldn't predict what might happen as the dimensions of the solid were reduced, eventually to the size of individual molecules. This was what the astronomers wanted to know. Huffman did predict, however, that as interest in smaller and smaller particles grew within the physics community, then the physicists would likely meet the chemists coming the other way.

As the astronomers told him about the problems they were having, Huffman became increasingly interested. These tiny grains were a source both of frustration and fascination. Frustration, because they got in the way: they absorbed or scattered light from more distant and more interesting objects in the universe. Fascination, because there were so few clues to what these grains were made of and because theories of stellar and planetary formation suggested that the large interstellar clouds of gas and dust were breeding grounds for new stars, complete with planetary systems. The grains were clearly involved in the early stages of planetary formation, and yet so little was known about them.

What was known was that the dust particles account for less than one per cent of the matter in interstellar space. Despite this, they have a tremendous power to dim the light from distant stars. On average, the visible light from a star some 3000 light years* away is dimmed by half due to the intervening dust, and each additional 3000 light years cuts the light intensity by half again. This dimming, or *extinction* of light—a combination of absorption and scattering—is not uniform across the electromagnetic spectrum. It is strongest for ultra-violet light, gradually falling off with increasing wavelength. In the visible region, blue light is dimmed much more than red light, causing an overall reddening of starlight as it passes through an intervening dust cloud on its way to earth.

The shape of the extinction curve gave astrophysicists some clues to the likely sizes of the dust grains. Electromagnetic theory predicted that the smaller the grains, the greater the reddening effect. The measured extinction curve in the visible region pointed to grain sizes of the order of 0.1 micrometres. The strong extinction in the ultra-violet indicated grains of the order of a few nanometres, whereas the much weaker extinction in the infra-red suggested particles as large as one micrometre.

These grains are much smaller than dust particles we normally encounter in the home (or in the laboratory), and are more realistically thought of as a kind of smoke rather than dust. Their density in interstellar space obviously varies, but in a region of space where we might expect on average to find one atom per cubic centimetre, the grains may be hundreds of metres apart. It has been estimated that a space the size of St Paul's Cathedral in London would contain just one dust grain.

But what really fascinated Huffman were the additional spectral features superimposed on the otherwise smooth extinction curve. These features are due to increased absorption of light by the grains and they provide the only clues to the nature of their composition. Apart from the 40-odd diffuse interstellar bands in the visible, there are absorption features in the infra-red at 3.1, 3.4, 6.0, 6.8, and 9.7 micrometres and a strong ultra-violet feature at 217 nanometres. Huffman set himself the difficult task of identifying the origins of these features. As funding was not so forthcoming for this kind of esoteric activity, it became an indulgence. Huffman would enjoy working at this particular hobby whenever he could spare the time from his more supportable research.

In the late 1970s, experimental progress was being sought in two ways. One way was to pick a material, which could be a single element or a compound of elements, and measure its optical constants (refractive indexes and extinction coefficients) as a function of wavelength. Scattering theory could then be used to calculate what the extinction curve should look like in the wavelength regions where the various interstellar bands were observed. To do this kind of calculation, it was first necessary to make some assumptions about the sizes and shapes of the dust grains, but correspondence between calculation and

*A light year is the distance travelled by light in one year. It corresponds approximately to 6000 billion miles.

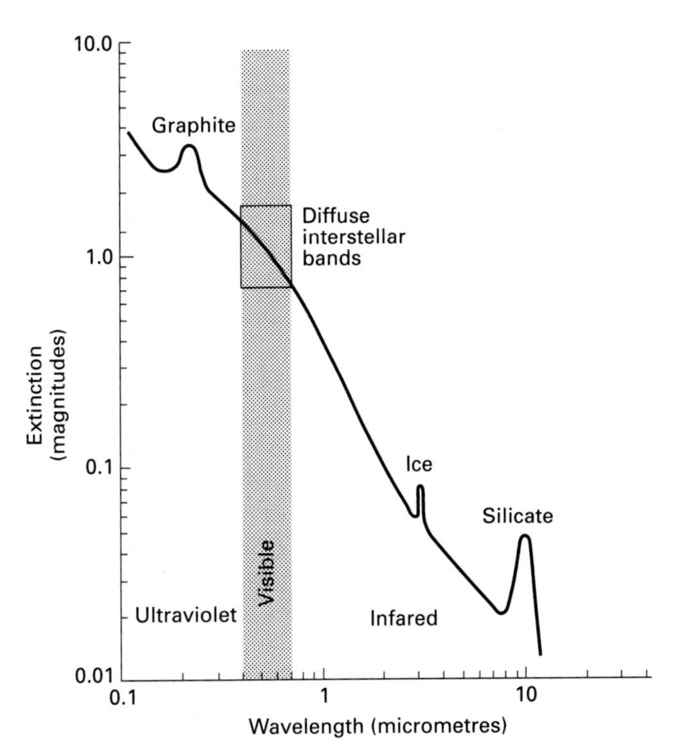

The extinction of light from distant stars increases smoothly from low energy (long wavelength) to high energy (short wavelength) radiation. Superimposed on this smooth background are distinct absorption features in the infra-red, visible and ultra-violet. This figure shows the infra-red bands at 3.1 and 9.7 micrometres and the ultra-violet 217 nanometre band. (the diffuse interstellar bands in the visible are not shown explicitly). The infra-red bands are thought to be due to interstellar ice and silicate particles, and the ultra-violet band is probably due to graphite particles. Huffman and Krätschmer worked together to try to engineer particles with the properties required to reproduce the exact shapes and intensities of the bands. Adapted, with permission, from Wynn-Williams, G. (1992) *The fullness of space,* Cambridge University Press.

observation could be taken to indicate that the material might be present in space.

An alternative (and complementary) approach was to generate and collect small dust particles produced directly from the material in the laboratory. The spectra of these particles could then be measured and compared with the interstellar features. Close correspondence between the two in terms of wavelength, width, and intensity provided evidence that the element or compound might be present in the interstellar medium, in the form in which it had been produced in the laboratory. Correspondence with the extinction curve calculated from the optical constants also gave some reassurance that the theory was being properly applied.

Kroto's task was in many ways much simpler. In the search for interstellar molecules like the cyanopolyynes, the microwave emission signals provided characteristic 'signatures' which could be identified with specific molecules provided their microwave spectra were known. Assignments could therefore be made with a good degree of confidence, a feature of the work that had appealed to Kroto from the beginning. However, the higher-energy infra-red, visible, and ultra-violet features did not give such clear signatures and were not so readily assignable. The diffuse interstellar bands in the visible region plagued the physicists just as much as they plagued astronomers and chemists.

If dust particles were responsible for some of these features, then a major part of the problem lay in the nature of the particles themselves. It was apparent that the calculated extinction curves were very sensitive not only to the type and composition of the material, but also to other factors such as the shape, size, aggregation, and crystallinity of the particles. The experimental challenge was to make particles in the laboratory with properties as close as possible to those they would be expected to have if they were formed in interstellar space.

The experimental philosophy that Huffman and Krätschmer adopted in 1977 for their collaborative research programme was relatively simple and straightforward. Because of the nature of their origin, interstellar dust particles could well be horribly complicated mixtures of elements, they could come in a variety of shapes and sizes, and could be of uncertain crystallinity. Possibly the spectral features detectable in the light from distant stars reflected this complexity. However, for reasons of economy of materials, time, energy, and thought, Huffman and Krätschmer reasoned that the experimentalist must start out with simple, pure systems. They could build complexity into their systems if the simple ones didn't fit, but it was important to start simple and get complicated only when absolutely necessary.

A second important aspect of their approach was to attempt only experiments that seemed possible and (preferably) easy. In 1977, there were many unknowns in interstellar spectroscopy, but some unknowns appeared more accessible to experiment than others. Huffman had become convinced that the 9.7 micrometre band was due to silicate particles, and that the 217 nanometre band was due to small particles of graphite. Huffman had already expended much research effort on the optical properties of these materials, and in 1977 offered Krätschmer a choice between the two. Krätschmer elected to work on the silicates.

They chose to look at the mineral olivine, a naturally occurring silicate which contains the elements magnesium, iron, oxygen, and silicon. The infra-red spectrum of crystalline particles of olivine did show a feature in the region of 10 micrometres but this did not correspond particularly well to the observed interstellar band. Huffman believed they could obtain much better correspondence with more glassy (more amorphous or non-crystalline) particles. A facility was available in Tucson which they could use to bombard olivine particles with high-energy argon ions, disrupting the crystal structure and making the particles more amorphous. There were parallels between this approach and Krätschmer's earlier work, a factor that influenced his decision in favour of the olivine study.

Over a period of six months, they bombarded the olivine crystals with high-energy ions and measured the optical constants of the resulting particles. The calculated extinction curve now fitted much better with the observed 9.7 micrometre interstellar band. This was confirmed by further studies carried out at about the same time by other astrophysicists. They all reached the same conclusion: floating in the space between the stars are billions of tiny bits of rock.

Krätschmer returned to Heidelberg and in the succeeding years maintained only loose contact with Huffman. Their collaborative work on olivine was published in 1979, and in the meantime Krätschmer embarked on his own experimental programme concerned with water ices. Ice particles had been suggested some time before as a possible explanation for the 3.1 micrometre band. As with olivine, crystalline ice (of the kind used to cool drinks here on earth) did not quite fit the bill, and Krätschmer therefore studied the spectra of amorphous ice, which in interstellar clouds was presumed to condense onto the surfaces of silicate particles. This work was not entirely conclusive, but Krätschmer was able to show that to explain the observed spectrum, any ice grains in the interstellar medium would need to be fairly large. In the densest clouds, some of the bits of rock are covered with a thick layer of ice.

By the summer of 1982, the problem of the 217 nanometre band was still not considered to be solved, and yet Huffman was still convinced that graphite particles were responsible. Together with his Tucson colleague Kendrick Day, Huffman had re-measured the optical constants of laboratory-produced graphitic 'soot' in 1973. They had used a simple carbon evaporation apparatus in which a high voltage electric current was passed through two contacted graphite rods in a low-pressure atmosphere of inert gas. Apart from the brilliant light emitted by the arc between the electrodes, copious quantities of carbon smoke were also produced and condensed as thin layers of soot on a small quartz plate. From subsequent measurements of the optical properties of the soot, Huffman had concluded that the extinction curve of the graphite particles they had made gave a peak at about the right wavelength in the ultra-violet, but the shape of the curve did not give a perfect match to the 217 nanometre interstellar band.

Over the years, Huffman had continued to chip away at the problem and had concluded that, as with olivine, factors which were not being controlled in the experiment such as the shapes, sizes, crystallinity, and especially the aggregation of the graphite particles formed in the evaporator could, perhaps, be accounting for the discrepancy. The observed interstellar feature was rather intense. Conventional soot gave a band which peaked in the right wavelength region but which was much weaker and broader. In contrast, crystalline graphite gave a more intense but narrower band. Huffman was beginning to think that he could improve the match by making graphite particles as small and as spherical as possible.

There was again no particular reason to believe that the real interstellar grains were composed only of carbon. They could also contain other elements such as hydrogen. The most logical sources of interstellar carbon were the clouds of smoke and soot produced by carbon-rich red giant stars such as IRC + 10°216. The very same stars were being cited by Kroto and other astrochemists as possible sources of the cyanopolyynes. Huffman was well aware that the interiors of these stars contained many elements other than carbon.

But in the spirit of the 'start simple' philosophy, Huffman came to the conclusion that further carbon soot experiments would still be worthwhile. He had some sabbatical leave lined up, and so renewed contact with Krätschmer to see if they could do these experiments together in Heidelberg. Krätschmer agreed, and Huffman secured funding from the prestigious Alexander von Humboldt Foundation to cover his expenses. He set off for West Germany with his family in September 1982.

Heidelberg is a nice place to do physics under any circumstances. This picture-postcard town, nestling on the banks of the river Neckar and overlooked by the gently sloping hills of the Odenwald, boasts Germany's oldest university and an imposing fourteenth-century castle, for five centuries the residence of the Palatine Prince Electors. The castle, the river, the Old Town, the mountains, forest, and vineyards all combine in apparent tribute to European culture and civilization. It is a romantic place, a source of inspiration for poets and writers from the Romantic Age to the present day.

The town is also steeped in a long history of academic study. Founded in 1386 by King Ruprecht I of the Palatine, Heidelberg University became a major centre for learning and scholarship in the early years of the nineteenth century, and has seen eight of its academic scientists honoured with Nobel Prizes in physics, chemistry, and medicine. With nearly 30 000 students in a population of 130 000 inhabitants, the spirit of the town is strongly influenced by aspects of student life.

The Max Planck Institute for Nuclear Physics is hidden among the trees of the Odenwald, reached by one of the many twisting roads that climb up into the hills to the south-west of the town. Driving along this road in the leafy, late summer sunshine of 1982, the Huffman family could not fail to note the contrasts with the Arizona desert. Heidelberg promised the obvious cultural and tourist attractions—the castle, the fine buildings, museums, theatres, pubs, restaurants, and long walks over the hillside, following in the footsteps of generations of philosophers. But Huffman's three children were also looking forward to experiencing a different kind of climate. It did not often snow in Arizona.

Krätschmer's laboratory was located on the second floor of the Cosmophysics Department, a short walk from the Institute's main gate. The equipment they were planning to use was housed in a small annexe off Krätschmer's main lab, which was not large but was littered with the usual assortment of experimental fall-out: piles of yellowing papers, glassware, and bits of apparatus in various stages of repair. Across the corridor was Krätschmer's office. This was also small and cramped and contained two desks, back-to-back: one for Krätschmer and one for the use of students or visitors working in his lab. Huffman settled in quickly.

Krätschmer had an apparatus similar to the kind that Huffman had used in his earlier experiments on carbon soot with Kendrick Day. This was an evaporator that had been built for the purpose of depositing thin coatings of carbon suitable for electron microscopy studies. It consisted of a large bell-jar mounted vertically and hinged so that it could be opened outwards. The jar was covered with a strong wire mesh to reduce the danger of flying glass in the event of an implosion under vacuum. On the vertical panel enclosed by the jar when sealed shut were two graphite rods mounted on copper electrodes. One of the rods was ground to a point at one end, and was contacted with the flat end of the other rod. The electrical resistance of the rods at the point of contact caused them to heat up when a current was passed between them. The inside of the bell-jar would be bathed in the brilliant light of the electrical arc, the temperature would climb to several thousand kelvin and fragments of graphite and carbon atoms would literally 'boil' off the surface and nucleate to form a vapour with an appearance not unlike cigarette smoke.

The bell-jar was evacuated using a combination of vacuum pumps which could be closed off by a gate valve. The pumps were tucked neatly beneath the bell-jar inside a rigid metal frame. The whole unit was compact and self-contained.

Inert gas such as helium or argon could be admitted to the bell-jar through a separate inlet. The inert gas served to cool the hot carbon vapour produced in the arc and promote the formation of soot particles. Because the physicists were concerned to make the particles as small as possible, it was important to keep the pressure of the gas low. High pressures meant more inert gas atoms and more collisions between these atoms and the growing soot particles. Too many collisions and the particles would tend to clump together and grow too large. The predictions of electromagnetic theory promised that such clumping would spoil any chance of correspondence with the 217 nanometre interstellar band.

Krätschmer was largely innocent of the finer experimental details, and was content to watch as Huffman tried various ways to optimize the conditions. They were joined by Norbert Sorg, Krätschmer's first postgraduate student, who was planning to incorporate the carbon soot study as part of his Diploma.

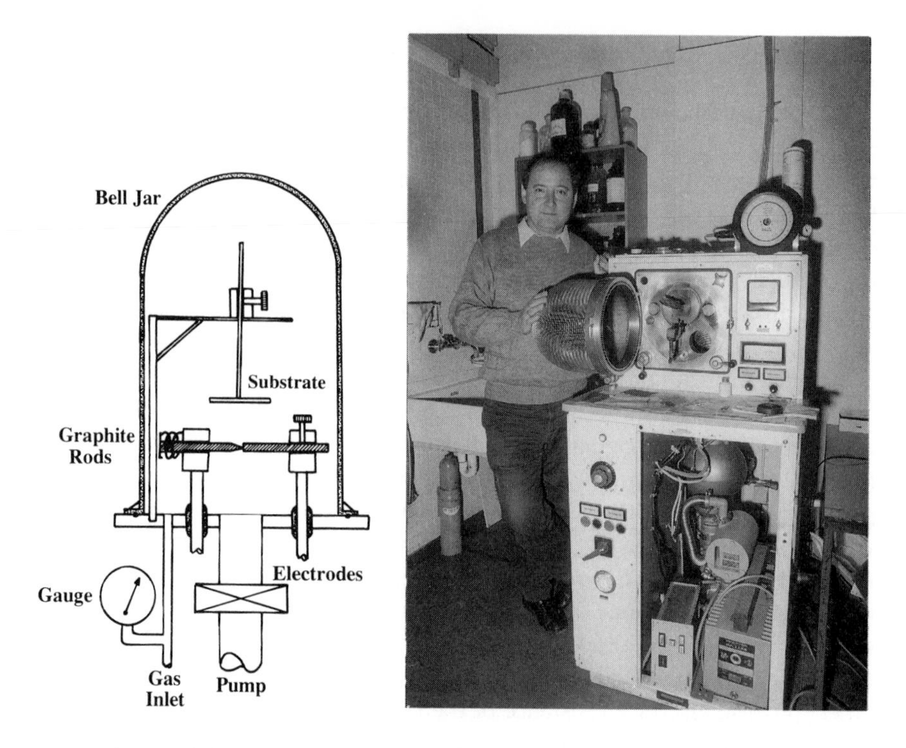

The Heidelberg evaporator. Passing a current through the contacted graphite rods generates a high-temperature arc. Carbon smoke produced by the evaporated graphite is collected on a substrate (a thin plate of quartz or other suitable material) and removed to a spectrometer for analysis. The photograph shows Wolfgang Krätschmer with the apparatus.

The experiments themselves were relatively quick and simple. Within a few seconds of operation, the arc produced a thin layer of soot which the physicists collected on a quartz plate fixed to a movable mount. They had to watch the process carefully. If the film was too thin then there would not be enough material present to measure the ultra-violet spectrum. Too thick a film would make the coated plate opaque to the light source in the spectrometer, preventing any measurement of the spectrum. Experience was really the only guide to the optimum film thickness.

After waiting ten minutes or so for the apparatus to cool, they would remove the coated quartz plate from the bell-jar and place it in a conventional ultra-violet/visible absorption spectrometer. They would then compare the resulting ultra-violet spectrum of the soot with the 217 nanometre interstellar band.

The first few experiments confirmed their suspicions. The aggregation of the graphite particles which constituted the soot was very sensitive to the pressure of inert gas in the bell-jar. They varied the pressure over a range from a few torr up to 20 torr.* This was about the maximum they felt they could tolerate and still avoid too much clumping of the particles. Even so, the ultra-violet spectra of the soot did not give them the perfect match they were looking for.

They concluded from this that the graphite particles were clumping together too much, even at low pressures of inert gas. Such was the sensitivity of the particles' optical properties that even a small amount of unavoidable clumping had a big effect on the extinction curve. But there was something odd about some of the measured spectra. With helium or argon pressures of 20 torr the physicists occasionally saw two—sometimes three—extra bumps in the spectra, superimposed on a broad band which peaked at around 220 nanometres and was characteristic of the ordinary carbon soot. These bumps appeared at 215, 265, and 340 nanometres. Sometimes they would be more noticeable, sometimes less noticeable. Sometimes they wouldn't be there at all.

Huffman thought this was very peculiar. As far as they knew, the conditions inside the bell-jar hadn't changed, and yet the bumps seemed to come and go unpredictably. Huffman had spent his entire career as an experimental physicist staring at spectra like these, and he was intrigued.

Krätschmer was generally downbeat. He had the feeling that the appearance of the bumps showed that something had gone wrong with the experiments. Every now and then a small piece of graphite would fall and become lodged in the gate valve, preventing the valve from closing properly and allowing oil vapour from the vacuum pump to enter the bell-jar. The oil vapour consisted of hydrocarbon compounds which would decompose in the arc and produce a terrible mess. When this happened, they had to strip down the apparatus and clean it out. Krätschmer believed that the bumps in the ultra-violet spectra were probably caused by contamination of the soot with some decomposition products of the oil vapour, or perhaps the grease they were using to provide a good vacuum seal around the rim of the bell-jar.

In any case, it was clear that their carbon soot experiments were not going to yield particles of the right size to provide a conclusive explanation for the 217 nanometre band. Huffman was not prepared to let the business of the mysterious bumps go that easily, but he agreed that they needed to try another approach if they were going to make any further progress on the problem they had set out to solve.

Somehow they had to try to avoid the clumping of the particles which they believed was causing all the trouble. One possible approach was to freeze the particles and trap them in a matrix of solid inert gas before they could grow too large. The spectra of such 'matrix-isolated' particles could then be measured and compared with the 217 nanometre band, as before. However, these were not easy experiments. They involved generating the graphite particles at high

*One atmosphere pressure is equivalent to 760 torr.

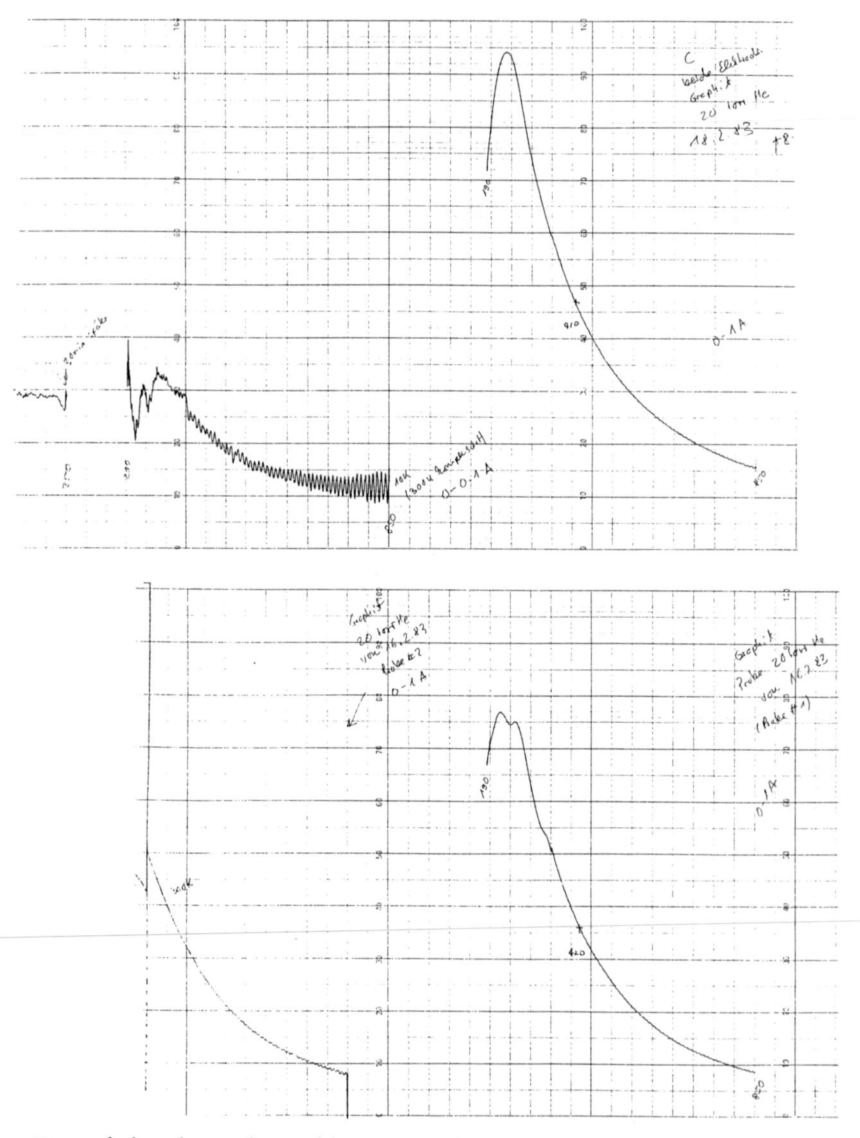

Two of the ultra-violet/visible spectra of carbon soot produced by evaporating graphite rods. The upper spectrum, recorded on February 18, 1983, shows the smooth curve expected of 'ordinary' carbon soot. The lower spectrum, recorded two days earlier, shows three odd 'bumps' superimposed on the curve. The pressure of helium was 20 torr in both experiments. The writing is Krätschmer's.

temperatures and at the same time producing a matrix of solid argon at 10 kelvin. The particles had to be produced at relatively high pressures (a few torr or so), whereas the matrix had to be formed under very low pressures.

When they measured the ultra-violet spectrum of the matrix-isolated particles, they found that the correspondence with the 217 nanometre band was really no better than before. Clumping was almost certainly taking place in the vicinity of the arc: the particles had already grown too big before they could be incorporated into the matrix.

Such is the lot of the experimentalist. However, the seasoned experimenter avoids the gloom and despondency that results from failed experiments by finding ways to turn them into successes. This is done by looking for something (anything!) positive from an otherwise unhelpful experiment. Huffman, Krätschmer and Sorg noticed that under certain conditions, the *visible* spectra of the matrix-isolated particles contained some very intense, sharp lines. These lines were actually nothing to do with graphite particles, they were characteristic of small carbon *molecules* such as C_3.

When they gently heated (or 'annealed') the matrix, they found that the lines due to C_3 gradually disappeared, to be replaced by a great many new lines at different wavelengths in the visible region. A careful study of these lines revealed that they were due to a variety of larger carbon molecules. As the temperature of the solid matrix increased, they reasoned, the C_3 molecules became more mobile, they migrated through the matrix and combined to form larger molecules, from C_4 through to C_9.

Huffman was not slow to note that after correcting for the distorting effects of the matrix, the wavelength of the line which they had assigned to C_7 corresponded approximately to that of the strongest of the diffuse interstellar bands. He had read Alec Douglas's 1977 proposal that the C_n polyynes might be responsible for the diffuse bands. Having failed to solve the problem of the 217 nanometre band, the physicists simply changed their target. This is how experimentalists overcome adversity.

Not that Huffman had given up with the carbon soot experiments. There was still this nagging mystery of the extra bumps which Huffman was inclined to think might be something important. They repeated the bell-jar experiments in an attempt to find out what was going on. Krätschmer started to refer to the bumps as 'camel humps', and he referred to the soot samples which gave the strange spectra as 'camel samples'. It was a terminology that stuck.

The German winter of 1983 gave way to spring (alas, despite some very cold weather, no snow fell to delight the Huffman offspring) and the physicists tried all manner of things to find out where the camel humps were coming from. They looked at the infra-red spectrum of the camel samples, but their spectrometer was old and not very sensitive: they saw nothing unusual. They tried heating the camel samples in air, and found that the humps disappeared. They reasoned that if the camel humps were really due to contamination of the soot by hydrocarbon compounds, then it might be possible to sublime them.

This involved heating the sample to drive off the low-boiling fraction as a vapour, which then condenses—in pure form—on a cold surface positioned above the sample. However, they had not collected enough of the soot to produce anything conclusive. They also tried to wash the contaminants out using a solvent. Lacking a training in basic chemistry (or having forgotten everything they had once learned), the physicists assumed that everything must be soluble in acetone, and so tried to dissolve out the contaminants in this solvent. The soot, and whatever else there might have been in it, remained stubbornly insoluble.

At no time during Huffman's sabbatical in Heidelberg did they try to raise the pressure of inert gas inside the bell-jar above 20 torr. But then there was no good reason why they should.

The support that Huffman had secured from the Humboldt Foundation to cover his expenses was rather lavish. It included use of a brand new 700 Series BMW, the like of which Huffman had never before experienced, and he thoroughly enjoyed roaring down the fast lanes of the German autobahns. It included two meetings at the Foundation premises in Bonn, with Huffman and his wife weekending at a hotel across from the German Bundestag. Huffman's wife remarked that the main topic of conversation at these gatherings of illustrious Humboldt Award-winners was often not science, but the excitement of driving the 700 Series BMWs that had been put at their disposal. The funding also covered expenses for visits to other Max Planck Institutes in Bonn (radioastronomy) and Stuttgart (solid state physics). Huffman decided to take some of the camel samples to Stuttgart to measure their Raman spectra.

Raman spectroscopy (named after the Indian physicist C. V. Raman) is based on the scattering of light from samples rather than its absorption. It yields spectra that feature the same kinds of vibrational energy states as infra-red spectra, but the two types of spectroscopy are to a certain extent complementary. Only molecules which possess or can develop a dipole moment as they vibrate can absorb infra-red light, but this restriction does not apply to light scattering. Consequently, the symmetries of some molecular vibrations make them 'inactive' in infra-red spectroscopy and they do not appear in the infra-red spectrum. These same vibrations may, however, be active in the Raman spectrum.

The Raman spectra of the camel samples seemed to confirm that there was something else in these samples apart from ordinary carbon soot. Huffman began to speculate that the something else might be a new form of carbon. There was a candidate. It was known as carbyne.

Carbyne had been suggested as a novel high temperature form of carbon, in addition to the familiar diamond and graphite forms, as a result of studying fused rock from a crater in Germany. Geoscientists Ahmet El Goresy and G. Donnay, at the Geophysical Laboratory of the Carnegie Institution of Washington in Washington, DC, had found thin layers of a carbon-based

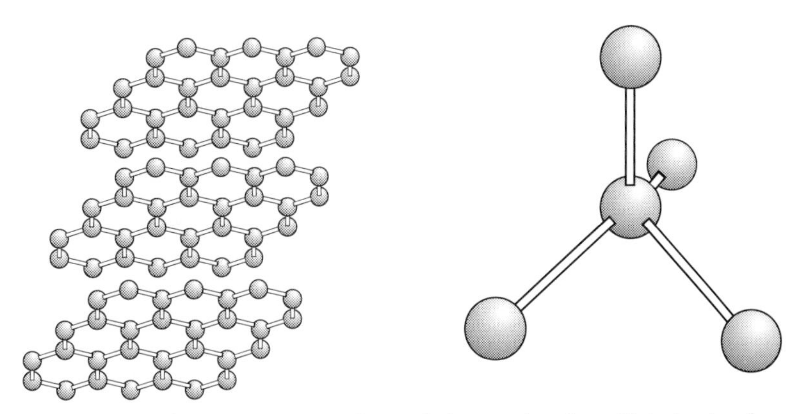

Graphite is the most common form of elemental carbon. Flat sheets of hexagons, looking like chicken wire fencing, stack one on top of the other spaced 0.335 nanometres apart. In contrast, the geometrical structure of the carbon atoms in diamond is quite different. Each atom in diamond is bonded to four others in a tetrahedral arrangement which gives the structure considerable intrinsic stength and hardness. Despite this, graphite is actually the more stable of the two forms.

substance in fused rocks from the Ries Crater in Bavaria.* These layers, which in reflected light appeared metallic grey to white in colour, alternated with layers of graphite, and were found only in rocks that had been subjected to shock, sufficient to fuse the graphite and to fuse the silicates to glass. They had described their observations in an article published in the American journal *Science* in July 1968.

Diamond and graphite, the two forms (or 'allotropes') of carbon, possess very different properties because they are constructed from different spatial arrangements of carbon atoms. Diamond has the familiar tetrahedral arrangement of atoms, which forms a structure of great intrinsic strength. Graphite is composed of flat layers of carbon atoms in a hexagonal array. These layers are arranged in planes of atoms resembling chicken-wire fencing, and the spacing between successive planes is 0.335 nanometres (billionths of a metre). The layers stack one on top of the other, and tend to slip and slide over one another.

Carbon atoms can join to one another through single, double, or triple bonds. In the case of diamond, four single bonds reach outwards from each carbon atom in the crystal lattice, pointing towards the corners of a tetrahedron. In graphite, each carbon atom is connected to three neighbours by a combination of two single bonds and one double bond. This structure looks much less elegant and less strong than the structure of diamond but, although diamond is the denser and harder of the two substances, graphite is in fact the more stable.

*This is an impact crater, similar to craters on the surface of the moon. In fact, the Ries Crater was used as a training ground for the Apollo astronauts.

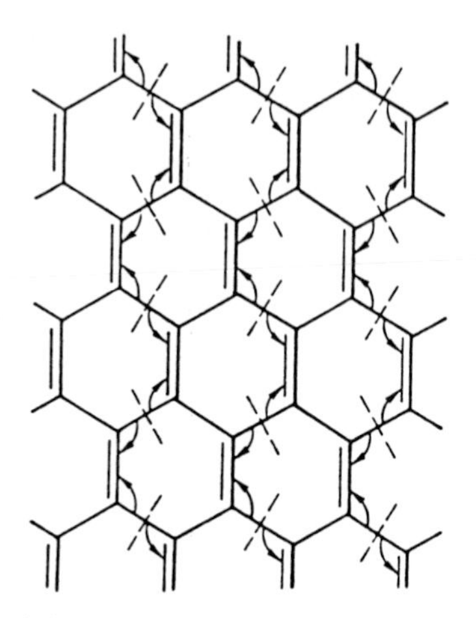

Whittaker proposed that at temperatures above 2600 kelvin, the chicken wire fencing structure of graphite transforms to the linear chain structure of carbyne.

Graphite can be converted into diamond at temperatures above about 3000 kelvin, but very high pressures are also required to force the atoms to take up the tetrahedral arrangement.

El Goresy and Donnay had proposed that carbon could have a third form, which became known as carbyne. There followed several reports proposing the existence of a number of carbyne forms. In 1978, A. Greenville Whittaker at the Materials Sciences Laboratory of the Aerospace Corporation in El Segundo, California, heated graphite to high temperatures and produced a mysterious 'white carbon', which he identified with carbyne. He suggested a mechanism by which graphite might be converted into linear carbyne chains at temperatures above about 2600 kelvin. In this mechanism, the single carbon–carbon bonds in graphite are broken and the electrons thus released combine with the double bonds to form triple bonds. The result is an array of linear carbon chains containing alternating single and triple bonds, identical to the (much shorter) carbon chains in the polyynes.

According to Whittaker, carbyne is supposed to be the stable form of carbon until the temperature is raised further to 3800 kelvin, at which point the bonds rearrange again to form diamond. These kinds of temperatures were being produced in the bell-jar experiments in Heidelberg, and it seemed logical to Huffman to propose that the carbon vapour might be condensing to form structures characteristic of carbyne.

Krätschmer was not so sure. He did not think that the experimental evidence for the existence of carbynes was all that convincing. There were arguments raging in the scientific literature over whether carbyne really existed. It had been questioned whether Whittaker's white carbon and the other proposed forms of carbyne might, after all, be layer silicates. Krätschmer was still of the opinion that the camel humps were due to 'some kind of junk'.

Eventually, Huffman too came to be persuaded that they must just be making junk. The carbon soot experiments had failed to provide a conclusive explanation of the 217 nanometre band, and appeared to have been dogged by an experimental artifact which produced the camel humps. They had no real desire to pursue the matter further. For one thing, no scientist likes to waste time on a difficult problem unlikely to produce anything more than the satisfaction of knowing just exactly what had gone wrong.

Fortunately, the matrix-isolation studies had proved to be quite fruitful, although not in the direction they had originally hoped. The carbon molecules they had observed certainly warranted further study, and Huffman and Krätschmer agreed to continue with these experiments. Krätschmer believed that there could be an element of truth in Douglas's proposal, and matrix-isolation appeared to be a reasonably inexpensive method for studying the spectroscopy of C_n molecules.

Huffman and his family returned to Arizona in the summer of 1983. Although the experiments hadn't worked out in quite the way he had planned, Huffman had enjoyed immensely his time in Heidelberg. In fact, Huffman and Krätschmer had just made the most important scientific discovery of their lives, but they didn't know it.

3

Welcome to the machine

The travel guides like to boast that Texas is a state of mind. Kroto certainly had plenty on his mind when he made the journey to Texas towards the end of February 1984. He had managed to scrape together enough travel money to get from Sussex to Austin, and a biennial conference on molecular structure. At the conference he had met Bob Curl, a fellow microwave spectroscopist he had got to know at a conference in England seven years before. At their first meeting, Kroto had invited Curl to visit him in Sussex. Now came the opportunity for Curl to return the favour, and he invited Kroto to visit his home and laboratory at Rice University in Houston.

Visiting America gave Kroto the opportunity to pursue two of his favourite non-scientific activities: travelling and browsing through bookshops. From the conference in Austin he had driven to Dallas, where he explored a number of half-price bookstores filled with books on graphic art, design, and architecture. These were subjects that had held him in thrall for longer than he had been a career scientist and the hours passed quickly as his attention wandered from one bookshelf to another. From Dallas he took Highway 45 towards Houston, eventually coming within sight of the white cement and sparkling chrome towers of the oil giants, which dominate the skyline. Here, on the outskirts of Space City in the Lone Star State, interstellar molecules were very much on Kroto's mind. He still occasionally dreamt about $HC_{33}N$.

A few miles out from the Downtown area, he pulled over and telephoned Curl for directions to his home. Curl lived on Bolsover, a stone's throw from the Rice campus which lies a few miles southwest of Downtown along Main St. Kroto followed Curl's instructions, and found his way to a quiet suburb. It was hard to believe that this was but a short distance from the centre of the fourth largest city in the United States.

Kroto enjoyed the warm hospitality of the Curl home and, relaxing after dinner, their conversation turned towards science. Curl was very enthusiastic about the work he was doing with his Rice colleague, Rick Smalley. Smalley had designed and built a new machine for looking at weird and wonderful molecules that could not be made by conventional means. Curl and Frank Tittel, another Rice scientist from the Department of Electrical Engineering, had joined forces with Smalley some two years before and had secured funding from the US Army to study small clusters of semiconducting materials such as silicon, germanium, and gallium arsenide.

This work was going extremely well. The machine had proved to be a powerful new tool, and Smalley was actively engaged in many other projects. Curl was particularly excited by some new spectroscopic results that Smalley and his colleagues had obtained for the molecule SiC_2, and he strongly urged Kroto to spend some time the next day talking to Smalley and looking around his lab.

The original Rice Institute was founded in 1891 with the aid of a $200 000 endowment from Texan businessman William Marsh Rice. Nine years later, Rice was murdered in New York City by his valet. It transpired that the valet was involved in a million-dollar conspiracy with a lawyer who was working for Rice's ex-wife. Together the lawyer and the valet had planned to get their hands on the estate of Rice's deceased second wife. Subsequent legal wrangling held up a further $10 million endowment for the Institute until 1912.

With this money the astronomer Edgar Odell Lovett, the Institute's first president, commissioned a firm of Boston architects to convert some bleak Texas prairie into a university campus in the spirit of the great Ivy League universities of Princeton and Harvard. Forsaking the Gothic style they had employed at Princeton and West Point, the architects Cram, Goodhue, and Ferguson designed a campus more suited to the sticky heat of Houston, drawing on and adapting a wide range of 'southern' and Mediterranean themes.

The result is an extremely pleasant mix of pastel colours, ornament, and cloistered solitude, seen at its most striking in the architecture of Lovett Hall. Rice's ashes are interred inside a bronze statue which sits imposingly in the centre of Academic Court, adjacent to Lovett Hall. Space has no premium in Texas and the Court, like the rest of the campus, embraces the Texan propensity for bigness. The remaining, widely dispersed campus buildings are connected by tree-lined roads or footpaths through well-tended lawns and neatly trimmed hedges.

Kroto made his way to Smalley's lab on the third (and top) floor of the Space Physics Laboratory, built in the early 1960s when funding from NASA was more plentiful. This is a novel building which has all the office and laboratory space concentrated in a large rectangular column in the centre of the structure while the stairs and corridors run around the outside. The corridors are surrounded by railings and a series of thin concrete pillars, affording little protection from the elements but then such protection is hardly needed in Houston. Despite appearances, however, there was no mistaking the fact that Kroto was entering familiar territory. Like experimental laboratory buildings the world over, the corridors doubled as storage space for gas cylinders (both empty and full) and large cryotanks of liquid nitrogen.

Now there is a myth that modern science is carried out by white-coated scientists inside pristine laboratories. This myth is derived largely from interminable television commercials promoting the efficacy of soap powders or drugs (usually on the basis that their ability to clean or cure has been scientifically or clinically 'proven' by the white-coats). The reality, at least for modern chemical physics, is often somewhat different. Smalley's lab is reasonably clean, but it is also profoundly untidy. Bits of equipment and tools litter the free workspace, dropped where they were last used to be found again (sometimes after a lengthy and irritating search: 'Who's got the five-sixteenths?'). For some, such untidiness reflects a lack of discipline and rigour. For others, it is a sure sign that there is plenty of exciting stuff going on.

Kroto surveyed the impressive array of equipment that Smalley had assembled. Smalley himself was pretty impressive, too. At a time when chemical physics in America meant increasingly big budgets and large groups, the academic scientist was being constantly pushed into a role as figurehead and fundraiser. Up to a point this was okay, but the figurehead inevitably becomes more and more distant from the work going on in the lab for which he is striving to find funds. All the hands-on stuff is then left to the army of postgraduate students and postdoctoral scientists, who in turn need maintenance grants which means more fundraising. At the extreme, the students and post-docs do all the work and the figurehead simply signs his name to the papers.

Smalley had obviously avoided all this. He had somehow managed to build up a large lab, staffed with good people doing some great science, but at the same time maintaining a hands-on familiarity with the apparatus. He could climb over this elaborate equipment and he could make it work himself if he wanted to.

Smalley was proud of what he had achieved. After spending four years as an industrial research chemist with Shell Chemical Company, he had gone to Princeton University to study for a Ph.D., which he completed in 1973. But it was as a postdoctoral research associate with Lennard Wharton and Don Levy at the University of Chicago that he had developed the know-how that was to define his life's work. He became a builder and user of big machines.

The machine particularly in question on the day of Kroto's visit was a second-generation cluster beam apparatus known lovingly as AP2 (pronounced 'app-two'). In contrast to the wide-open spaces of the Rice campus, AP2 was squeezed into one corner of the lab, filling almost all the available space from floor to ceiling. The centrepiece was a large cylindrical stainless steel chamber mounted on top of an equally large vacuum pump. Radiating outwards from the chamber were tall benches on which the lasers and optical devices were fixed. Light from the lasers was passed into the chamber through quartz windows fitted into steel flanges. Because of the sheer size of the vacuum pump, these flanges were about six feet off the ground, requiring the lasers also to be mounted at this height. To work with AP2, the researchers literally had to climb on top of it using a stepladder. Standing on the floor, the view was one of a mish-mash of copper tubing, supports, cables, and pump hoses—a visual cacophony in metal and plastic.

From each of the many electronic control and detection devices there flowed one or more screened cables in black, grey, or white. Some of these plastic streams gathered to a torrent as they wound their way up and over the apparatus before descending from the ceiling to make their connections with the rack of amplifiers, counters, discriminators, and analogue-to-digital converters that stood alongside. Like the puzzles in a child's comic book that challenge you to find which cartoon character has hooked the fish, so the students had to take considerable pains to ensure that the outputs were connected to the right inputs through the cable spaghetti. Many of these cables carried enough voltage to be quite deadly, others carried timing pulses calibrated in billionths of a second.

Controlling AP2 and receiving and storing the digital data was the task of a personal microcomputer which sat on a wooden bench next to the bank of electronic equipment. In front of the computer was a wooden stool and a small oscilloscope mounted on a pedestal and angled upwards so that its screen was clearly visible. Nobody was at work on the machine that day, and the mood in the lab was one of uneasy calm.

Using the ladders to climb up onto AP2, Smalley beckoned Kroto to follow. From above, the machine began to make a little more sense. Off to one side sat the long, squat cream-and-brown Quanta-Ray laser. This was a neodymium–yttrium–aluminium garnet (Nd:YAG) laser, capable of delivering high-energy light pulses in the infra-red and the visible. Further around sat the large, square, sky-blue Lumonics excimer laser, which produced intense pulses of invisible ultra-violet light. Kroto could now see how the light pulses from these lasers could be combined using mirrors and directed into the chamber along the same optical axis.

Smalley clambered over AP2, explaining the function of each bit and describing how he and his group had used it in their recent study of SiC_2. Kroto was fascinated by what he saw and heard.

The basic idea behind Smalley's work was to generate clusters of atoms—unusual molecules ranging from very small (two atoms) to very large (50 atoms or more)—to characterize them, and then to do some spectroscopy on them. In itself this was not particularly novel: a number of groups around the world had set up to do these kinds of experiments. But Smalley wanted to study clusters formed from the atoms of refractory materials—metals such as chromium and vanadium and semiconductors such as silicon, germanium, and gallium arsenide. Such unconventional studies required an unconventional apparatus.

By coincidence, the interest in the spectroscopy of SiC_2 was partly astrophysical in origin. Absorption bands in the blue–green region of the spectrum observed in the light from carbon-rich stars as long ago as 1926 were identified in the mid-1950s as belonging to SiC_2. This identification was achieved by generating SiC_2 in the laboratory and measuring its absorption spectrum. A partial analysis of this spectrum suggested that SiC_2 had a linear structure.

The problem was that this novel molecule had to be generated in a conventional manner. A hot oven was used in which graphite and silicon were co-vaporized. Silicon and carbon atoms in the vapour combined at high temperatures to form a range of small molecules and clusters. Now increasing the temperature means increasing the frequency (and force) of the collisions between the atoms and molecules present in the vapour. The molecules acquire more and more energy, moving higher and higher up the ladder of energy states. The more molecules present in the higher energy states, the more lines appear in the spectrum. When there are just too many lines in the same region of the spectrum, the result is usually an unresolvable blob. Any information contained in the pattern of lines is lost.

The blue–green absorption bands of SiC_2 correspond to a transition in which an electron is excited in the molecule and its pattern of 'electron waves' is changed. Associated with this electronic transition are sets of vibrational and rotational transitions, the patterns of which yield information about the structure of the molecule. The scientist analysing the SiC_2 spectrum in the 1950s had measured part of the pattern of vibrational energy states, and had based his assignment of the structure on this. A measurement of the pattern of rotational energy states would have been much more informative, but this was impossible because at the temperature of the experiment there were simply too many rotational lines in the spectrum to be resolved in his spectrometer.

What was needed was a machine which produced large quantities of exotic molecules like SiC_2, allowed discrimination between the molecule of interest and any other molecules formed in the process, and which allowed the molecule's rotational spectrum to be measured. Studying the rotations required a very low-temperature environment, so that the molecules would all be dragged into the lower rotational states—the bottom rungs on the ladder—thereby giving fewer lines in the spectrum. This was obviously at odds with the need for high temperatures. No technique applied between the mid-1950s and 1983 had been able to overcome this problem.

What was needed was AP2. With this machine, Smalley could use light instead of heat. He used pulses of green light with a wavelength of 532 nanometres from the Nd:YAG laser to blast atoms from the surface of a solid target, in this case a rod of silicon carbide, which was mounted inside the chamber. This laser could deliver pulses of light with energies of the order of 60–70 millijoules (thousandths of a joule) per pulse. Now this may not sound like much (the nutritional information on a Cornflakes packet reveals that 100 grammes of cereal yields 165 *thousand* joules of energy). But the energy from the laser was being delivered in a pulse lasting only five billionths of a second, giving peak powers of the order of ten million watts. This is equivalent to the power of one hundred thousand 100 watt light bulbs. You can do a lot of damage with this kind of power, especially when it is focused to a spot no more than a millimetre in diameter.

And damage was certainly what Smalley was inflicting on the solid targets inside AP2. The laser pulses completely disrupted the surface, flinging atoms into the space above and generating a plasma of ionized atoms and electrons with temperatures easily exceeding 10 000 kelvin, much hotter than the surface of the sun. To ensure more-or-less uniform conditions from one laser pulse to the next, the solid target was rotated so that each pulse struck a different part of the surface, preventing the formation of deep pits.

The laser pulse was timed to strike the target an instant after the opening of a valve which released a slug of helium gas into the chamber. The gas had a pressure typically three times that of normal atmospheric pressure. The space above the target was flooded with helium and the ions and electrons blasted from the target formed a plasma which was swept away towards a 'cluster zone'. This was where the exotic molecules would form. Collisions between helium atoms and the ions produced by the laser pulse helped to recombine the ions and electrons to form neutral atoms once again. These atoms then combined together to form clusters with a range of sizes, anything from a few atoms to 50 or more.

Beyond the cluster zone was a tiny aperture through which the newly formed clusters and helium atoms would squeeze before expanding into another vacuum chamber. Forcing the atoms and molecules through the aperture increased the number of collisions, encouraging the formation of more and larger clusters.

Unlike molecules, helium atoms possess no internal energy states of vibration or rotation, and helium gas therefore has a very low heat capacity (a low capacity to store energy). In a collision between a molecule and a helium atom, energy may be transferred only into the atom's translational 'degrees of freedom', which means that the atom leaves the encounter with a greater velocity. Having passed through the aperture into a vacuum, the compressed gas comprising the clusters and helium atoms expanded rapidly. Collisions between the clusters and helium atoms transferred energy out of vibration and rotation of the clusters and into translational motion of the helium atoms. As the atoms accelerated to supersonic velocities, the clusters cooled dramatically.

Conventional notions of temperature are of no use here. The transfer of energy out of cluster rotation is more efficient than for cluster vibration (the clusters are dragged down to the bottom of the ladder of rotational energy states more efficiently than for the vibrational states). This means that the rotations are 'cooled' more than the vibrations, with typical 'temperatures' of the order of a few kelvin for rotation and 100 kelvin for vibration.

The relatively peaceful existence of the atoms in the surface layers of the solid target had been rudely interrupted, to put it mildly. They had been burned in the fire of a ten-thousand-degree plasma, compressed in a pulse of helium gas and forced to form uneasy coalitions in unusual cluster molecules, and then they were cooled to temperatures approaching absolute zero.

And it was not over yet. The newly formed clusters were subjected to a pulse of ultra-violet light from the excimer laser, timed to fire a few millionths of a second after the pulse from the Nd:YAG laser. The excimer laser pulses contained photons with sufficiently high energies to knock electrons out of the clusters, leaving them with a positive charge.

The light pulses from the excimer laser ionized the clusters just as they were passing into an electrostatic field between two charged metal grids. The resulting charged clusters were accelerated by the field down a 1.5-metre tube, much like electrons are accelerated down a television tube. Electrostatic energy was converted into kinetic energy of the accelerated ions, with the smaller, less massive clusters moving faster than the larger, more massive ones. The clusters, consisting of a wide range of masses which until now had all moved through the apparatus together in the same pulse of gas, began to spread out as they travelled down the tube. The clusters arrived at the end of the tube in order of size; those containing few atoms arriving first with those containing many atoms following. At the end of the tube was an electron multiplier, a device for counting the voltage signals produced from the impact of the charged clusters. Because detection is based on discrimination of the clusters by mass, which in turn depends on the time taken for the clusters to travel down the length of the tube, this kind of detection apparatus is called a time-of-flight mass spectrometer.

Each pulse of helium gas and each shot of the Nd:YAG and excimer lasers would produce a collection of cluster ions of different sizes which would be discriminated by mass and detected by the time-of-flight mass spectrometer. The solid target would be rotated to expose a fresh part of the surface, and the whole process would be repeated, with the new set of signals from the spectrometer being added to the previous set. This process would be repeated about ten times per second, and data from typically 1000 repeats would be accumulated to build up an adequate spectrum of cluster sizes. Here was an apparatus that could be used to generate exotic new molecules, cool them to very low temperatures, and then tell you the number of atoms in each new molecule you had made.

The scientists' cleverness did not end here. Once the different clusters had been characterized by mass spectrometry, further manipulations were necessary to do spectroscopy on one (or more) of them. With a silicon carbide target, the cluster beam apparatus produced large quantities of SiC_2 molecules which were detected as SiC_2^+ ions in the time-of-flight mass spectrometer. The trick now was to use AP2 to do some spectroscopy on SiC_2.

In fact, this was a relatively straightforward task. The SiC_2 molecules could only be satisfactorily detected if they were first ionized by the excimer laser. It turned out that this was possible only with very high laser pulse intensities, or

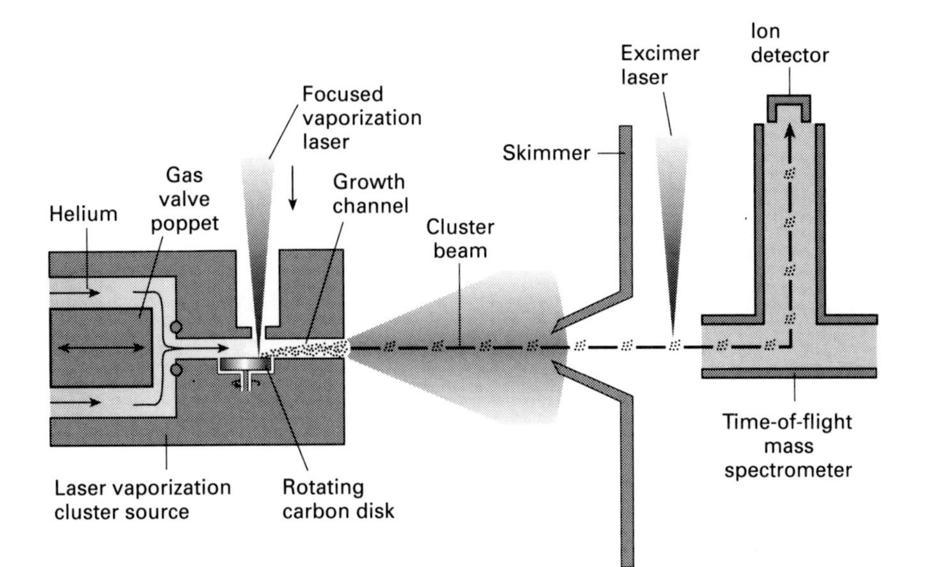

A schematic diagram of AP2. Pulses from the Nd:YAG laser are focused onto a rotating target. The laser is timed to fire just as a valve is opened, flooding the chamber with high-pressure helium gas. The plasma created by the laser is carried 'downstream', where the ions and electrons recombine and the atoms begin to form clusters before being expanded through a small nozzle to supersonic velocities. Further clustering occurs as the gas expands and cools. The middle of the expanding cone of gas is skimmed off and ionized using light from the excimer laser. The ions are deflected and accelerated down a 1.5-metre tube, where they spread out in order of mass with the lightest clusters reaching the ion detector first. By calibrating the 'time-of-flight' of the clusters (typically tens of microseconds), it is possible to deduce their relative masses and hence the cluster sizes (number of atoms per cluster). Adapted from 'Fullerenes' by Robert F. Curl and Richard E. Smalley.

with lower pulse intensities if the SiC_2 molecules were already electronically excited. The detection was therefore sensitive to the excitation of the SiC_2 molecules in the pulse of helium gas as it passed through the apparatus. In experiments carried out just a few months before, Smalley and his colleagues D. L. Michalopoulos, M. E. Geusic, and Patrick Langridge-Smith had used yet another laser to excite the SiC_2 molecules.

This was a dye laser. It differed from the other two in that its wavelength was continuously variable (or 'tunable') over a small region of the visible spectrum. The SiC_2 molecules could only absorb light from this laser when its wavelength or frequency matched the gap between two rungs on the molecule's ladder of energies. As the wavelength of the dye laser was varied, the SiC_2 molecules would absorb at some wavelengths and not others. Only SiC_2 molecules that had absorbed light from the dye laser would then be further excited and ionized

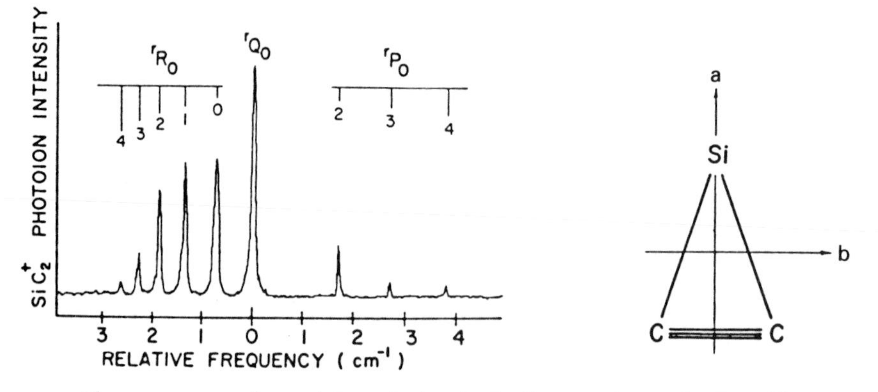

The resonance-enhanced two-photon ionization spectrum of ultracold SiC_2 molecules. By cooling the molecules in a supersonic expansion, their rotational energies are dragged down to the lowest rungs on the ladder, giving a spectrum with very few lines. Analysing the spectrum allowed Smalley and his colleagues to estimate a rotational temperature for the SiC_2 molecules of just five kelvin. The intense line in the centre appears at a wavelength of 498 nanometres.

Detailed analysis of the spectra of SiC_2 enabled Smalley and his colleagues to deduce that this molecule must have a triangular shape. The axes a and b are inertial axes about which the molecule rotates. A third, c, axis lies perpendicular to the plane of the figure. Adapted, with permission, from Michalopoulos, D. L., Geusic, M. E., Langridge-Smith, P. R. R., and Smalley, R. E. (1984). *Journal of Chemical Physics*, **80**, 3556.

by the excimer laser. This is the same as saying that only molecules that absorbed light from the dye laser would be ionized.

By monitoring the signal in the time-of-flight mass spectrometer corresponding to SiC_2^+ and varying the wavelength of the dye laser, the scientists determined the absorption spectrum of neutral SiC_2 molecules in the cluster beam apparatus. This was a spectrum of ultra-cold molecules, most of them occupying the lowest rungs on the ladder of vibrational and rotational energy states producing a few easily resolved lines. The whole process relied on the absorption of two photons (one to produce excited SiC_2, the other to ionize it) and an enhancement of the probability of ionization when the dye laser was tuned to, or in 'resonance' with, a transition in SiC_2. This is referred to in the scientists' jargon as resonance-enhanced two-photon ionization.

It was a very powerful technique, one that had already produced a string of high-quality research publications for Smalley and his colleagues and one which promised much more still to come. Applying the technique to SiC_2, the group at Rice University had measured the rotational part of the absorption spectrum and concluded that SiC_2 is not linear, as previously supposed, but triangular. This result was of interest to Kroto since it was consistent with some calculations he had carried out with his Sussex colleague John Murrell. Despite this interest, however, Kroto's mind was elsewhere. As Smalley clambered over AP2 and explained its intricacies, Kroto could think of only one thing.

The machine produced atoms in a very high-temperature environment, just like the surface of a red giant star. It crowded them together with helium atoms so that they would form clusters, just as might be supposed to occur at the surface of a red giant star. It then cooled them to very low temperatures, just like the expanding outer atmosphere of a red giant star. Replace the silicon carbide target with graphite, and AP2 would be a small-scale laboratory version of IRC + 10°216. Would long-chain carbon molecules be produced? With a little hydrogen and nitrogen, would it be possible to produce very long chain cyanopolyynes? Such as $HC_{33}N$?

Kroto said nothing to Smalley, but raised the subject with Bob Curl later that evening. Curl was immediately enthusiastic. He was also well aware of Alec Douglas's 1977 proposal that linear carbon chains might be responsible for the diffuse interstellar bands. With a graphite target, he felt that they should be able not only to see if long chains were indeed formed in Smalley's apparatus, but also to do some spectroscopy on them. Using resonance-enhanced two-photon ionization, they should be able to measure the absorption spectra of the carbon chains and see if they matched up with the diffuse bands. Curl agreed that AP2 seemed perfectly suited to the study of Kroto's dream astromolecules.

Smalley himself was not so enthusiastic. He agreed to the idea in principle but was in the middle of a heavy programme of research on clusters formed from the atoms of semiconductor materials. It was not going to be possible to divert from this programme to accommodate a new study of carbon clusters. In any case, Smalley was not convinced that there was anything really new to be learned from doing such a study. Carbon is a ubiquitous element. It pervades interstellar space and is abundant on earth. The chemistry of carbon compounds is fundamental to living organisms and constitutes a whole sub-branch of chemistry. Chemists and materials scientists had been studying carbon for centuries. What more was there to discover? Smalley's programme of research on semiconductor clusters seemed much more likely to produce interesting new results.

Smalley did not rule out the possibility that they could some day take a look at carbon clusters, but he did not commit himself to a schedule. The experiments would have to await a convenient slot in the programme.

Kroto had not been back in Sussex very long when one of his colleagues, Tony Stace, drew his attention to the manuscript of a research paper that was circulating amongst members of the small community of British cluster chemists. Stace was working on very large clusters of inert gas atoms—such as argon and krypton—doing experiments that shared many common features with those of the group at Rice University. He had been sent the manuscript by Pat Langridge-Smith, who, since completing his postdoctoral studies with

Smalley in Houston, had returned to take up a lectureship at the University of Edinburgh. In turn, Stace passed the manuscript to Kroto. It had been prepared by three scientists at Exxon's Corporate Research Science Laboratory at Annandale in New Jersey, and it described the results of some experiments they had done on carbon clusters formed by the laser vaporization of graphite.

Kroto was quite irritated. These were the very experiments he had tried to persuade Smalley to do with his cluster beam apparatus at Rice. Exxon scientists Eric Rohlfing, Donald Cox, and Andrew Kaldor had used a virtually identical experimental arrangement. They had blasted graphite with a Nd:YAG laser, entrained the carbon atoms in a pulse of high-pressure helium, clustered them together, expanded the clusters in a supersonic beam of gas and then ionized, discriminated, and detected the clusters in a time-of-flight mass spectrometer. That the technique was identical to Smalley's was no real surprise: the Exxon cluster beam aparatus was, in fact, a clone of AP3—a third-generation machine built by Smalley's team in Houston. The Rice scientists had made a copy of this machine and had shipped it to New Jersey in 1982.

Rohlfing, Cox, and Kaldor had produced and detected carbon clusters ranging in size from two atoms to 190 atoms. The distribution of clusters with less than 30 atoms followed a more-or-less expected pattern, similar to the one seen in a carbon arc by von Hintenberger and his colleagues in 1963. In this distribution, carbon clusters containing odd numbers of atoms were more prominent than those containing even numbers of atoms. Clusters with 3, 11, 15, 19, and 23 atoms produced ion signals in the mass spectrometer that stuck out among their neighbours, although generally the signals diminished with inceasing size beyond C_{11}.

This kind of distribution had been seen many times before. It was consistent with theoretical predictions which said that for ionized linear carbon chains, ions with an odd number of atoms are generally more stable than ions with an even number of atoms. The more stable the cluster, the more likely it is to survive the ride in the cluster beam apparatus. The distribution tailed off towards C_{33}, the object of Kroto's desires, but the signals reappeared at C_{38} and grew rapidly in magnitude to a maximum at C_{60}, beyond which they diminished again. No cluster was picked out as particularly special. There certainly seemed to be nothing special about the signal due to C_{60}.

Now the nature of the distribution had changed completely. Instead of an odd–even intensity alternation with the most intense signals coming from the odd clusters, above C_{38} there appeared to be *no signals at all* from clusters with odd numbers of atoms. The odd clusters simply weren't there. This was a completely unexpected result; one that demanded an explanation. The Exxon scientists didn't really have a good explanation, but they speculated that the large clusters might be symptomatic of a new form of carbon. Using the same line of reasoning that Donald Huffman had followed as he puzzled over the camel humps in Heidelberg two years before, Rohlfing, Cox, and Kaldor wondered if they could be seeing carbyne.

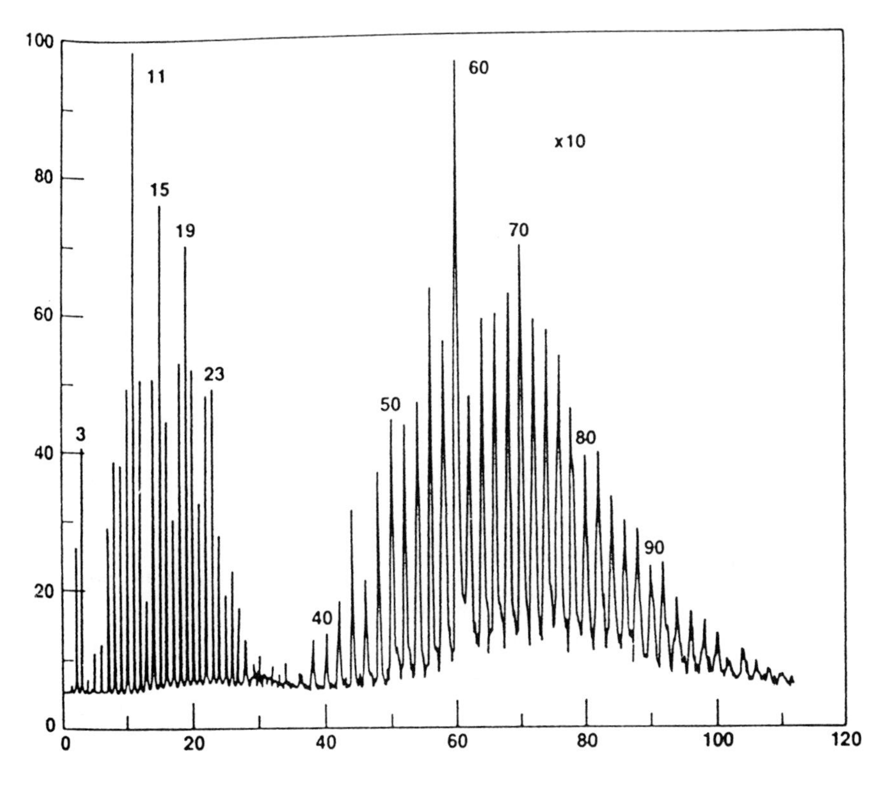

The cluster distribution reported by Rohlfing, Cox, and Kaldor. In this spectrum, the size of the positive ion signal is related to the number of ions detected (note the × 10 magnification for ions larger than C_{30}^{+}). The smaller clusters show a distribution not so dissimilar from that seen by von Hintenberger, Franzen, and Schuy in 1963. In contrast, the larger clusters show a distinctly different distribution, with no signals at all from clusters with an odd number of atoms. Adapted, with permission, from Rohlfing, E. A., Cox, D. M., and Kaldor, A. (1984). *Journal of Chemical Physics*. **81**, 3322.

Whittaker had suggested that above 2600 kelvin the bonding between the carbon atoms in graphite rearranges to give carbyne. These kinds of temperatures were being exceeded in the Exxon group's laser vaporization experiments and so it seemed logical to propose that the vaporized atoms were combining to give carbon clusters with carbyne structures. This fitted with the observation of only even carbon clusters above C_{38}. The clusters would be formed from a basic subunit containing two atoms, —C≡C—, and so there would always be an even number of atoms. The Exxon scientists considered other possiblities, but it seemed that only carbyne could explain the preference for even numbers.

Of course, the Exxon team had done the basic experiments that Kroto had proposed to Curl and Smalley. But they had not done everything that could be done. The results described in the manuscript were fascinating (although Kroto did not believe the carbyne explanation*), but they did not address the key questions concerning the role of these clusters in interstellar space. Rohlfing, Cox, and Kaldor had vaporized graphite that had been treated with potassium hydroxide and had succeded in generating some novel potassium–carbon clusters. Specifically, it appeared that they had made clusters of the form $K(C{\equiv}C)_nK$, the two ends of a long-chain polyyne molecule occupied by potassium atoms. But they had not tried hydrogen and nitrogen to see if long-chain cyanopolyynes could be produced. They had not used the resonance-enhanced two-photon ionization technique to measure the absorption spectrum of the carbon clusters they had made, and so could not discover if these molecules were responsible for the diffuse interstellar bands. There was still every reason to repeat and extend these experiments if ever the opportunity arose. Curl wrote to Kroto in May 1984 to say that they were still looking for a slot on AP2 to try the carbon cluster experiments.

But Smalley again saw things a little differently. The Exxon paper was published in the *Journal of Chemical Physics* in October 1984, and its appearance only confirmed Smalley's original view that he should not get involved in this kind of work. For sure, the paper described some unexpected and interesting new results, but Smalley's commitments lay elsewhere and he saw no reason to enter into needless competition—to tread on the toes of the Exxon group.

Andrew Kaldor presented the Exxon group's results in a poster session at the third International Symposium on Small Particles and Inorganic Clusters (ISSPIC), held at the Free University in Berlin in the summer of 1984. The poster was an abbreviated description of the work, giving the background, apparatus, experimental results, and conclusions in large print pinned on a board and displayed at eye level. Time was set aside during the conference for participants to view the many posters displayed. Such sessions had become the hallmark of national and international conferences, and reflected the simple fact that there was not enough time to allow all the new science worthy of discussion to be given in formal, verbal presentations by the participants.

The Exxon poster attracted a great deal of attention during the first session on Monday, July 9 and Kaldor, wearing a broad grin, was kept busy answering the many questions put by the group of keenly interested scientists that had gathered around it. This group included two physicists—Donald Huffman and Wolfgang Krätschmer. Having had some success in Heidelberg with their matrix-isolation experiments, the physicists had resolved to study the small carbon molecules C_3–C_9 in greater detail. They too were fascinated by the Exxon spectra, which showed carbon clusters up to C_{190}. They realized that their own work on carbon clusters was hardly begun.

*Kroto did not believe in carbyne at all.

Huffman and Krätschmer had delivered a short paper on their matrix-isolation expriments to the symposium, but a similar report on their carbon evaporation experiments had been rejected by the organizers as being beyond the scope of the conference. They had chosen not to mention the curious camel hump phenomenon. Not only did they not wish to discuss something they couldn't properly explain, they had in any case already reconciled their observations as the result of some kind of junk. Despite the fact that Kaldor's poster provocatively suggested that the carbon cluster beam might represent a unique state of matter, Huffman made no connection between these results and his own speculations that carbyne might be responsible for the camel humps. Nobody remarked on the slight prominence of the C_{60} and C_{70} signals in the Exxon spectra.

There followed a series of letters and telephone calls between Houston and Sussex, but it was over a year later, in August 1985, that Kroto received word from Curl that a slot on AP2 had finally come vacant. In fact, the students had already done some preliminary experiments with graphite. Should Curl just send Kroto the results or did he want to come to Houston to participate in the experiments? Kroto did not need to think before answering. He booked himself on the next available flight to Houston.

Back at Rice, Smalley informed his students Jim Heath and Sean O'Brien. He was still sceptical about the whole thing and saw it as a slightly irritating disruption of his smoothly running research programme. He was determined that they should spend no more than a few weeks on this silly game with graphite.

4

The Lone Ranger

Graphite was put into AP2 for the first time on Friday August 23, 1985. Jim Heath and Sean O'Brien were joined in the lab by Yuan Liu, one of Frank Tittel's graduate students, and Qing-Ling Zhang, who was a Curl student. Together Liu and Zhang repeated the Exxon cluster experiments that had been done some 18 months earlier. At this stage they were not particularly concerned about the results they were getting, but just wanted to get a 'feel' for the carbon system. The first few time-of-flight mass spectra they obtained did appear to be consistent with the spectrum reported by Rohlfing, Cox, and Kaldor in their paper in the *Journal of Chemical Physics*. AP2 had no difficulty producing the unusual 'bimodal' distribution, with the odd-atom clusters prominent for small cluster sizes switching to a distribution containing only even clusters above about C_{38}.

However, one spectrum, obtained shortly after midday, later proved to be somewhat different. The students had set the time-of-flight mass spectrometer to show C_{64} full-scale. With this setting, the spectrum they measured showed the signal corresponding to C_{60} off-scale, but nobody knew by how much. When this spectrum was later replotted, the C_{60} signal was revealed to be some 20 times the size of its nearest neighbour, C_{62}, and significantly larger than anything the Exxon group had reported. The signal due to C_{70} was also prominent.

Liu made notes in the AP2 'log book'. This was a ruled, pale brown hard-cover notebook decorated with the academic seal of Rice University in which the students entered a running commentary on their experiments. With the C_{60} signal off-scale, there seemed to be no difference between the spectrum they had measured and that reported previously by the Exxon group. Liu therefore wrote that they had observed the 'same distribution as E.A. Rohlfing.' What had really been captured in the data was something quite extraordinary: AP2 was already revealing some of carbon's longest-held secrets.

They also looked to see if resonance-enhanced two-photon ionization was possible with the smaller carbon clusters. While they saw no enhancement of the ion signals for C_2–C_6, they did observe some enhancement for clusters with more than ten atoms as they tuned the wavelength of the dye laser. Heath continued these experiments the next day and confirmed that resonance-enhanced two-photon ionization would work with carbon clusters in the range C_{14}–C_{25}, giving them a chance to put Alec Douglas's proposal to the test.

The stage was set. On August 26 the students knew that Kroto was coming to join them and so they returned to their study of germanium and silicon while they waited for him to arrive.

Kroto arrived at Houston's Intercontinental Airport during the afternoon of Thursday, August 29. He was once again staying with Curl, and early the next morning the scientists walked the short distance from Curl's home on Bolsover and made their way to Smalley's office.

The office doubled as a meeting room. It had white painted walls free of decoration but for a few notices and the odd spectrum. Curtains at one end of the room appeared to be permanently drawn and so the only form of lighting was the intense artificial brightness from the ceiling. The wall at the other end of the room was covered with shelves on which Smalley had accumulated his personal collection of issues of the *Journal of Chemical Physics*, the *Journal of Physical Chemistry*, *Physical Review Letters*, and other scientific periodicals. Like all academic scientists with their own journal subscriptions, Smalley had a problem. So many scientific papers were now being published that the journals rapidly ate up shelf space. Journals that couldn't be fitted in lay piled on the floor awaiting a decision on their fate.

In front of the shelves sat Smalley's desk, which was littered with papers and issues of journals left open at crucial places. In front of the desk was a low, marble-topped coffee table, a worn, hessian-covered, three-seater settee and a collection of padded office chairs in brown and blue. This was Smalley's debating chamber. It was here that he would periodically gather the group together to discuss the research, draw conclusions, and evolve plans for the future.

Smalley was a tough research adviser. During these often lengthy group meetings, he would buttonhole his colleagues and his students and force them to defend some conclusion they had reached or decision they had made, chipping away at their reasoning until they could defend themselves no longer. He would force his exposed victims to admit that their rock-solid explanations were, perhaps, not so solid after all, from which it would usually be clear what kinds of further experiments were necessary to nail the truth. He would close a bout of aggressive intellectual fencing by declaring: 'That's the experiment we need to do!'

The students were frequently irritated by Smalley's apparent lack of concern for what he saw as irrelevant experimental detail (although he knew the detail well enough), preferring instead to concentrate on the bigger picture. He was sure of what could be achieved using the machines he had created; it was up to the students to make the experiments work. He certainly made their lives difficult, but they were under no illusions—top-flight academic research in this area of chemical physics was no picnic. Nevertheless, Smalley's aggressive attitude made some students cautious and defensive. They were often unwilling to share their hard-won data until they were sure they had first understood it.

Into this arena stepped Harry Kroto. With Smalley, Curl, Heath, and O'Brien gathered around the coffee table, Kroto gave an informal seminar on everything he knew about carbon, interstellar molecules, the cyanopolyynes, the diffuse interstellar bands, stars, and soot. He told them of his dreams about $HC_{33}N$. The seminar lasted more than two hours. Kroto and Smalley then closeted themselves in Smalley's office and drew up plans for the experimental programme.

After a long discussion, they lunched together at the *Goode Company*, a Mexican restaurant on Kirby Drive, a short car ride from the Space Physics building. In America, and especially in Houston, nobody went anywhere without taking the car. Perhaps Houston was no longer the 'unbelievably torrid effluvial sump' of Tom Wolfe's *The right stuff*, but at midday in August the weather was typically very hot and humid, and it simply did not pay to stray for too long beyond the protection of an air-conditioned environment.

Kroto spent that Friday afternoon talking to the students in the lab, getting to know them and trying to establish a rapport. It was one week after Liu and Zhang had carried out those first preliminary experiments with graphite. The spectrum showing that there might be something rather interesting to be discovered about C_{60} and C_{70} remained tantalizingly hidden from view, stored as a data file on floppy disk.

The carbon cluster experiments couldn't be immediately restarted, so on Saturday Kroto spent some time looking around the City and searching for bargain books. He was not disappointed. There were many bookshops, large and small, some specializing in rare or antique books. Finally, on Sunday, September 1, 1985, Kroto at last joined Heath to begin their programme of experiments on carbon clusters.

Both Heath and O'Brien were expert at handling AP2 and making it work. That first day, and on subsequent days, Kroto sat down in the lab and listened carefully as Heath or O'Brien explained the mechanics of the apparatus in great detail, asking detailed questions in return. Kroto wanted to understand exactly what happened in the cluster source so that he could relate this to the experiments he had in mind.

For Kroto, working with AP2 proved to be an exhilarating experience. When the machine was operating, the Nd:YAG laser would fire at a rate of about ten times per second, producing a high-pitched whine and brilliant flashes of green light, some of which was scattered to form ghostly images on the walls of the laboratory. To this was added the rapid, monotonous click-click-click of the excimer laser, and the noise from the pumps and the cooling fans built into the data-gathering electronics.

Kroto was happy to leave the practicalities of the experiments to the students, and positioned himself in front of the computer monitor which displayed the time-of-flight mass spectra as they accumulated, just as he had watched the oscilloscope screen with Lorne Avery and Takeshi Oka at Algonquin Park in 1977. From this vantage point, he would make suggestions for the next experiment based on the results of experiments just completed. The students had been more used to cranking the handle on AP2 and thinking about what the results meant at the end of the session. The interactive approach of a seasoned experimental scientist, forcing them all to be creative and hypothesize about what the results might mean as they went along, was both novel and highly beneficial.

Heath and Kroto carried out their first experiments together during the early part of Sunday afternoon, using helium as the 'carrier' gas. Initially, they saw nothing that had not already been seen in the Exxon experiments. Heath was intrigued by the different times of arrival of the small and the large clusters at the ion detector, implying that they were perhaps not being cooled and thermalized as effectively as they could be. He decided to add a physical extension to the nozzle known as an 'integrating cup', which had the effect of prolonging the time available for collisions and so promoting the complete thermalization of the clusters.

The spectrum they obtained with the integrating cup in place showed that the C_{60} signal was slightly affected by the change, increasing to about eight times the size of the signal for C_{62}. But AP2 wasn't working well and all the signals were very weak.

The next morning, the scientists discussed at length the results of the previous day's experiments. The dominant topic was the observation that only even clusters were formed for cluster sizes bigger than C_{38}, just as the Exxon group had reported in their paper published a year before, and the possible explanations. Kroto gave his reasons why he did not believe the carbyne explanation. As they debated different alternatives, they tried to picture in their minds what was going on inside AP2. They conjured up images of atoms and random bits of graphitic chicken-wire fencing blown off the surface of the target by the vaporization laser. These atoms and bits of fencing swirled, collided, and stuck together as the tide of helium atoms washed over them and pushed them into the cluster zone, to produce . . . what?

Smalley had by this time done a lot of thinking about the structures of the kinds of tiny bits of solid matter that AP2 could create. One of the basic aims of the collaborative work with Curl and Tittel was to learn more about the properties of semiconductor materials by studying the behaviour of microscopic fragments. During this work, the problem of tying up 'dangling bonds' had been a central theme.

Like all microscopic fragments, a piece of graphitic chicken wire fencing has edges. Carbon atoms on the edges cannot tie up all their electrons in bonds to other carbon atoms, simply because at the edges there are no other atoms with which to bond. These 'untethered' electrons represent reactive sites, and are referred to as 'dangling bonds'.

A plane of carbon atoms in a graphite-like structure of hexagons has dangling bonds at its edges, making it more reactive and so less stable. In bulk graphite, these dangling bonds are stabilized by picking up hydrogen atoms. Similarly, the tetrahedral arrangement of atoms in diamond has dangling bonds which also pick up hydrogen. But unless it was deliberately added, there was no hydrogen or any other reactive gas in the cluster zone with which the growing carbon clusters could tie up any dangling bonds. These bonds would act as sites for chemical attack, and experience had shown that, if they existed, the most stable structures were the ones that didn't have any dangling bonds.

The only obvious way to tie up these bonds completely was to form some kind of closed structure. A closed ring of carbon atoms was just such a possibility. Could the polyyne chains bend around on themselves so that the dangling bond at one end tied up the dangling bond at the other? Just as Rohlfing, Cox, and Kaldor had argued, the basic —C≡C— unit would at least explain why only clusters containing even numbers of carbon atoms were found in the second distribution.

The experimental programme recommenced in the late afternoon. Instead of helium, the researchers now switched to pure hydrogen as the carrier gas and looked for evidence for the formation of hydrogen-containing clusters in the mass spectra. Any gas could be used to provide the right kinds of clustering conditions, but their aim in using a reactive gas was to see if they could get hydrogen atoms to stick to the ends of the long carbon chains, much as the Exxon scientists had done with potassium atoms. They found that the signals from the large, even clusters grew even more pronounced. But they also found long chains which they ascribed to molecules of general structure $H(C{\equiv}C)_nH$, with n between 6–20. This was confirmation that some long-chain polyynes could be formed under the kinds of conditions thought to prevail in the outer atmospheres of carbon-rich red giant stars.

There were many other things to be checked. They had to be sure that what they were seeing was indeed a reflection of the chemistry that was going on inside the apparatus. With a large and complex machine, producing output that seemed far removed from the behaviour of the molecules themselves, it was all too easy for them to be fooled into seeing what they wanted to see. They checked the effects of changing the voltage on the deflector plate which propelled the ions down the flight tube in the time-of-flight mass spectrometer. They checked the dependence of the ionization signals on the power of the Nd:YAG and excimer laser pulses. They checked the effects of changing the wavelength of the excimer laser. They checked the effects of changing the timing of the firing of the two lasers. They checked the effects of changing the timing of the firing of the Nd:YAG laser in relation to the opening of the valve which released the slug of high-pressure helium or hydrogen gas into the vaporization chamber.

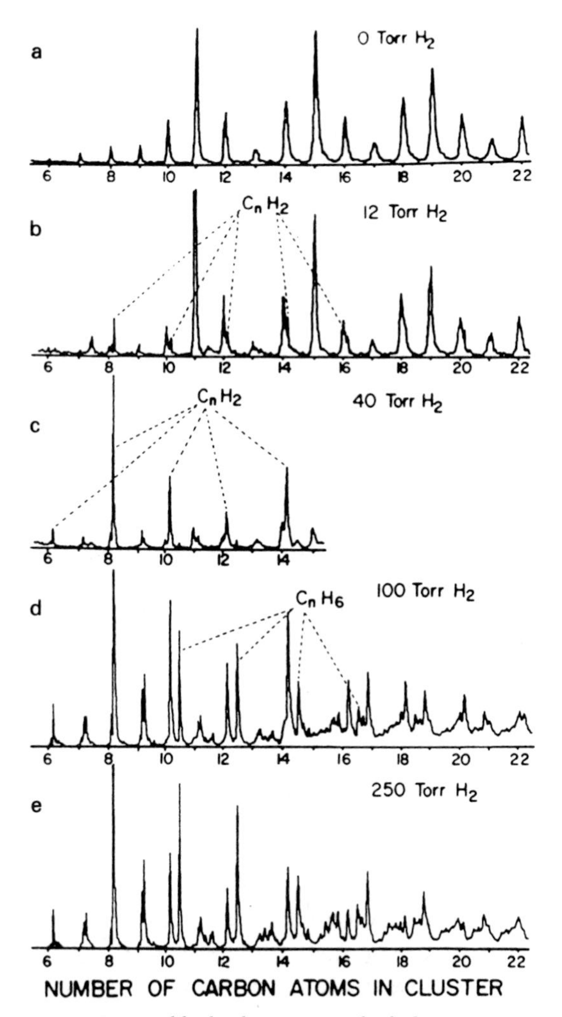

Introducing a reactive gas like hydrogen into the helium carrier gas inside AP2 produced distinct $C_{2n}H_2$ clusters, presumed to have linear chain structures of the type H—(C≡C)$_n$—H. In (a), no hydrogen has been added and only bare carbon clusters are seen, as in this spectrum showing clusters in the range C_6–C_{22}. In (b) and (c), increasing the proportion of hydrogen in the carrier gas increases the intensity of 'satellite' signals corresponding to $C_{2n}H_2$ clusters. In (d) and (e), still higher proportions of hydrogen produce clusters $C_{2n}H_2$, $C_{2n}H_6$, and $C_{2n}H_{10}$. Adapted, with permission, from Heath, J. R., Zhang, Q., O'Brien, S. C., Curl, R. F., Kroto, H. W., and Smalley, R. E. (1987). *Journal of the American Chemical Society*, **109**, 359. Copyright (1987) American Chemical Society.

The mass spectra produced in these experiments showed that the even clusters tended to survive intact, suggesting that they were much less reactive towards hydrogen than the smaller clusters. This was consistent with the idea that these large clusters had structures in which the dangling bonds were somehow tied up. Although there was nothing particularly special about the signals due to C_{60}, Kroto and Heath had noted its odd behaviour of the day before. Heath decided to enter a comment on it in the AP2 log book.

While all this was going on, Smalley and Curl were occupied largely by other matters. Both had many other responsibilities, as well as other research programmes and students to look after. Kroto often returned with Curl to his home for lunch, and at regular intervals both Smalley and Curl would look in on all the activity in the lab to see how things were going. Most of the discussion within the group took place in Smalley's office. At convenient points, Kroto would collect together the computer print-outs from each experimental run, make several copies, and bind them together into a set of notes. These would then form the focus for the group discussions.

Smalley was still less than enthusiastic about the work, although he had become rather intrigued by the behaviour of the large clusters. Curl was a bit disappointed that Kroto seemed more intent on trying to make the polyynes and cyanopolyynes in the cluster apparatus. Curl's hopes for these experiments lay in the measurement of the absorption spectra of the pure carbon clusters using resonance-enhanced two-photon ionization, and the possibility that these might correspond to the diffuse interstellar bands. For Kroto, the spectroscopy was secondary. He didn't think that the two-photon ionization experiment was going to be easy, and wanted to tackle the easier experiments first. In any case, he felt there were flaws in Douglas's 1977 proposal which made it less attractive.

As the work progressed, Kroto got to know Smalley's students. Together they lived and breathed the experiments, sometimes working late into the night. It turned out that Heath was an avid reader and collector of books, with an impressive collection of first editions of the works of Rudyard Kipling. When the experiments allowed, Heath and Kroto would sneak away from the lab to browse around the bookstores in the 'Village', a small shopping area near to the Rice campus bounded by University Boulevard, Kirby Drive, Morningside and Tangley.

Heath was also an outstanding musician, playing guitar, piano, and other instruments. In his student days, Kroto had developed a proficiency with the guitar that had allowed him to play occasionally at various folk clubs. But Kroto's major non-scientific preoccupation had always been graphic art. In fact, had he been able to obtain some decent career advice about graphics and design or architecture at his school in Lancashire, the young Kroto would probably not have become a scientist.

Kroto also developed a strong friendship with O'Brien, based around another shared passion: religious debating. O'Brien—a devout Catholic—and

Kroto—a devout atheist—would often spend hours locked in intense theological discussion as enjoyable for both participants as their scientific studies. At the end of their long sessions in the lab, which would sometimes not conclude until two or three in the morning, Kroto, Heath, and O'Brien would make their way to the *House of Pies*, a 24-hour pie and coffee shop on Kirby Drive. Over endless cups of coffee, they talked about art, books, music, science, and religion.

This kind of exhausting schedule was not particularly unusual for the three scientists. Getting a sophisticated laser experiment to work successfully demands total commitment to the complex (and often temperamental) equipment on which it is based. Although the lasers were supplied by commercial companies they still required nursing from time to time. The pumps, valves, data collection electronics, and computer were all subject to occasional faults which would cause the experiment to grind to a halt. When the experiment was up and running properly—the lasers delivering pulses with the right characteristics and firing when they should, the valves opening, pumps creating a good vacuum, electronics and computer functioning—you didn't stop just because you hadn't had dinner yet.

Yuan Liu spent all of Tuesday fixing some bugs in the computer software, and the experiments resumed just after five o'clock on Wednesday afternoon. Now the researchers tried nitrogen as the carrier gas to see if they could form carbon chains with nitrogen atoms on the ends. These experiments were unproductive, but when they switched back to helium shortly after six o'clock, they obtained a very surprising result. The mass spectrum was now dominated by signals corresponding to C_{60} and C_{70}. The Exxon spectrum had shown the C_{60} and C_{70} signals just creeping head and shoulders above their neighbours. The experiments of the last few days had shown much the same thing, although the C_{60} peak had behaved erratically. Now, Kroto, Heath, O'Brien, Liu, and Zhang were looking at a spectrum with a C_{60} peak at least a factor of 30 larger than C_{62} (the C_{60} signal was again off-scale), with C_{70} also very prominent. In the AP2 log book, Liu wrote: 'C_{60} and C_{70} are very strong!'. They were so surprised by this spectrum that they immediately re-measured it. Of course, AP2 had shown this kind of thing some days ago, but now the scientists were sitting up and taking notice.

Kroto bound this spectrum up with the others they had collected that day, and they discussed the results with Curl and Smalley in the group meeting the following morning. On his own copy of the spectrum Kroto marked in the top left-hand corner the date, the nature of the experiment ('He/C_n repeat') and the observations 'C_{60} huge' and 'C_{70} also'. On the spectrum itself he drew an arrow pointing to the huge peak and wrote 'C_{60}^+?'. What was going on?

This strange behaviour sparked some lively discussion. Smalley was particularly taken with one 'Krotoism'. Kroto had developed the habit of referring to the clusters as 'wadges' of carbon atoms. Smalley picked up the

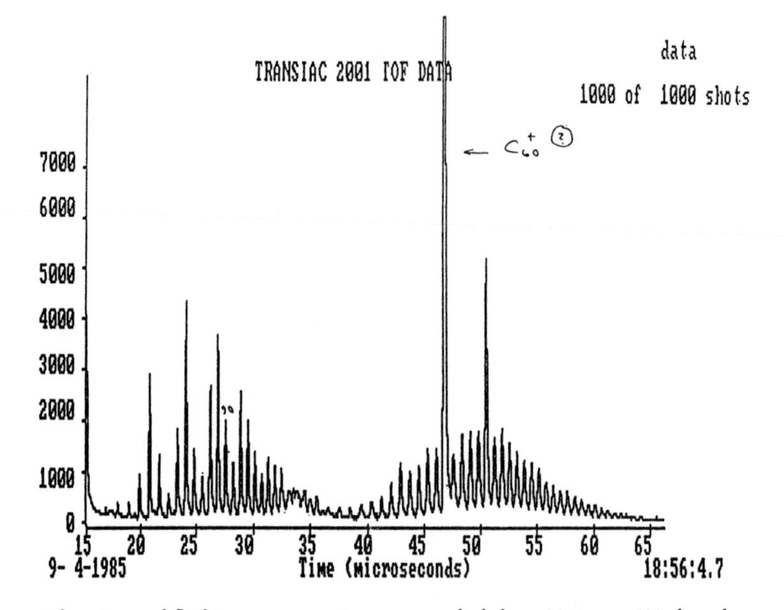

The time-of-flight mass spectrum recorded by AP2 on Wednesday September 4, 1985 and annotated by Kroto. This spectrum was not converted from flight time (measured in microseconds) to cluster size, but the large signal at about 47 microseconds corresponds to C_{60}^{+}. The second largest signal corresponds to C_{70}^{+}. It was this spectrum that had the scientists sitting up and taking notice.

habit and started to refer to C_{60} as the 'mother wadge'. Its omnipresence in the time-of-flight mass spectra led Kroto to call it the 'God wadge'. There were many such exchanges, some drawing inspiration from the characteristically zany humour of *Monty Python's Flying Circus*, whose programmes had been a source of great amusement to Kroto and which were endlessly repeated on the American public television networks. But C_{60} was not entirely alone. The signal jumped about unpredictably, but the much smaller C_{70} signal would usually follow it faithfully. In a rather appropriate homage to the legends of the Lone Star State, Kroto later referred to C_{60} as the Lone Ranger, and C_{70} as his faithful sidekick, Tonto.

Gathered around the coffee table in Smalley's office, the scientists scrambled for an explanation. Clearly, C_{60} was prominent because it was somehow more stable than the other clusters formed in AP2. Again they collectively conjured up images of the mayhem created by the pulse of light from the vaporization laser. How could such randomly produced atoms and fragments of chicken-wire fencing combine to form a very stable cluster with exactly 60 atoms? And why 60? It didn't matter that this had nothing to do with the cyanopolyynes; the scientists had been confronted with one of nature's puzzles, and they were determined to solve it.

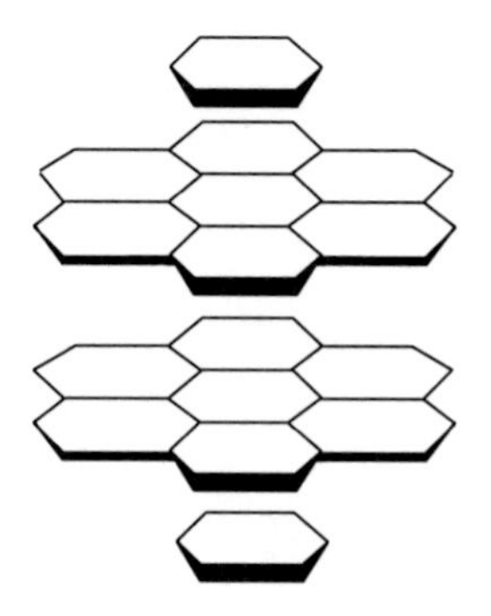

The flat sandwich structure for C_{60} suggested by Kroto. The structure has four layers of graphite chicken wire in a C_6:C_{24}:C_{24}:C_6 pattern. It was not regarded as a particularly strong candidate because it has dangling bonds all the way around the edges of the layers. The scientists were agreed that the most stable structure for C_{60} would be the one with no dangling bonds.

Perhaps the stability of C_{60} itself offered some clues. Once formed, it seemed to be resistant to physical or chemical attack, surviving inside AP2 perhaps at the expense of the other clusters. This implied that it must have a special structure.

Kroto wondered if it might be possible to form a stable 60-atom cluster out of four bits of chicken wire fencing stacked one on top on the other. In the middle of this sandwich structure, there would be two flat planes of seven hexagons, each with 24 carbon atoms. Top and bottom would be two hexagons, giving the structure an overall spherical shape and making 60 atoms in all. There was still the problem of the dangling bonds all around the edges of each layer in the sandwich, but Kroto speculated that these might not be quite so reactive if the layers were reasonably close together.

It was a neat idea, but they were all agreed that the most stable structure would be the one with no dangling bonds, if only they could find it. Of course, a ring was one such possibility but, try as they might, they could not think of any compelling reason why a ring of 60 atoms would be so much more stable than any of its larger or smaller cousins.

By the same reasoning, they wondered if it could be possible to bend a sheet of chicken wire fencing around on itself, so that the dangling bonds along one edge would tie up those of another, and so on all the way around. The result would look like some kind of closed cage—perhaps a sphere—made up of hexagons. Was such a structure possible? This was all somehow reminiscent of the 'geodesic' domes designed by the eccentric inventor and architect Richard Buckminster Fuller, which had enjoyed some popularity in the 1950s and 1960s.

Kroto himself had vivid memories of walking around inside one of these structures at Expo '67 in Montreal. This structure—a nearly three-quarter sphere—had housed the American Pavillion and at the time was the largest of Fuller's domes. Kroto was then a postdoctoral scientist, coming to the end of a

The geodesic dome designed by Buckminster Fuller for the American Pavillion at the Montreal Expo in 1967. The dome is a three-quarter sphere, 200 feet high and 250 feet in diameter. © 1994 Allegra Fuller Snyder. Courtesy, Buckminster Fuller Institute, Santa Barbara.

three-year period of research in Canada and America before returning to England to take up a lectureship at Sussex University. He remembered walking around inside the dome, pushing his small son Stephen in a pram and looking at the exhibits. As far as he could recall, the dome had been constructed from hexagons, just like the hexagons in the graphite chicken wire. Was it possible that C_{60} could be a geodesic sphere?

None of the scientists was sufficiently familiar with the design principles behind Fuller's geodesic domes to be sure how they were put together. However, Kroto did remember that he had once constructed for his children a model geodesic 'stardome', a spherical map of the night sky, which he had purchased as a cardboard kit. He thought this dome also had pentagonal faces, as well as hexagonal ones, and wondered if it had 60 vertexes. His family still had this model somewhere, stored away in an old Xerox carton.

They continued to bounce different ideas around, but without some further information about the nature of the geodesic spheres they couldn't really reach a conclusion. They decided to resume the experiments, but to keep a watchful eye or two on the behaviour of C_{60}, hoping to find some further clues.

Now a special note would go into the AP2 log book whenever the C_{60} signal was observed to be particularly strong. They began with experiments using nitrogen, and this time saw that for the smaller clusters, nitrogen-substituted carbon chains were now being produced. The mass spectra revealed the

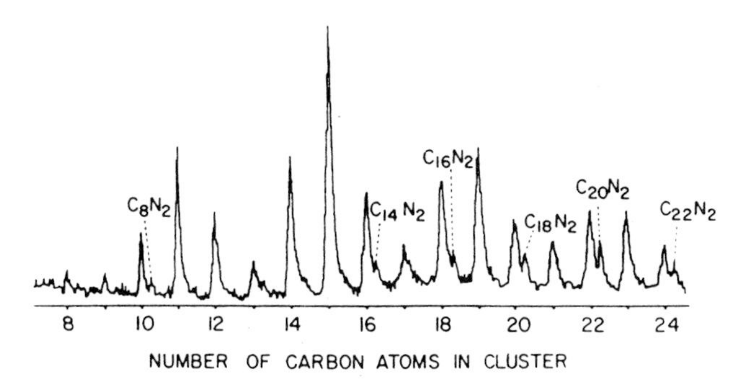

Introducing nitrogen into the helium carrier gas inside AP2 produced clusters of general formula $C_{2n}N_2$, presumed to have the linear chain structures N≡C—(C≡C)$_n$—C≡N. Combined with the results obtained in the experiments with hydrogen, this spectrum provided strong support for the idea that long-chain cyanopolyynes are formed in the outer atmospheres of carbon-rich red giant stars. Adapted, with permission, from Heath, J. R., Zhang, Q., O'Brien, S. C., Curl, R. F., Kroto, H. W., and Smalley, R. E. (1987). *Journal of the American Chemical Society*, **109**, 359. Copyright (1987) American Chemical Society.

presence of clusters such as $C_{20}N_2$, to which they ascribed structures of the general form N≡C—(C≡C)$_n$—C≡N.

At last Kroto had his answer. The results that Smalley's cluster beam apparatus had produced in only four days of experimentation were enough to confirm the possibility that he had first raised almost four years before in his lecture to the Faraday Division of the Royal Society of Chemistry. Under the right conditions, carbon chains would bond with hydrogen and nitrogen to produce the molecules H—(C≡C)$_n$—H and N≡C—(C≡C)$_n$—C≡N, a short step away from the cyanopolyynes H—(C≡C)$_n$—C≡N. The experiments confirmed that a cool carbon star like IRC + 10°216 could be a source of long-chain cyanopolyynes like those that had been detected in interstellar space in the late 1970s and early 1980s. If Kroto could produce these molecules and somehow measure their microwave spectra, he could begin to look for them in space and, perhaps, extend the list of interstellar molecules. Despite the excitement over the odd behaviour of C_{60}, he had not forgotten his dream about $HC_{33}N$.

The experiments with nitrogen were followed on Friday morning by experiments with oxygen. Now the aim was to see if they could find evidence for the formation of clusters of general formula C_nO. They could see no further signals in the time-of-flight mass spectrum that would have been indicative of clusters containing oxygen atoms, except one possibility next to the C_{60} signal which they tentatively assigned as $C_{60}O$.

At the group meeting in Smalley's office on Friday afternoon, they discussed what to do next. The experiments on the reactions of the smaller carbon chains with hydrogen and nitrogen already looked very promising and could probably be wound up with a little more effort. Kroto had got what he had come to Houston for, but they were all agreed that the surprising behaviour of C_{60} seemed to hold additional promise, unconnected with the purpose of the experiments that Kroto had originally proposed. The strange behaviour of C_{60} was something that was, perhaps, worth pursuing. They talked again of sandwich structures, rings, and geodesic domes. Smalley suggested that Kroto find out more about the domes by checking some books about Fuller out of the university library. However, this was impossible: as a temporary visitor, Kroto did not possess a library card.

Kroto was due to return to England the following Tuesday, so Curl suggested that if someone was willing to work over the weekend, it might be worthwhile doing further experiments to find out just how special C_{60} was. Both Heath and O'Brien agreed. Kroto had already decided to journey to Dallas for the weekend—to visit art galleries, bookstores and in particular to seek out half-price books. He left Houston later that evening.

5

Buckminsterfullerene

The challenge was to find out why the C_{60} signal was behaving so erratically. It was almost as though the cluster had a mind of its own, coming and going as the mood took it. Of course, molecules don't have minds; they are forced to obey the laws of chemistry and physics and respond in predictable ways when subjected to the experimenter's mechanical questions. There must have been something erratic about one or more of the experimental variables, such as the laser pulse energies, the age of the graphite disk, or the exact time delay between opening the valve and firing the vaporization laser. These were aspects of the experiments that were less well controlled than the others. If for some reason the formation of C_{60} was very sensitive to one or more of these variables, the corresponding signals in the mass spectrometer would be difficult to reproduce. If the students could find out which variables were responsible and fix them, the signal would be expected to become much more reproducible. They might be able to find ways to make the C_{60} signal even stronger.

O'Brien set to work alone that Friday evening. He first checked to see if the C_{60} signal was in any way sensitive to the age of the graphite target (and the amount of violent abuse it received from the vaporization laser). If something was being formed on the surface of the disk in the pit caused by one experiment, perhaps to be picked off as the disk was rotated around to the same spot in subsequent experiments, then careful monitoring of the signal should have revealed a connection. O'Brien studied the spectra produced following one, two, and three passes over a single disk. Although in these experiments the ratio of the C_{60} and C_{62} signals was consistently large (about 30 or so) it did not seem to depend on the extent of wear and tear of the target.

The next thing to look at was the effect of varying the power of the vaporization laser. O'Brien fixed a new graphite disk into the apparatus and began by looking at the effects on the C_{60} signal of changing the laser power manually. He noted that the energy threshold for the dramatic rise in the signal was slightly lower than that for the other clusters. Under the conditions of his experiments, he found that the signal would begin to increase when the laser pulse energy was increased above about 40 millijoules. Signals for the other clusters would not begin to increase in the same way until the laser pulse energy was increased to 50 millijoules. At 40 millijoules per pulse, he observed the C_{60} signal to be some 30 times larger than that for C_{58}. This was in itself a very striking result. At 100 millijoules per pulse, the C_{60} signal was only twice the size of that for C_{58}.

But there was clearly more to it than this. Despite carefully controlling the Nd:YAG laser pulse energy, the C_{60} signal continued to behave unpredictably. The pulse energy certainly seemed to be having some effect but it was just as clearly not the whole answer. Nevertheless, O'Brien shut down AP2 with a feeling of satisfaction: he had ruled out some of the more obvious of the possible explanations, and had seen the first signs that the size of the C_{60} signal relative to its neighbours could be controlled and adjusted.

Heath took up the challenge the next day. As if to demonstrate his purposefulness of mind, Heath wrote 'Why C_{60}^{+}?' at the top of a fresh page of the log book. O'Brien stopped by early in the morning to explain what he had done the previous evening, then left. Heath repeated the experiments that O'Brien had done and confirmed that varying the laser power did have some effect on the relative size of the C_{60} signal.

Heath then turned his attention to the effects of varying the helium gas pressure above the graphite target as it was blasted to atoms by the vaporization laser. He could control this by changing the pressure of the initial helium charge (the 'backing pressure') prior to opening the valve which released the gas into the vaporization chamber. A quick way of making this change was simply to shut off the helium tank which supplied the gas to the apparatus, and watch what happened to the time-of-flight mass spectrum as the pressure dropped.

Here, at last, was an important clue. The distribution of clusters changed dramatically as the pressure fell. The higher pressures favoured the larger (greater than C_{40}) clusters. This was not entirely unexpected: the greater the number of collisions (and so the higher the gas pressure), the more favourable the clustering conditions.

Heath then varied the helium backing pressure in a more systematic manner, and watched the effect on the ratio of the signals corresponding to C_{60} and C_{62}. These experiments were very revealing. With low backing pressures, the ratio of C_{60} to C_{62} was relatively small. Increasing the pressure considerably increased the size of the C_{60} signal compared to C_{62}. So, the formation of C_{60} depended very sensitively on the clustering conditions, and was favoured over its neighbours by high helium pressures. O'Brien's results from the previous evening could now be understood in terms of the effects of high laser powers on the pressure of gas in the clustering zone. With very high laser powers, the gas absorbed more energy and expanded, decreasing the local density of helium atoms in the region where collisions with the growing carbon clusters were apparently crucial for the formation of C_{60}.

The other, related variable to change was the time delay between the opening of the valve and the firing of the vaporization laser. If C_{60} was indeed being favoured by the presence of a large number of helium atoms with which to collide, then this timing should be important. Heath did the experiments with growing excitement. With a long delay between opening the valve and firing the laser, such that most of the slug of helium had passed over the target before it could be vaporized, the C_{60} signal was virtually identical in size to C_{62}. In fact, the resulting spectrum was very similar to the one that had been published by Rohlfing, Cox, and Kaldor. With a shorter delay, such that the laser fired when the helium pressure passing over the target was at its peak, the size of the C_{60} signal increased to about 30 times that of C_{62}, giving a spectrum similar to the ones that AP2 had produced several times before. This was why the C_{60} signal had been so erratic. It was very sensitive indeed to the clustering conditions inside AP2, which depended on the complex interplay of experimental variables such as the laser power, backing pressure, and timing. Heath reasoned that if C_{60} was favoured by increasing the number of collisions with helium atoms, then it should also be favoured by prolonging the time available for such collisions to occur.

Most of the clustering took place in the cluster zone, where carbon atoms and molecule-sized fragments of graphite, already cooled to room temperature by collisions, would combine before being expanded through the nozzle in a supersonic jet. If this cluster zone were to be made physically longer, then there would be opportunity for further collisions, possibly encouraging more C_{60} clusters to form. There was nothing wrong with Heath's powers of reason. On Sunday, September 8, he fitted an extension to the nozzle and the size of the C_{60} signal was immediately restored to 30 times the size of the signal for C_{62}. He began to realize that they were dealing here with something totally new and unexpected, and very, very important.

Heath was now definitely on a roll. He looked once more at the reactions of the carbon clusters formed in the extended nozzle with oxygen, systematically increasing the proportion of oxygen in the helium carrier gas. He began to see evidence first for $C_{60}O$ and $C_{60}O_2$, and then as he increased the proportion of oxygen he saw evidence for a wide variety of oxygen-containing clusters in the mass spectrum. He also looked again at the reactions with nitrogen and with hydrogen.

Kroto returned to the laboratory that evening, and Heath showed him the results he had obtained. Now Kroto became very excited. Heath had shown him that the C_{60} signals were not due to some experimental artifact of AP2: they were controllable and reproducible. A totally new kind of carbon molecule—perhaps even a new form of carbon altogether—had been unexpectedly created in the cluster beam apparatus.

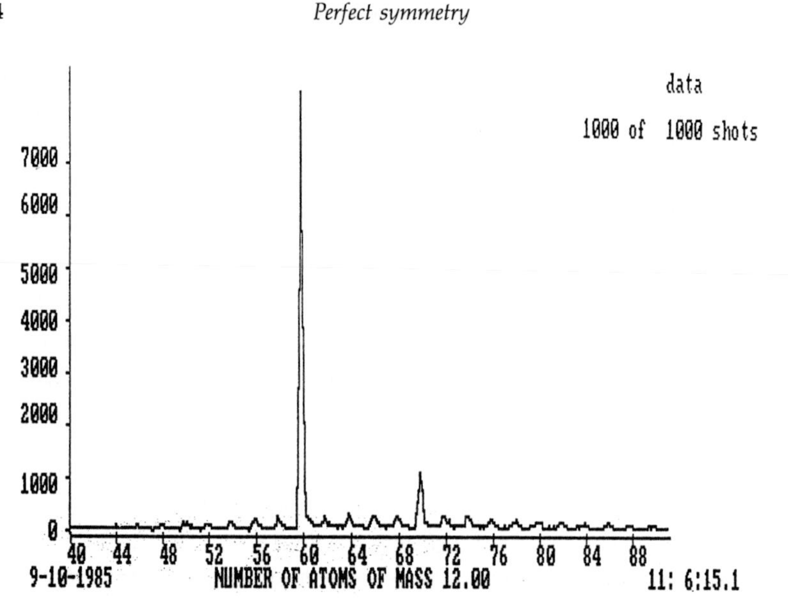

This 'flagpole' spectrum was recorded on Tuesday, September 10. With conditions inside AP2 suitably optimized for cluster formation by adjusting the timing of the vaporization laser and extending the cluster zone, the result was a striking mass spectrum in which the signal due to C_{60} stands virtually alone. Also prominent is the ever-faithful C_{70}. This result confirmed the observation that the scientists had made the previous Wednesday: there must be something remarkably special about C_{60}.

Could the C_{60} signal be made even stronger? Heath fitted the integrating cup back inside AP2, thereby creating the most favourable clustering conditions possible. Early on Monday morning, just before the group was scheduled to meet again in Smalley's office, Heath carried out a quick experiment with the integrating cup in place. The mass spectrum he obtained was truly incredible. The C_{60} signal was now 40 times that of C_{62} and, with the exception of a little C_{70}, stood virtually alone. The other clusters had all but disappeared from the mass spectrum. It didn't matter that this had nothing to do with the cyanopolyynes, or $HC_{33}N$. The Lone Ranger had come into his own.

Smalley called it the 'flagpole' result. The mass spectrum which before had shown signals corresponding to a broad range of carbon clusters had turned into a spectrum with one unmistakable feature: a flagpole without a flag. Gathered together around the coffee table in Smalley's office, the group was electrified by this spectrum. Clearly, there was something unique about C_{60} that made it stand out in this way. Sixty was obviously some kind of 'magic' number. Their task now was to find out why.

The scientists again reviewed what they knew. By increasing the backing pressure, optimizing the timing and extending the cluster zone, the C_{60} signal had been enhanced at the expense of all the other clusters in the group from

C_{40}–C_{120}, with the exception of the ever-faithful C_{70}. All these changes to the experimental arrangement had simply increased the opportunity for collisions to occur between the growing clusters and the helium atoms in the carrier gas, prior to their expansion in the supersonic jet.

The conclusion was fairly obvious. Given the opportunity, the carbon clusters preferred to grow towards C_{60} (with a few forming C_{70}). The other clusters which were seen in the spectrum when the conditions were not optimum were clearly less stable than C_{60} for some reason. With further collisions, these other clusters would either grow further or break up, but every time a C_{60} cluster formed it remained, immutable. There must therefore be something special about the *structure* of C_{60} that made it especially stable, something that made it reluctant or unable to grow larger, and resilient when bombarded with helium atoms.

Now their discussion became very intense. What was the structure of this thing? Was it made of flat graphitic sheets? A ring of carbon atoms? A geodesic dome? O'Brien was beginning to grow weary of this seemingly endless debate about carbon, and was eager to return to the experiments on gallium arsenide.

Kroto, Heath, and O'Brien went back to the laboratory to do some more experiments. This time they looked at the reactions of the smaller carbon clusters with water—H_2O—and 'heavy' water—D_2O—using the integrating cup. The results of these experiments gave them their strongest evidence yet for the formation of polyynes. Reactions between the carbon clusters and H_2O had produced signals in the time-of-flight mass spectrometer that they could ascribe to $H{-}(C{\equiv}C)_n{-}H$, just as they had seen in the experiments with hydrogen. Reactions with a mixture of H_2O and D_2O gave them the anticipated substitution patterns, $H{-}(C{\equiv}C)_n{-}H$, $H{-}(C{\equiv}C)_n{-}D$, and $D{-}(C{\equiv}C)_n{-}D$.

But they couldn't resist an occasional peek at C_{60}. Varying the size and shape of the integrating cup produced mass spectra in which the C_{60} signal was stronger than ever: now more than 40 times larger than its C_{58} neighbour.

Kroto and Smalley discussed the research papers they should write based on the work they had done on carbon clusters over the last week or so. Smalley agreed that they had enough data on the reactions of the carbon chains with hydrogen and nitrogen to demonstrate that long chain cyanopolyynes could in principle be formed in the expanding shell of gas and dust surrounding a carbon-rich red giant star like IRC + 10°216. They would consider these results and perhaps write them up in a paper for submission to the *Astrophysical Journal.*

Smalley also agreed that the C_{60} story—and particularly the flagpole result—was itself deserving of publication. However, as an advisory editor for the journal *Chemical Physics Letters,* Smalley felt that anybody refereeing such a paper would automatically return it, asking the authors to say something about possible structures. He was therefore reluctant to publish anything until they had managed to come up with one or two suggestions. Kroto admitted that this was probably true, but despite their intense discussions they were no closer to a structure.

This Fuller dome, built for the Union Tank Car Company in Baton Rouge, Louisiana, was opened in October 1958. It housed a car rebuilding plant and, with a diameter of 384 feet and a height of 116 feet, was at the time the largest clear-span structure ever built. It appears to be constructed entirely from hexagons. Reproduced from Marks, R. and Fuller, R. B. (1973). *The Dymaxion world of Buckminster Fuller*, Anchor Press/Doubleday, New York. © 1960 Allegra Fuller Snyder. Courtesy Buckminster Fuller Institute, Santa Barbara.

Smalley resolved to go to the university library and try to find out what he could about Fuller's geodesic domes. He withdrew the book published in 1960 by Robert W. Marks, *The Dymaxion world of Buckminster Fuller*, which contained many illustrations and photographs, and included a long section on the domes. The group gathered again in Smalley's office.

Now it is uncanny, but the one photograph of a Fuller dome that caught Smalley's attention was virtually the only one that did *not* blatantly show the all-important design feature that would have helped them to solve the C_{60} puzzle almost immediately. Studying almost any other photograph would have given the scientists the clue they were looking for. Instead, they scrutinized the photograph of a hemispherical dome that had been built in 1958 in Baton Rouge, Louisiana, for the Union Tank Car Company. It seemed to be made entirely of hexagons.

This photograph appeared to confirm the reasonableness of the idea that a closed structure could be formed out of chicken wire fencing. But the Union Tank dome was a large one with many hexagons, and there was still no clue as to why 60 should be such a magic number. The scientists continued their discussion as the day wore on, reaching no firm conclusions.

Kroto was due to make his return flight the next day, and so had invited the team he had worked with so intensely to a farewell meal. Together Kroto, Smalley, Heath, and Heath's wife Carmen made their way to the *Goode Company* on Kirby Drive. Curl and O'Brien had already made other plans for that evening.

There was only one topic of conversation over dinner. What was the structure of C_{60}? Could it be a geodesic sphere? What was so special about the number 60? They went over the same ground again and again, searching for the clue that would unlock the secret of this most mysterious of structures: the clue that would unmask the Lone Ranger. They talked about flat structures and round structures, dangling bonds and Fuller's domes. They drew out their ideas on paper napkins provided by the restaurant, arguing one way and then another. Kroto again mentioned the cardboard stardome that was tucked away somewhere in his Sussex home, and the fact that it had pentagonal faces. But they were getting no nearer to the answer.

After the meal, the group parted company. Kroto headed back to the lab, determined to have one last go at solving the puzzle before his flight home the next day. He searched the lab for the copy of Marks's book that Smalley had checked out of the library, but could not find it.

Back at Curl's home, Kroto again raised the question of the stardome with Curl, and wondered again if he should call his wife, Margaret, to see if she could find it and maybe describe its shape and count its vertexes. However, Curl dissuaded him. It was by now the early hours of Tuesday morning in Sussex, and it seemed pointless for him to disturb his wife's sleep on what Curl viewed as some wild-goose chase. Curl thought it too far-fetched a coincidence for the model stardome to have 60 vertexes. Kroto did not press the point. He was indebted to Curl for his hospitality and his efforts to get the carbon cluster experiments off the ground. And, to an Englishman at least, long-distance telephone calls were expensive.

Heath and his wife went back to the lab to shut down AP2. They too were determined to find the solution, and on their way home they stopped off at a grocery store to buy toothpicks and a large supply of *Gummy Bears*. These sweets are variations on the more traditional *Jelly Babies*, in the shape of cartoon characters popular among small children.

At home they sat down and attempted to put together a sphere constructed out of 60 *Gummy Bears* and toothpicks. The *Gummy Bears* represented the carbon atoms and the toothpicks represented the bonds that connected them together. But it was no good. Starting with the *Gummy Bears* formed into a pattern of hexagons, just as they would appear in a layer of graphite, they tried to form a closed structure, but it would not work. The structure just would not hang together. They tried easing the strain by including some three-membered rings, but that was no good either. They concluded that a closed structure could not be made out of hexagons. Something, somewhere, was missing.

Smalley also headed for home, which was some way out of the City in the Meadows district. Like Kroto and the Heaths, he had also become possessed with the problem and determined to find an answer. He began by sitting at his home computer, trying to draw three-dimensional shapes on the screen. He persisted with this for several hours, muttering curses under his breath when things did not work out. Eventually, he abandoned the high-tech approach in favour of paper, tape, and scissors. He took a pad of legal paper and, with ruler and pencil, drew out a series of identical hexagons, with sides about an inch long. He then cut these out and attempted to stick them together with *Scotch* tape to form a closed structure.

But this was not working either. The result of sticking a few hexagons together along their sides was a flat plane, just like the flat planes of carbon atoms in the structure of graphite. Smalley could force the hexagons to distort more towards a curved surface only by overlapping them at the edges, forcing the structure to contort upwards. This was cheating; there was no way that such overlapping could occur in a carbon cluster, but by now Smalley was getting desperate. With a little more overlapping with each successive hexagon, he managed to produce something that looked like a salad bowl. This, at least, was beginning to look promising. He continued to build up the structure by adding hexagons around the edge, but now he had a problem. To the edge of this paper salad bowl he could add five more hexagons, but there simply was not room for a sixth. This made no sense. He could not leave a gap, but neither could he force fit another hexagon into the space. Even with cheating, it was not possible to make a sphere out of hexagons. In frustration, he screwed up his paper structure and tossed it in the waste bin.

It was now past midnight, but Smalley could not rest. He retrieved a beer from the freezer in the kitchen and, as he sipped it absent-mindedly, he recalled bits of the discussion he had had with his colleagues that day. He remembered Kroto talking about the model stardome he had once built for his children out of a cardboard kit. What was it that Kroto had said about this dome? That was it. Pentagons. Kroto had said that this model had pentagonal faces as well as hexagonal ones.

There are no pentagons in the structure of graphite, but there are plenty of other examples of organic molecules containing pentagonal rings of carbon atoms. Smalley sat down at his desk once more. This time he cut out some pentagons as well as hexagons, taking care to ensure that the sides of both shapes had the same length.

Now he started again, this time with a single pentagon. To each of its five sides he taped a hexagon. This time he did not need to cheat. The paper structure automatically curved upwards to form the shape of a bowl. Smalley quickly cut out some more pentagons and hexagons. He was beginning to think that he might have found something. He had started with one pentagon

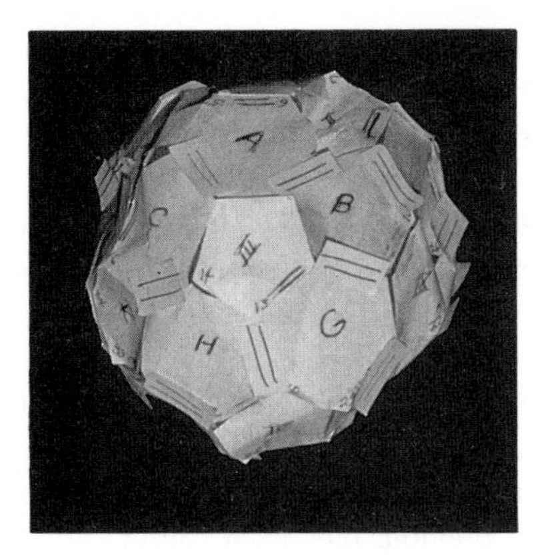

Smalley's paper model of C_{60} (pictured left). By cutting out paper pentagons and hexagons and sticking them together, Smalley discovered that it is possible to produce a perfectly spherical structure with exactly 60 vertexes. Kroto and Curl added the sticky labels, representing double bonds, to check if the carbon bonding requirements could be met for every one of the molecule's 60 atoms.

Kroto's polyhedral stardome (pictured right), which lay tucked away in an old Xerox carton 6000 miles away in his Sussex home, was later found to have the same structure.

surrounded by five hexagons, and found that this pattern could be repeated as the structure slowly took shape.

To the bowl he had added another five pentagons and five hexagons, a pentagon alternating with a hexagon all the way around the edge of what had now become a hemisphere. No cheating and no awkward gap. He counted the vertexes, the corners of the pentagons and hexagons which would correspond to the positions of the carbon atoms in a carbon cluster. He counted 40. But then, he actually had more than a hemisphere, because he was adding complete pentagons and hexagons some of which partly protruded above the line of the 'equator'. He worked out that the true hemisphere contained 30 vertexes, exactly half the magic number 60. His heart started to beat a little faster.

He hurredly added more pentagons and hexagons, sticking to the pattern of surrounding each pentagon with five hexagons. After adding two more rows, the structure was almost complete. The gap remaining was exactly the shape of a pentagon. That was it. Twelve pentagons and 20 hexagons, and he had the shape of a perfect sphere. It was so beautiful it had to be true. This had to be the structure of C_{60}.

As far as Smalley or any of the others were aware, nobody had ever proposed the existence or even the possibility of a perfectly spherical cage of 60 carbon atoms. What he held in his hand was a model of a totally new concept in molecular structure. It had no dangling bonds and was physically quite stable. Despite its flimsy construction, when he dropped his paper structure on the floor, it bounced. Add more atoms or take atoms away and the perfect symmetry would be lost. Sixty was a magic number because that was the only number of atoms that could combine to form a sphere.

During his long car journey to Rice the next morning, Smalley telephoned Curl to tell him what he had discovered. He left a message on Curl's answerphone telling him that the closed solution worked and that Curl should ask the other members of the group to gather in Smalley's office. Entering his office, Smalley tossed the paper ball onto the coffee table around which Kroto, Curl, Heath, and O'Brien were sitting. Kroto was ecstatic: it was a beautiful structure. He also felt vindicated. There was no doubt in his mind that the paper model had the same shape as the stardome he had described the previous day. Heath and O'Brien were also enthralled. It was Heath's birthday.

Curl too was enthusiastic, but he was also a pragmatist. If this really was the structure of C_{60}, then it should be possible to satisfy carbon's bonding requirements for every one of its 60 atoms. It was not immediately obvious that this could be done. Borrowing a packet of sticky labels from Jo Anne Timmins, Smalley's secretary, Curl and Kroto set about finding out. Each carbon atom in the structure would be bonded to three others, requiring that some of the chemical bonds should be single and some double bonds, just as in the structure of graphite. They marked the sticky labels with two parallel lines to signify a double bond, and started to stick these to Smalley's paper model. A few moments of breathless activity, and it was done. It checked out. The single bonds alternated with the double bonds throughout the structure, so that each carbon atom was connected to its neighbours by two single bonds and one double bond. The model had passed its first test. Curl was convinced.

So symmetrical a structure just had to be known to those who studied these things. Smalley telephoned William Veech, the chairman of Rice University's Mathematics Department, and described the structure to him in detail. Veech called back after consulting with his colleagues. Curl took the call. Veech said that there was a number of ways he could explain this to them, '. . . but what you've got there boys, is a soccer ball'.

How could they have been so blind? Now it had been pointed out to them, it was all too perfectly obvious. The structure of C_{60} was not only the most wonderfully symmetrical molecular structure they had ever contemplated, it was also absurdly commonplace. A modern soccer ball is 20 white leather hexagons and 12 black leather pentagons stitched together, with each pentagon surrounded by five hexagons. It has 60 vertexes; 60 points where the corners of the pentagons and the hexagons meet along the seams. How many times had each of them looked at a soccer ball without really registering these simple facts? Heath raced to a sporting goods store in the Village and returned with a real soccer ball, just so they could admire it. O'Brien raced to the university bookstore, bought up its entire stock of molecular modelling kits and began to assemble what was to become the first of many, many models of spherical C_{60}.

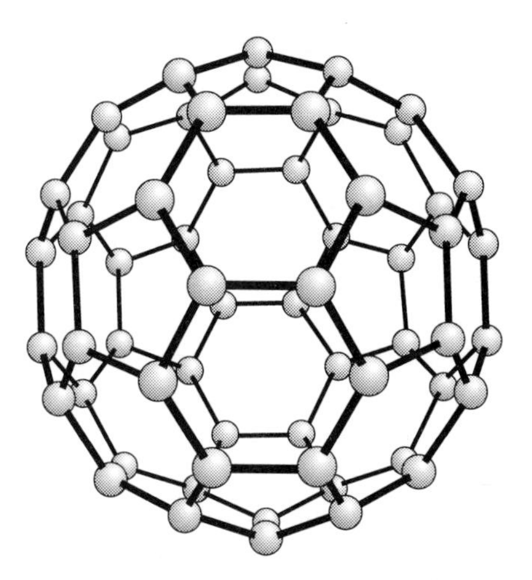

A molecular model of soccer ball C_{60}. Each carbon atom takes up a position at the vertex of a polyhedron with 12 pentagonal faces and 20 hexagonal faces, with each pentagon surrounded by five hexagons. A modern soccer ball has the same pattern of pentagons and hexagons. This perfectly symmetrical, all-carbon structure represented a whole new concept in molecular architecture.

Their discussions now reached fever pitch. The possibilities seemed endless. They had accidentally discovered a completely new kind of carbon molecule; perhaps a new form of carbon, the third after diamond and graphite. They had found it in a cluster-beam apparatus under conditions which were from the beginning believed to mimic those of the outer atmospheres of carbon-rich red giant stars. Did this mean that C_{60} might be abundant in interstellar space? Could C_{60} explain the diffuse interstellar bands? This would truly be a most elegant solution to the last great problem in astronomy.

If C_{60} were present in interstellar dust, it might be expected to provide an important surface for catalysing chemical reactions in space. If the earth condensed from a swirling cloud rich in C_{60}, or if C_{60} came to the primordial earth in a shower of meteorites, perhaps it had acted as an important catalyst in the formation of life-sustaining molecules.

But there was much, much more. What they had on their hands was more than a new molecule—they had a whole new concept in molecular architecture. If C_{60} could be made in large quantities, it could be used as a starting material in a whole new branch of 'round' chemistry. Atoms could be added to the outside of the cage, but it also possessed a unique site for atoms *inside* the cage. Perhaps putting atoms inside the cage would subtly influence the chemical and physical properties of C_{60} in ways that had never before been accessible to the chemist.

It was all too much to contemplate in a hurry. Kroto decided to delay his return to England so that they could discuss the possibilities and draft a short paper to describe what they had done. There was one other thing they had to do: they had to give their structure a name. Suggestions tumbled out. As the new molecule contained double bonds, convention demanded that its name end in 'ene', as in butene or benzene. What would be most appropriate? Ballene? Spherene? Soccerene? Of course, footballene was out of the question: the name would have been understood outside the United States but would have simply confused most Americans.

Kroto thought they should recognize and acknowledge Buckminster Fuller's architectural vision. It was their discussion of Fuller's geodesic domes that had put them on the right track, although it had taken Smalley's moment of inspiration to provide them with the structure. It was clear that C_{60} *was* a geodesic sphere. Kroto suggested they call their new molecule 'buckminsterfullerene'.*

Smalley and Curl were not too thrilled with this choice; it seemed far too long and cumbersome. It also seemed ridiculous. Could they really call it that? However, they agreed with its sentiment and, as they could not think of a better alternative, they eventually accepted Kroto's arguments. Smalley sat at his office computer and started to type the paper into his wordprocessor. He began by typing the title 'C_{60}: Buckminsterfullerene', and checked to see if everybody approved. They did.

They worked on the text of the paper for the rest of the day, completing it the following morning. Heath printed out some representative time-of-flight mass spectra to illustrate what they had found, and added a schematic drawing of the cluster beam apparatus showing the valve, the rotating graphite disk, the beam of the vaporization laser, and the integrating cup. To force home their point about the novelty of the structure, they also included a photograph that Kroto had taken of the soccer ball Heath had purchased the day before, placed on the grass in front of the Space Physics building.

What had begun as a rather esoteric exercise aimed at providing circumstantial evidence for the formation of long-chain cyanopolyynes in the outer atmospheres of carbon-rich red giant stars had turned into the most important discovery of their scientific careers. Kroto, Heath, O'Brien, Curl, and Smalley had all made their individual contributions, but the discovery was undeniably a team effort. Although Liu and Zhang had been involved in some of the experiments and had shared in the general excitement, they had not contributed to the intense debate that had finally brought forth the answer. Their turn would come later.

The five scientists commemorated the event by arranging a team photograph in front of the Space Physics building. Curl stood at the back, clutching the soccer ball. The others knelt in front, both Heath and O'Brien with one hand resting on a model of buckminsterfullerene constructed from

*Kroto recalls this moment vividly, but Smalley believes they were already using the name buckminsterfullerene towards the end of the previous week, before the soccer ball structure had been discovered (see Sources and notes, p. 269).

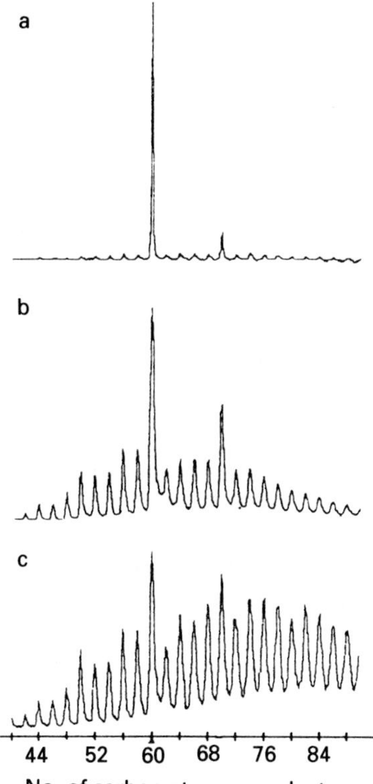

The time-of-flight mass spectra as they appeared in the *Nature* paper, published in November 1985. In (c) the pressure of helium passing over the graphite target during laser vaporization was about 10 torr. A broad range of even-numbered clusters were formed with C_{60} and C_{70} only slightly prominent, similar to the spectrum reported by Rohlfing, Cox, and Kaldor in 1984. In (b) the helium pressure is 760 torr (one atmosphere) and the C_{60} and C_{70} signals are much more prominent. The spectrum in (a) is the flagpole result. Adapted, with permission, from Kroto, H. W., Heath, J. R., O'Brien, S. C., Curl, R. F., and Smalley, R. E. (1985). *Nature*, **318**, 162. Copyright (1985) Macmillan Magazines Limited.

the molecular modelling kit that O'Brien had purchased. It was a proud moment. Together they were the ultimate five-a-side soccer team and they were on top of the world.

For posterity, Kroto also took a photograph of Smalley's paper model. He asked Heath if he could photograph the *Gummy Bears* model which, although it had not been a success, nevertheless represented an important milestone in the team's efforts to come to terms with the structure of C_{60}. Heath had to admit that the model no longer existed. Unlike Smalley's model, the carbon atoms in Heath's had been edible. Since putting it together on Monday evening, it had been slowly consumed by another graduate student.

The ultimate five-a-side soccer team. The 'bucky' pioneers are: Bob Curl (standing) and (kneeling, from left to right) Sean O'Brien, Rick Smalley, Harry Kroto, and Jim Heath. Curl is holding the soccer ball that Heath purchased from the sporting goods store in the Village. O'Brien and Heath have their hands on the first of many, many models of C_{60} constructed from a molecular modelling kit.

The following morning they discussed what to do next. The obvious thing to try was to see if they could somehow make C_{60} with an atom encapsulated inside the cage. Kroto, Smalley, Curl, and O'Brien agreed that they should try iron atoms, as there was a clear analogy with the structure of the molecule ferrocene which consists of two pentagonal hydrocarbon molecules with an iron atom sandwiched in the middle. Iron was also one of the most abundant metal atoms in interstellar space, and Smalley wondered if the diffuse bands might not be due to molecules of C_{60} with iron atoms trapped inside. Heath argued instead for lanthanum, on the basis that it forms a compound LaF_3 in which a lanthanum atom is associated with 12 fluorine atoms in the lattice structure. In essence, the lanthanum atom in this compound sits happily inside a ball of electrons, the kind of environment that Heath believed it would encounter inside the C_{60} cage. They reached a compromise and agreed to try iron first, then lanthanum.

By mid-afternoon the same day, O'Brien had tried experiments with ferric chloride adsorbed on a graphite target, searching the mass spectra for evidence of the formation of an FeC_{60} complex. Heath was in the lab but not really

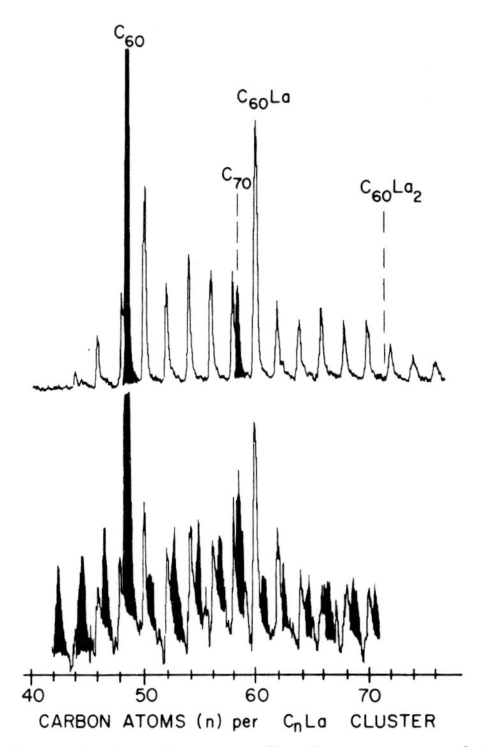

Evidence for the production of a range of lanthanum-encapsulating carbon clusters was obtained at Rice shortly after the discovery of buckminsterfullerene. In these mass spectra, the signals corresponding to the 'bare' carbon clusters are indicated as the blackened peaks. All the other signals correspond to lanthanum-containing clusters. In the upper trace, irradiation with high-energy pulses from the ionization laser has fragmented many of the bare clusters, leaving only C_{60}, C_{70}, and a range of lanthanum-containing clusters. Adapted, with permission, from Heath, J. R., O'Brien, S. C., Zhang, Q., Liu, Y., Curl, R. F., Kroto, H. W., Tittel, F. K., and Smalley, R. E. (1985). *Journal of the American Chemical Society*, **107**, 7779. Copyright (1985) American Chemical Society.

involved (he had a grudge against iron and really wanted to try lanthanum). Despite strenuous efforts, O'Brien drew a blank. He repeated these experiments later that evening, but again without success. The experiments with iron were finally abandoned and Heath got his chance to try lanthanum.

The time had finally come for Kroto to return to England. Smalley drove him to the airport. Kroto was airborne long before the plane took off from Houston on its non-stop flight to London. As he was lifted over the sprawling Texas landscape, he reflected on the events of the last two hectic weeks. If Texas was a state of mind, then it had given them all opportunities beyond their wildest imagination.

By Thursday afternoon, the scientists had packaged up their short paper on buckminsterfullerene and dispatched it by Federal Express to the Washington office of the journal *Nature*. Heath and Liu began a new experimental programme using lanthanum adsorbed onto the graphite target in AP2. By the evening they had seen signals in the mass spectra due not only to LaC_{60}, but also lanthanum complexes with all the large even clusters. They could not yet be certain that the metal atoms resided inside the clusters, but this seemed logical. This meant that all the other even clusters must have geodesic 'fullerene' structures, with C_{60} the most stable because of its perfect symmetry. The experiments with lanthanum had given them the first evidence for a whole new range of organometallic compounds: the first in a new chemistry. Nothing like this had *ever* been seen before. Clearly, the real adventure was just beginning.

The buckminsterfullerene paper was received at the Washington office of *Nature* on Friday. As is usual, the manuscript was passed to two academic referees for comments on its worth as a scientific publication and its suitability for *Nature*. One referee remarked that:

> I think it would be useful if the authors were to refer to some earlier work published in Nature by A. Douglas, 269, 130, 1977 on the carriers of the difused interestellar absorption. Also, while I cannot give the exact reference, the publication by W. Kratschmer, N. Sorg and D.R. Huffman.

PART II

From symmetry to substance

nature

INTERNATIONAL WEEKLY JOURNAL OF SCIENCE

Volume 347 No. 6291 27 September 1990 £2.50

A NEW FORM OF CARBON

UNDERSTANDING ANTARCTIC OZONE DEPLETION

The cellular defect behind cystic fibrosis

6

Form and geometry

It would, perhaps, take a bold commentator to remark that in the science of the late twentieth century, there is little potential for truly original thinking. If we can avoid confusing originality with creativity (no less admirable), then there is a case to be argued in favour of this seemingly rash statement. Analyse any apparently original idea from the recent history of science and you will generally find that it had been expressed before, probably in a different context and possibly in a completely different discipline. We rightly applaud (and honour) the creative act behind a particularly insightful observation, and that applause is no less deserved though the observation may be recycled or even merely restated. Truth itself has its time and place, and may be more readily accepted depending on who is expressing it.

With this kind of territory comes an occupational hazard. Discovering something in science that you believe is completely original, something you believe nobody has ever seen or thought of before, is an incomparable experience. If such moments are the stuff of scientists' dreams, then the nightmare is getting it wrong. In between lies the acute sense of anti-climax that comes with the further discovery that you are not the first. We could state Murphy's Second Law* as follows: Somebody, somewhere, has already had your 'original' idea.

Smalley's 'eureka' experience, as he taped the last pentagon to his paper structure, will no doubt remain with him forever and subsequent events cannot deny its validity on a personal level. But the fact that through that experience he was actually *rediscovering* a possibility with a relatively long history of some twenty years inevitably tarnishes its memory.

Of course, the soccer ball structure of buckminsterfullerene was itself commonplace, but that was not the point. For all that Smalley, Kroto, Curl, Heath, and O'Brien knew, nobody had ever proposed that an all-carbon molecule could be shaped like a soccer ball. In molecular terms, this was a unique

*Murphy's First Law: If it can go wrong, it will.

structure. The scientists enjoyed their moment of glory, before feeling the full force of Murphy's Second Law.

It was Kroto's colleague Martyn Poliakoff at Nottingham University who drew their attention to the fact that their notion of hollow cages of carbon atoms was predated by almost twenty years. David Jones, a rather eccentric British chemist who has hopped back and forth between jobs as an industrial scientist, consultant and academic at the University of Newcastle upon Tyne, has made an illustrious second career out of giving free rein to his extremely fertile imagination, concocting crazy ideas with just enough plausibility to make them amusing and endlessly entertaining. Using his pseudonym Daedalus, he described these crazy ideas for many years in the columns of the popular British science magazine *New Scientist*, and continues to do so today for *Nature*.

In an article published in *New Scientist* in November 1966, Daedalus speculated on the properties of huge graphite 'balloons', formed from ordinary graphite by introducing defects into the flat layers of hexagons so that they would buckle, curl up and close on themselves.* In his book *The inventions of Daedalus*, published in 1982, Jones elaborated on this idea and further suggested that the defects could be pentagons.

Defects were needed because, Jones concluded, without them he could not make a layer of hexagons form into a closed structure. As he noted in his book, this follows from a simple formula derived in 1752 by the great Swiss mathematician Leonhard Euler. This formula states that the number of vertexes (V) of a polyhedron plus the number of its faces (F) is equal to the number of its edges (E) plus two. That there must be some kind of relationship between V, E, and F becomes apparent when we recall that the number of vertexes determines the number of edges which determines the number of faces.

We can make a guess at a structure and feed its values for V, E, and F into Euler's formula. If we find that the two sides of the equation are equal, then it is possible (but not automatically proven) that the structure could be formed into a polyhedron. If, however, the values for V, E and F do not yield an equality then we must conclude that the structure *cannot* be a polyhedron. For example, if we try to make a polyhedron out of a 60-atom piece of graphitic chicken-wire fencing, Euler's formula tells us that we will fail.

Such a (hypothetical) polyhedron must have 60 vertexes ($V=60$), corresponding to the 60 atoms. Each vertex must be connected to three others and so each will have associated with it three 'half-edges' (one complete edge joins two vertexes). The total number of edges must therefore be 90 ($E=90$). If the faces are all hexagonal, then each edge must constitute one-sixth of each of two faces (one on either side), making a total of 30 faces ($F=30$). We see immediately that $V+F$ (90) does not equal $E+2$ (92) and so we cannot make a C_{60} polyhedron out of a piece of graphite.

*This article is reproduced as the Prologue.

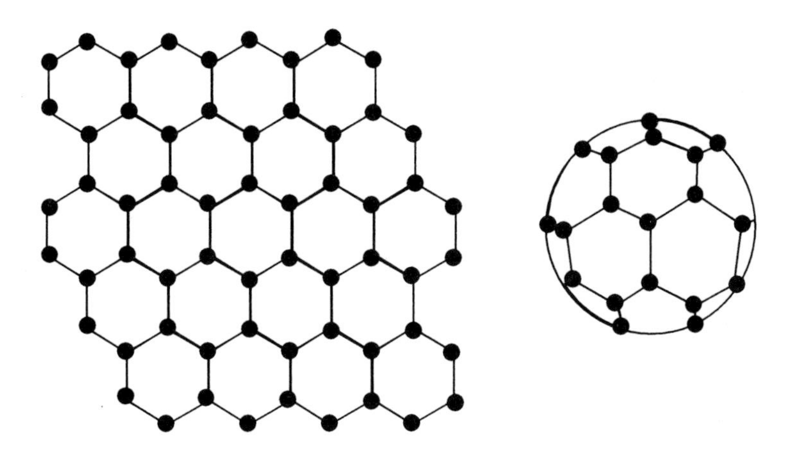

Don't be fooled by appearances. According to Euler's formula, the 60-atom sheet of graphite on the left *cannot* be folded into a closed sphere, no matter how convincing the appearance of the two-dimensional picture of the sphere on the right.

In fact, if we generalize this to structures with any number of vertexes, we find that it is mathematically impossible to obtain an equality in Euler's formula if the faces are all hexagonal. If Smalley, Kroto, Curl, Heath, or O'Brien had known of Euler's famous formula, then they would not have wasted time trying to construct closed cages out of hexagons: it just can't be done.

As Jones suggested, introducing some pentagonal faces into the structure along with the hexagons distorts it from its flat shape and offers the possibility that we can close it up to form a cage. Euler's formula tells us exactly what we need. If we look at the relationship between V, E, and F for the case of a polyhedron with both hexagonal and pentagonal faces, we can quickly deduce that to obtain equality we need 12 pentagonal faces—no more and no less. This result is quite independent of the number of hexagonal faces. Provided we have 12 pentagons, we can construct polyhedra with any number of hexagons (with, as it turns out, a single exception—we cannot make a polyhedron from 12 pentagons and one hexagon).

Nature long ago learned Euler's formula for herself. As D'Arcy Thompson described in his classic book *On growth and form*, first published in 1917, single-celled micro-organisms known as radiolarians possess polyhedral skeletons which appear to be composed entirely of hexagons. But closer inspection reveals that this is not so:

> But here a strange thing comes to light. *No system of hexagons can enclose space*; whether the hexagons be equal or unequal, regular or irregular, it is still under all circumstances mathematically impossible. So we learn from Euler: the array of hexagons may be extended as far as you please, and over a surface either plane or curved, but *it never closes in*. Neither our *reticulum plasmatique* nor what seems the very perfection of hexagonal symmetry in *Aulonia* are as we are wont to conceive them; hexagons indeed

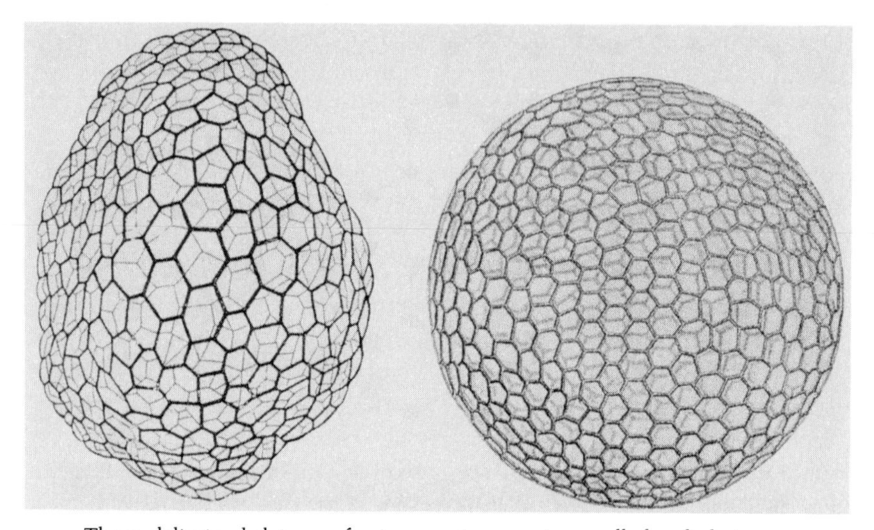

These delicate skeletons of microscopic organisms called radiolaria are natural examples of polyhedral structures composed predominantely of hexagons. These drawings are reproduced, with permission, from D'Arcy Thompson's *On growth and form* (republished in 1961 in an abridged edition by Cambridge University Press). Carnoy's drawing of *Reticulum plasmatique* is on the left, Haeckel's drawing of *Aulonia hexagona* on the right. Again, don't be fooled by appearances. The skeleton of *Aulonia hexagona* is *not* composed only of hexagons.

predominate in both, but a certain number of facets are and must be other than hexagonal. If we look carefully at Carnoy's careful drawing we see that both pentagons and heptagons are shown in his reticulum, and Haeckel actually states, in his brief description of his *Aulonia hexagona*, that a few square or pentagonal facets are to be found among the hexagons.*

The idea of introducing pentagonal defects to create hollow molecules from graphite came to Jones after reading this passage in Thompson's book and studying Ernst Haeckel's sketch of *Aulonia hexagona*. This beautifully symmetrical structure so wonderfully illustrates the notion of a hollow molecule that Jones reproduced Haeckel's drawing in his own 1982 compendium.

Jones had solved the basic problem of how to get from graphite to a closed structure. As he was primarily concerned with very large molecules—consisting of hundreds of thousands of carbon atoms—he stopped short of speculating on the possibility of soccer ball C_{60}. His purpose, after all, was to bridge the 'curious discontinuity' between the densities typical of gases and those typical of liquids and solids by inventing a 'vague fifth state of matter'.

*Thompson, D.W. (1961). *On growth and form* (abridged edition, Bonner, J.T. ed.). Cambridge University Press.

Jones regards his alter ego Daedalus as a kind of court jester to science. Looking back on his hollow carbon molecules after the discovery of buckminsterfullerene, he regards this invention as Daedalus's finest hour.

Haeckel's drawing of *Aulonia hexagona* also evokes the geodesic domes that were eventually to provide the breakthrough in understanding the structure of C_{60}, as well as its name. In 1917, the year D'Arcy Thompson's book was first published, the basic patterns of nature were also of some interest to a young inventor from New England. It was in this year that Richard Buckminster Fuller embarked on a study of idealized models of these basic patterns. This was a study that was eventually to lead to the construction of the first of his geodesic domes 35 years later.

Fuller had reached the conclusion that it is misleading to think that the macroscopic structures built by man—houses, office blocks, bridges—are built out of materials. On the surface this might seem to be the case, but the materials themselves are composed of microscopic structures—crystals, molecules, and atoms. He therefore believed that a study of these micro-structures would allow him to deduce some fundamental relationships governing the patterns of energy distributions in man-made macro-structures: relationships involving tension and compression.

Not having to hand a detailed theory of the inner workings of atomic nuclei, he chose to develop his own model based on 'energy spheres', idealized fields of energy in which all the forces are in perfect equilibrium. He then assembled these spheres in close-packed arrangements and examined the resulting structures. By starting with a single sphere, he found that he could fit a layer of exactly 12 spheres around it. A second layer has 42 spheres, a third 92.

From close-packing he deduced a basic structure which he called the *vector equilibrium*. This structure actually comes from studying the forces acting on the close-packed arrangement with one layer of 12 spheres surrounding a central sphere. Because of the inevitable voids, the final structure is not a larger sphere but a polyhedron with 14 faces—six square and eight triangular. All the sides of this polyhedron have the same length, and this length is equal to the distance between each vertex and the centre of the structure. If these lengths are taken to represent the magnitudes and directions of energy forces (vectors), then all the vectors are in balance. The forces tending to pull the structure apart are exactly balanced by the forces tending to pull it in on itself.

The reasoning that led Fuller from the vector equilibrium to the geodesic dome is convoluted and entirely typical of this rather eccentric inventor. The logical flow of ideas took him from the vector equilibrium to the icosahedron, octahedron, and tetrahedron. These are three of the five regular (or 'Platonic') solids, described in Plato's *Timaeus* and assigned as basic units of the elements water, air, and fire. Studying these most fundamental of three-dimensional geometric shapes led him to select the triangle as a basic arrangement of energy vectors. The final step in Fuller's intellectual journey came when he projected the

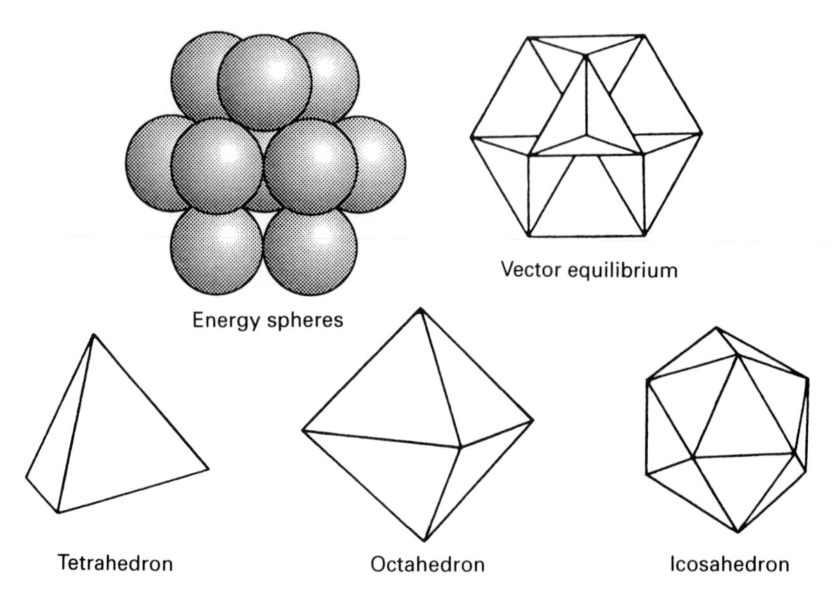

From close-packing to the vector equilibrium and the regular solids. Fuller discovered that packing 12 spheres uniformly around a central sphere generates a structure which, when translated into an arrangement of vectors, represents a unique balance of forces. The vector equilibrium is a polyhedron with 14 faces, six square and eight triangular. All sides have the same length, equal to the distance from each vertex to the centre. Fuller believed the vector equilibrium to be a fundamental structure, comparable to the regular (or 'Platonic' solids), of which three—the tetrahedron, octahedron, and icosahedron—are shown here.

faces of the vector equilibrium or the regular solids onto the surface of a sphere. When this is done the edges of the structures become arcs of great circles, or geodesics. Fuller called this general system of relationships *energetic geometry*.

At every step along the path from energy spheres to geodesics, Fuller had remained faithful to his original objective. He had begun with spherical building blocks representing idealized energy fields and had arrived at the projections of triangulated energy vectors onto the surfaces of spheres. What he had found was a means of translating the balance of forces embodied in the vector equilibrium and the Platonic solids into a spherical structure of arbitrary dimension.

What he had was a principle that would allow him to design and construct buildings with remarkably even distributions of load, tension, and stress. The geodesic dome offered tremendous prospects as the most efficient and economical of buildings, capable of a strength-to-weight ratio impossible in more conventional structures.

The key design element was the triangle—the greater the degree of triangulation of the polyhedron, the more uniform the distribution of energy points on the surface of the sphere on which it is projected. Pentagons and

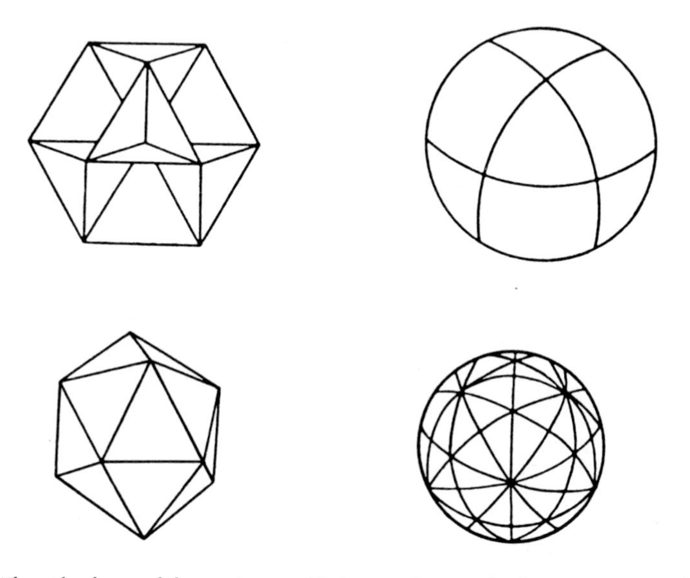

When the faces of the vector equilibrium or the icosahedron are projected onto the surface of a sphere, the edges are translated into the arcs of great circles, or geodesics. The balance of forces represented by the polyhedral structures is thus converted to a spherical form. This is Fuller's *energetic geometry*.

hexagons (and Euler's formula) emerge naturally from such considerations but were themselves not parts of Fuller's original design concept. Smalley, Kroto, Curl, Heath, and O'Brien can perhaps be forgiven for not immediately spotting the role of pentagons in Fuller's geodesic domes.

In fact, the first practical application of the principle that Fuller had devised involved the reverse process: the projection of the surface of a sphere onto a polyhedron. This was Fuller's *Dymaxion Map*, which he patented in 1946. All two-dimensional maps of the world involve some compromise, some distortion of the true three-dimensional surface which make parts of the world seem larger (or smaller) than they really are. By projecting a three-dimensional world map onto the faces of an icosahedron, such distortion is minimized and evenly distributed over the whole surface area. Unfolding the icosahedron to produce a flat projection gives the Dymaxion Map. It was the first two-dimensional map to give an accurate representation of the world's land masses.

The first of Fuller's geodesic domes was constructed for the Ford Motor Company in 1952, the concept was patented in 1954, and large-scale production began in earnest two years later. The Union Tank dome, whose appearance so misled the scientists at Rice, was built in 1958. At the time it was the world's largest clear span structure, stretching 384 feet and covering a floor area of 115 558 square feet. Yet the whole thing weighed only 1200 tons—two ounces for every cubic foot of space enclosed. It cost less than $10 per square foot.

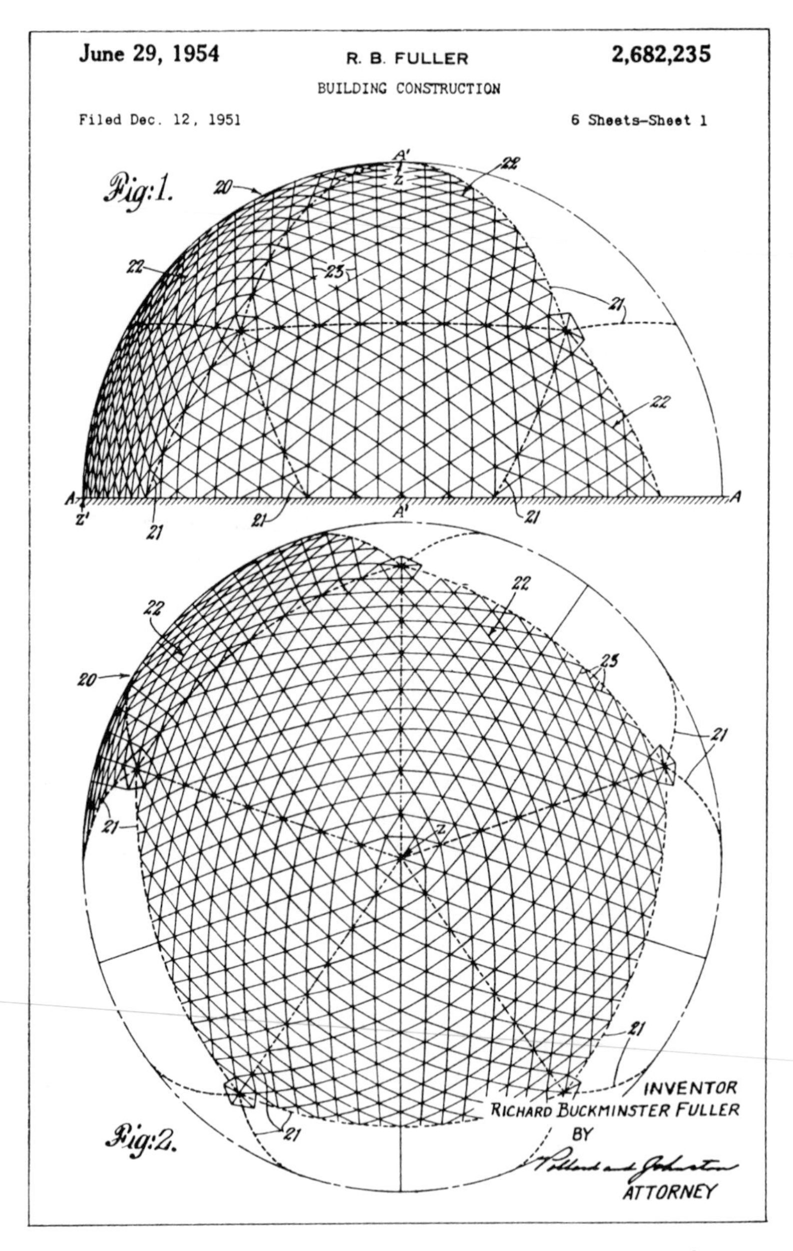

This page from Fuller's basic geodesic dome patent (June 29, 1954) shows how the triangular faces of a polyhedron are mapped onto a sphere, producing a network of points over which the building's load, tension, and stress is evenly distributed. Note how the geodesics (and the pentagons) emerge naturally from the process. © 1960 Allegra Fuller Snyder. Courtesy Buckminster Fuller Institute, Santa Barbara.

The geodesic domes came to symbolize American ingenuity, creativity, and industry. They were a much-needed success for their inventor. Fuller had suffered a long series of disappointments, both in business and in the practical application of his inventive genius. He had reached a turning point in 1927 when, broke and living in a Chicago tenement with his wife, a newborn child, and an Al Capone hit man for a neighbour, he had contemplated suicide. Instead of taking his own life, he abandoned the idea of earning a conventional living and dedicated himself to an intellectual programme that he was later to call *comprehensive anticipatory design science*.

This programme produced a string of inventions that captured the public imagination but never really took off. His failures had led people to perceive him as a rather accident-prone eccentric, a perception encouraged by his use of a rather convoluted and fanciful language. It was all to easy to dismiss Fuller as a crackpot.

But there was no denying the touch of genius that lay behind the domes. After walking around inside a geodesic dome constructed for the 1959 World's Fair in Moscow, Nikita Khrushchev is reported to have remarked: 'I would like to have J. Buckingham Fuller come to Russia and teach our engineers.' A geodesic dome was an obvious choice to house the American exhibition at the Montreal Expo in 1967. This three-quarter sphere, 250 feet at its widest and 200 feet high, housed exhibitions on several levels connected by escalators and stairs. The dome required no internal supports or struts, allowing visitors breathtaking views from every vantage point. Like many others visiting the exhibition, a young English post-doctoral research scientist by the name of Harry Kroto was entranced by what he saw.

Fuller's thoughts about the patterns of forces in structures formed from energy spheres had led him to the geodesic domes. He had developed some new design principles for macro-structures by studying the principles that governed the patterns of force in hypothetical micro-structures. That his geodesic domes should serve as a basis for rediscovering these principles in the context of a new form of carbon micro-structure has a certain symmetry that Fuller would have found pleasing, if not very surprising.

The soccer ball structure did not feature in Fuller's work, although a triangulated version, with 92 vertexes, 180 faces and 270 edges does appear in a chart in which he set out some of the basic interrelationships that characterized his energetic geometry. Daedalus had invented the concept of the hollow molecule but had stopped short of soccer ball C_{60}. Was it the case, then, that the idea of a spherical molecule consisting of exactly 60 carbon atoms was still very much an original concept? The answer was no. It does not do to underestimate the power of Murphy's Second Law.

In the late 1960s, Japanese chemist Eiji Osawa was, like many other organic chemists of the period, fascinated by the structures and properties of aromatic molecules. These molecules are so named because the first few of the class that

A view of the inside of Fuller's geodesic dome at the Montreal Expo. Note the appearance of the pentagon in the structure, towards the top of the picture just right of centre. © 1994 Allegra Fuller Snyder. Courtesy Buckminster Fuller Institute, Santa Barbara.

were isolated are rather sweet-smelling substances. They possess cyclic structures which can be drawn with alternating single and double carbon–carbon bonds. In reality, the electrons which form the double bonds do not remain fixed, but are spread out or 'delocalized' over all the bonds in the ring, reducing the overall energy of the molecule.

What happens is that electron waves on adjacent carbon atoms, which are usually termed 'orbitals' and pictured as lobes pointing out into space, overlap to form the bonds that lie between the atoms and hold them in their molecular framework. These are called sigma bonds. There is another set of orbitals at right angles which may also overlap to form so-called pi bonds which lie above and below the plane of the molecular framework. A 'double' bond therefore consists of one sigma and one pi bond.

In a molecule with alternating single and double carbon–carbon bonds, the double bonds do not remain fixed in position, but their pi bond electrons become delocalized and wander over all the carbon atoms in the structure to give a more uniform distribution with a lower energy. In benzene, this delocalization produces a pi bond that extends over the whole ring, with electron orbitals pictured as doughnuts sitting above and below the plane formed by the ring of atoms. This is what makes the molecule 'aromatic'. The consensus of opinion was that the kind of electron delocalization required to

The pi-electron orbital system of benzene. Each carbon atom in the hexagonal molecular framework has a pi-electron orbital which points above and below the plane of the hexagon. Each orbital contains a single electron. These overlap to produce continuous, doughnut-shaped orbitals which allow the electrons to become 'delocalized' over the whole molecule, reducing its energy. This delocalization has consequences for the physical and chemical properties of benzene, properties that are summarized by its classification as an 'aromatic' molecule.

produce an aromatic molecule is possible only when the molecule has a flat structure, with benzene as the archetype. Osawa was not so sure.

In 1966, two organic chemists at the University of Michigan, Wayne Barth and Richard Lawton, announced that they had succeeded in synthesizing a molecule composed of six rings; a central pentagon surrounded by five hexagons with the empirical formula $C_{20}H_{10}$. The systematic name for this molecule, based on certain rules of nomenclature, is dibenzo[*ghi,mno*]fluoranthene, but Barth and Lawton coined the simpler 'trivial' name *corannulene* from the Latin *cor*, meaning 'heart' and *annula*, meaning 'ring'. Corannulene challenged the existing preconceptions of aromaticity.

On the one hand, it could be argued that corannulene should have a flat structure, allowing full delocalization of its pi-bond electrons, maximizing its aromatic character and increasing its stability. On the other hand, it was clear that the central pentagonal 'defect' would create considerable strain in a flat structure, strain which could be relieved if the structure distorted to a bowl shape, as Smalley discovered for himself with a paper model many years later.

The result was the compromise that the chemists had anticipated. Corannulene is an aromatic molecule but it does not have a flat structure. Crystallography showed quite clearly that the molecule forms a shallow bowl. Osawa studied this shape intently. Was it possible to extend this to a sphere, and so obtain a wholly novel three-dimensional aromatic molecule?

During an idle moment, Osawa pondered on this problem as he watched his young son playing soccer. The ball his son was playing with was decorated with the usual black pentagons, and Osawa recognized the bowl-shaped

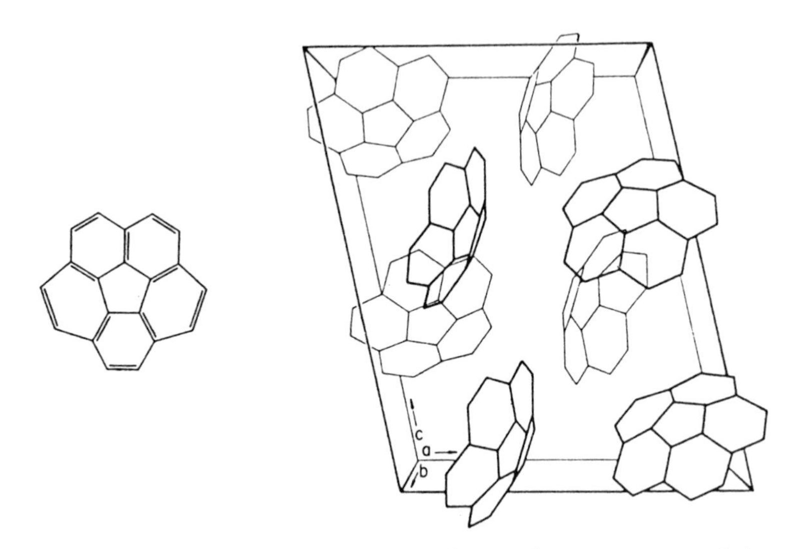

The molecule corannulene (left) consists of a central pentagon surrounded by five hexagons. To minimize the strain inherent in its geometric structure, the molecule needs to distort into a bowl shape. On the other hand, to maximize the extent of delocalization of its pi-bond electrons (and hence minimize its energy) it needs to remain flat. The reality is a compromise, as shown by the shapes and positions of molecules in crystalline corannulene derived from X-ray diffraction studies (right). Adapted, with permission, from Barth, Wayne E. and Lawton, Richard G. (1971). *Journal of the American Chemical Society*, **93**, 1730. Copyright (1971) American Chemical Society.

structure of corannulene in the pattern. So there it was. Extending the corannulene-type structure all the way to a sphere appeared to be at least geometrically possible.

Osawa learned that this spherical structure is a *truncated icosahedron*. Cut off each corner of the icosahedron and the result is a polyhedron with 32 faces: 20 hexagonal and 12 pentagonal. What he had found was a molecule shaped like a truncated icosahedron with a carbon atom at each of its 60 vertexes. He described this novel molecule in an article published in the popular Japanese chemistry journal *Kagaku*, and expanded on it in his book *Aromaticity*, co-written with Z. Yoshida and published in 1971.

Osawa had had the *idea*, but was a long way from being able to realize it in the form of a synthesis. There was, however, one check he could make to discover if such a molecule might be stable in principle. In the 1930s, the German theoretician Erich Hückel had devised a simple approximate method for calculating the energy levels of molecules with delocalized pi bonds. The approach is so simple (and yet so effective and illustrative), that Hückel theory and its more modern variants now form a fundamental part of the undergraduate chemistry curriculum. Osawa carried out some Hückel calculations based on the geometry of soccer ball C_{60}, and concluded that it should indeed be stable.

However, soccer ball C_{60} still remained no more than an elegant structure and an interesting theoretical notion. Although the theoretical calculations indicated that the molecule should be stable, nobody in Japan who read Osawa's book could see an easy way to make it. It was an idea before its time, and Osawa eventually abandoned it in favour of more practical pursuits.

In the meantime, the concept of soccer ball C_{60} was independently rediscovered by Soviet theoreticians D. A. Bochvar and E. G. Gal'pern. In a paper published in the Proceedings of the Academy of Sciences of the USSR in 1973, they presented the results of calculations very similar to the ones that Osawa had carried out. For Bochvar, C_{60} was one of a large number of highly speculative carbon structures that were of interest, ranging from carbyne, carbon molecules shaped like polyhedra, and exotic infinite networks. As with Osawa, C_{60} was little more than a rather appealing theoretical curiosity.

The Hückel calculations were repeated again in 1981 by Robert Davidson at du Pont's Central Research and Development Department in Delaware. His results were entirely consistent with the previous calculations. He noted that the results were very characteristic of aromatic molecules and, if it could ever be synthesized, soccer ball C_{60} would represent the first example of three-dimensional aromaticity.

In the early 1980s, organic chemist Orville Chapman believed his discipline was in great disarray. It had become inward-looking and self-absorbed, with organic chemists showing little interest or enthusiasm for other branches of science. What Chapman needed was something that would revitalize his science. Looking out from his office window over the sunlit campus of the University of California in Los Angeles, he sought guidance from a higher authority. 'If God would give me the grace to make one molecule,' he asked himself, 'what would that molecule be?'

The answer was soccer ball C_{60}. In July 1981, Chapman arrived in Germany to begin a two-month period of study leave at the University of Erlangen. He was met at Frankfurt airport by his former postdoctoral associate, François Diederich, who had since returned to the Max Planck Institute for Medical Research in Heidelberg. Over lunch, Chapman told Diederich of his idea for a spherical carbon molecule and of his plans to take time during his stay in Erlangen to write proposals to obtain funding for work on the development of synthetic routes to C_{60}. To Diederich, it all sounded incredibly new and exciting.

Chapman failed to secure funding from the US National Institutes of Health, but did succeed with the National Science Foundation, and the first of five graduate students began work shortly afterwards. During the period 1981–85, Chapman talked extensively to his colleagues in the organic chemistry community about soccer ball C_{60}, delivering lectures at various institutes on the subject. On the whole, the community was sceptical that such a molecule could be made.

By September 1985, the idea of soccer-ball C_{60} was certainly not original. It had been conceived first by Osawa in the early 1970s and independently rediscovered by at least four other groups. In fact, the soccer ball concept had been applied to a molecular system even earlier than this. J. L. Hoard and his colleagues at Cornell University and the University of Pennsylvania showed that the 60-atom truncated icosahedron forms an instrinsic part of the crystal structure of beta-rhombohedral boron, a crystalline form of boron stable at high temperatures. Another icosahedral structure of 24 atoms is bonded inside the B_{60} cage, producing a B_{84} unit which is repeated throughout the crystal lattice.

Such is the degree of specialization in modern science, and the sheer impossibility of keeping track of an ever-expanding scientific literature, that Kroto, Heath, O'Brien, Curl, and Smalley were completely unaware of the earlier proposals made by scientists in other branches of chemistry. However, if they were right in their interpretation of the results of their carbon cluster experiments, then they were certainly the first to propose that they had actually made soccer ball C_{60} in the laboratory. Now all they had to do was prove it.

7

The fullerene zoo

Kroto hardly had time to catch his breath before he was back on an airplane bound for Houston. Although he had been present in Smalley's laboratory only as a visitor, he had been directly involved in the birth of buckminsterfullerene and Smalley accepted that he had every right to help bring up the baby. The agreement they reached between them required Kroto to be an active participant in further research on buckminsterfullerene, not a passive observer from across the Atlantic. This meant being with the team at Rice as often as allowed by his other research commitments and his university teaching and administrative duties.

The result was an excruciating round of international travel, with periods of three-to-five weeks working in Sussex punctuated by periods of two-to-three weeks in Texas. However, Kroto was relaxed about the arrangements: he enjoyed travelling and repeated visits to Houston would at least allow him to remain acquainted with the city's bargain bookstores. He was back in Houston on October 3, only three weeks after leaving.

Smalley had submitted a short paper on the results they had obtained at Rice with lanthanum-impregnated graphite to the *Journal of the American Chemical Society*, extending the list of authors to include Zhang, Liu, and Tittel. The paper summarized the evidence from the experiments on AP2 that had been initiated by Heath. The mass spectra revealed the presence of a large number of LaC_n complexes, with C_{60} and C_{70} the only 'bare' (metal-free) clusters. The spectra also revealed a tendency for the C_n clusters to pick up only one lanthanum atom, consistent with the idea that the atoms were being incorporated at unique binding sites *inside* the cages and therefore suggesting that all the clusters might have closed cage structures. It was straightfoward to show that no C_{60} complex with two lanthanum atoms could form, because there simply is not enough room inside the cage.

Their interpretation of these results was taking them quite a long way beyond their original discovery that C_{60} is special. If they were right, then

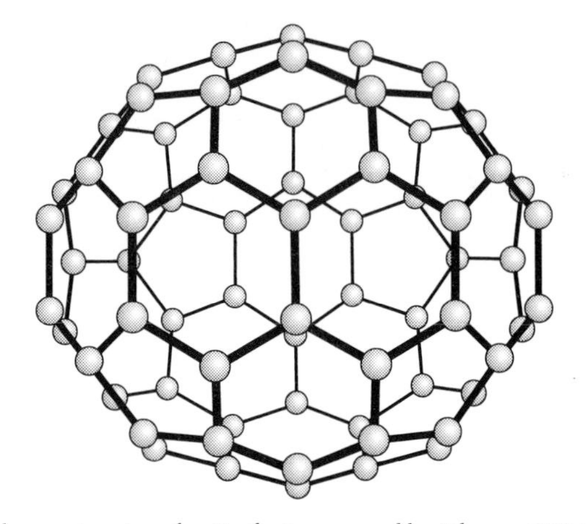

The closed cage structure for C_{70} first proposed by Zhang, O'Brien, Heath, Liu, Curl, Kroto, and Smalley.

formation of closed-cage molecules above a certain size was a general phenomenon in contradistinction to everything known about the properties of graphite. They presumed that these molecules would possess polyhedral structures of pentagons and hexagons much like buckminsterfullerene itself but obviously in all cases less symmetrical. As an example, they described a possible structure for C_{70}. This consisted of two halves of C_{60} with an extra belt of 10 carbon atoms around the middle, producing a kind of 'squeezed' sphere.

By the time Kroto arrived back in Houston in October, O'Brien and Zhang were busily investigating the reactivity of the large (C_n, $n > 40$) clusters. Heath had taken on the challenge of trying to use the laser vaporization technique to make C_{60} in bulk quantities, allowing a much more detailed (and leisurely) study of its properties through conventional chemical analysis.

The purpose of the original experiments begun in August had been to react long-chain carbon molecules with hydrogen and nitrogen to make new molecules related to the cyanopolyynes. This was an approach that could be stood on its head and repeated for the larger clusters. Rather than demonstrate how new molecules could be made by reacting the clusters with various gases, the scientists wanted to show that C_{60} *refused* to react with even the most reactive gases.

They reasoned as follows. If C_{60} did possess a closed-cage structure, it would have no dangling bonds and so would be relatively resistant to chemical attack. Certain small molecules such as nitrous oxide (NO), sulfur dioxide (SO_2), and oxygen (O_2) are known for their reactivity towards a class of chemical species called radicals—molecules with one or more unpaired electrons (or dangling bonds). Such is their reactivity that NO, SO_2, and O_2 are sometimes referred to as radical 'scavengers'. It is a standard procedure in gas-phase chemistry to add scavengers in large concentrations to a system in which radicals are thought to be important intermediates leading to the formation of certain products. If radicals are indeed important in the reaction, the scavengers will rapidly mop them up, inhibiting the formation of the products.

The earlier experiments on AP2 had shown that the larger clusters were relatively unreactive towards O_2. O'Brien's new experiments confirmed that the same pattern of behaviour was obtained with NO and SO_2. While the results continued to highlight the special stability of C_{60}, the general lack of reactivity of all the other large clusters supported the idea that they too had closed-cage structures. Smalley complained for several weeks that these experiments were failing to reveal any kind of reactivity. In the end, he was forced to accept O'Brien's results. It looked as though they had found not only the most remarkable of molecular structures, but a veritable zoo of closed-cage molecules ranging in size from 40 atoms all the way to 100 and more. These later came to be known collectively as 'fullerenes'. In contrast, the small (C_n, $n < 40$) clusters reacted very rapidly with NO and SO_2, consistent with the idea that these clusters possess completely different radical structures—long chains with dangling bonds at either end.

The group discussions in Smalley's office were no less energetic than before. Their attention now became focused on the question *how?* By what mechanism were C_{60} and the other fullerenes being formed inside AP2? Kroto, Heath, and O'Brien ransacked the well-stocked Rice University library for clues. They set themselves the task of finding out as much as possible about the formation of small carbon particles, injecting what they had learned into the group discussions.

The experimental data on C_{60} appeared to them to suggest that it was a survivor. Out of all the closed cage molecules produced inside AP2, only C_{60} could survive certain clustering conditions. That C_{60} became prominent under the most favourable clustering conditions did not necessarily mean that such conditions were prerequisite for its formation. On the contrary, perhaps all the other clusters were instead reacting and growing further, acquiring more and more carbon atoms and eventually becoming tiny soot particles. These would be small in terms of the dimensions typical of ordinary dust particles, but would be very large and heavy in molecular terms and so would not be expected to register in the mass spectrometer. They reasoned that the spherical symmetry and stability of buckminsterfullerene meant that it could not grow to become a soot particle and so survived the clustering process. Eventually, all the other clusters would grow too large, producing a lot of soot and leaving a little C_{60}.

This all seemed to be tied up somehow with the process of soot formation. On their many forays through the scientific literature, Kroto, Heath, and O'Brien had discovered enough about the current understanding of the mechanism of soot formation to convince themselves that there could be a connection. Soot is often found in the form of small spherules between one and five millionths of a metre in diameter which combine to give more complex structures. From their speculation and intense discussion in Smalley's office there emerged a rather intriguing possibility.

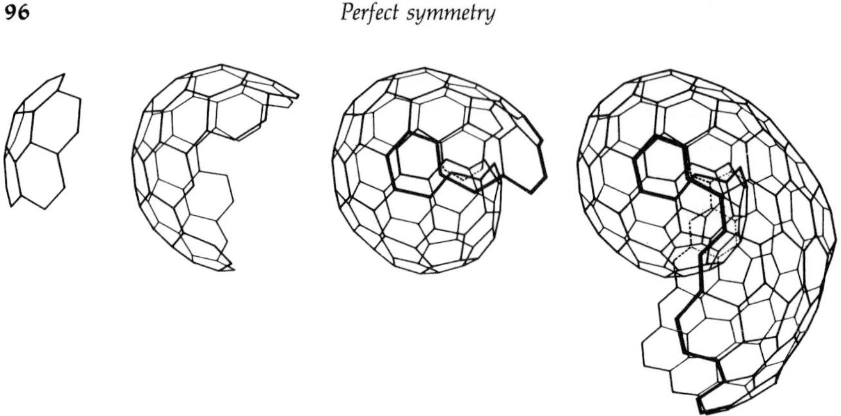

The icospiral nucleation mechanism. Beginning with a basic corannulene unit—one pentagon surrounded by five hexagons—the structure grows to a near spherical shape but does not close. Instead, the growing edge folds over the top of itself and spirals like a nautilus. The group at Rice proposed that such structures seed the formation of soot particles. Adapted, with permission, from Kroto, Harold (1988). *Science*, **242**, 1139.

They pictured the process this way. The instability associated with the dangling bonds all the way around the edge of a small fragment of graphitic chicken-wire fencing can be reduced by introducing pentagons, enabling the structure to close up on itself, as Daedalus had proposed in 1966. As carbon atoms add to the growing fragment, a leading edge is produced which continues to incorporate more pentagons. However, rather than automatically close up to form a sphere, the leading edge is more likely to grow and curve over the top of itself, spiralling like some nautilus and continuing all the way to a full-size soot particle.

They called this process the 'icospiral nucleation mechanism'. It was appealing, if rather fanciful. If it was true, it meant that the formation of soccer ball C_{60} was very much an exception rather than a rule, relying on the chance closure of the growing structure. They estimated that the chance of forming C_{60} in the chaos created by the vaporization laser and subsequent supersonic expansion inside AP2 was one in ten thousand, or even one in a million. To a certain extent, this made them feel a little more comfortable. To produce such a beautifully ordered structure out of chaos seemed to go against the grain of the second law of thermodynamics, which demands that in a spontaneous change, order tends to *decrease*. Surely, making buckminsterfullerene inside AP2 had to be a chance event.

Still, the implications were profound. The mechanism suggested that where there is soot, there is also soccer ball C_{60}. Could it be that C_{60} is produced in the smoke from every candle flame? There was more. Huge carbon-rich red giant stars like IRC + 10°216 are pumping vast quantities of smoke and soot into the interstellar medium. If buckminsterfullerene was being produced along with the soot, then it must be one of the most ubiquitous molecules in the universe, and certainly amongst the oldest.

While they were all enthusiastic about the mechanism, Curl urged caution. Scientists had been studying the formation mechanisms and properties of soot for many years and were unlikely to greet such a bold new idea with open arms. The others paid scant attention to Curl's words of warning. Having staked their scientific reputations on an outrageous claim for soccer ball C_{60}, making another outrageous claim seemed of little consequence. They wrote up their proposals for the soot mechanism together with the reactivity results in a paper submitted to the *Journal of Physical Chemistry* at the end of November 1985. The *Nature* paper had just appeared in print.

They still had no hard evidence that the C_{60} they were making inside AP2 had the soccer ball structure. All they had was a prominent peak in a mass spectrum corresponding to a cluster of 60 carbon atoms, a spherical paper model constructed from 12 pentagons and 20 hexagons and some circumstantial evidence based on reactivity comparisons. This wasn't much. Connecting the C_{60} signal with the soccer ball required a little imagination and an awful lot of faith. To argue that the soccer ball was so beautiful a structure it just had to be true was unlikely to to cut much ice with a community of scientists trained to be sceptical.

What they needed to do was make C_{60} in sufficient quantities to enable a thorough chemical analysis. Although Heath was working hard on this task, they might as well have been asking for the moon. The scientists had so far made C_{60} in only the most minute quantities, caught for a fleeting instant, ionized in the beam of a powerful laser, and detected with one of the most sensitive of analytical instruments, the mass spectrometer. They had come to the conclusion that forming C_{60} was an exception rather than a rule, and they estimated that they were making at most a few tens of thousands of C_{60} molecules at a time. This was more than enough to detect with the sophisticated instrumentation of AP2, but it was certainly not enough to see and hold in your hand.

But they still had their dreams. Smalley and Kroto both imagined what buckminsterfullerene might look like in solid form or in solution if it could ever be produced on a large scale. Smalley talked of a 'yellow vial'. Kroto thought that buckminsterfullerene might be pink. They were not sure that they would ever get the chance to see who might be right.

For Kroto, there was one key measurement that would provide a most dramatic proof that C_{60} had the soccer ball structure. This was a measurement of its nuclear magnetic resonance (NMR) spectrum.

Atomic nuclei consist of protons and neutrons, but the nature and properties of a chemical element are determined only by the number of protons in its nuclei. Atoms with the same number of protons but different numbers of neutrons have different masses but belong to the same element. Such atoms are called isotopes. The most common isotope of carbon is carbon-12 (six protons and six neutrons), but carbon-13 (six protons and seven neutrons) occurs naturally on earth in small amounts.

NMR spectroscopy relies on the fact that some atomic nuclei, including carbon-13, behave like tiny magnets and will become aligned when placed in a magnetic field. They can then be made to 'flip' or 'resonate' between different alignments by absorbing electromagnetic radiation at specific frequencies in the radiowave region. The precise frequencies which cause the nuclei in a molecule to resonate depend on the extent to which the nuclei are 'screened' from the magnetic field by electrons in neighbouring atoms, and hence on the chemical environment of the nuclei.

When produced from graphite with carbon-13 in natural abundance, a few C_{60} molecules would on average be expected to have one carbon-13 nucleus sitting somewhere in their structures. In a less symmetrical structure, different NMR resonance frequencies would be observed, related to the different possible locations (and hence chemical environments) of the carbon-13 nucleus in the structures. But in soccer ball C_{60}, all the carbon atoms have exactly the same chemical environment. No matter where a carbon-13 nucleus sits in such a structure, it will produce the same characteristic NMR resonance frequency. If it could ever be measured, the NMR spectrum of buckminsterfullerene would consist of just one line. Kroto believed this was the ultimate 'one-line proof'. If only they could make enough C_{60}.

If the experimentalists were constrained by the apparent limitations of nature, the theoreticians were constrained only by their imaginations. The proposal that there could exist a whole new kind of all-carbon molecule presented the theoreticians with a golden opportunity, and they had a field day.

Since the discovery of quantum theory in the early 1920s, theoretical chemistry had undergone almost 60 years of development and refinement. On the one hand, the application of quantum theory from first principles (so-called *ab initio* theory) is feasible only for relatively simple molecules. On the other hand, those 60 years had been spent fruitfully developing all manner of approximate theoretical descriptions varying in their sophistication and complexity: Hückel theory was just one of many different possible approaches. Soccer ball C_{60} fell from the torrid environment of AP2 into the laps of theoreticians with access to an elaborate collection of tools designed for calculating its properties. For many theoreticians, it was an opportunity just too good to miss.

Osawa, Bochvar and Gal'pern, and Davidson had already applied Hückel theory to C_{60} and their results had indicated that the molecule should be quite stable. Unaware of these earlier calculations, O'Brien began to work on his own shortly after the revelations in Smalley's office on September 10. These calculations took over a week to complete, but by the end O'Brien had independent confirmation that the theory supported the argument in favour of the soccer ball structure.

At exactly the same time, Anthony Haymet at the University of California in Berkeley was carrying out identical calculations. He had discovered soccer ball C_{60} all over again, unaware of the previous work and of the experimental work going on at Rice. His own calculations confirmed that C_{60} should be stable.

As the experimental evidence began to emerge that there might be a whole family of closed-cage fullerenes, some theoreticians looked back at Euler's formula and other mathematical relationships that exist between polyhedra with hexagonal and pentagonal faces. That these elegant structures could be connected with the large carbon clusters opened the door to a fascinating new world of carbon chemistry, at least on paper. If the large carbon clusters did indeed have polyhedral structures then the preference for even numbers could be immediately understood, since it is impossible to construct a polyhedron with an odd number of vertexes. But then there was a huge number of possible polyhedra, so why was C_{60} the most stable? Furthermore, it emerged that there are 1760 different ways of combining 60 atoms to form a closed-cage molecule, so what was it about the soccer ball structure that was supposed to be so special?

The closed-cage dodecahedral molecule C_{20}—called dodecahedrene—consists of just 12 pentagons (no hexagons) and, because it is highly strained, it is unlikely to be stable. It also suffers another disadvantage. A Hückel calculation of the energy levels of polyhedral C_{20} reveals that it has two unpaired electrons and the structure is therefore said to have an 'open' electronic shell. The most stable molecules are those with closed electronic shells (not to be confused with closed cages, which refer to the molecules' geometric structures).

Curl had insisted on checking the bonding of Smalley's paper model as a rudimentary check on its *electronic* structure because he knew that any unpaired electrons 'left over' would be destabilizing. It very quickly became obvious that in trying to decide if a particular polyhedral molecule is likely to be stable, it is important not only to look at the likely strain inherent in its geometrical shape, but also to count up its electrons and discover if the molecule has a closed electronic shell. Haymet's calculations, in line with all the others that had been carried out thus far, showed that soccer ball C_{60} has a closed electronic shell and supposedly gains a dramatic stabilization from the delocalization of its pi bond electrons. This stabilization more than offsets the small amount of strain which is in any case evenly distributed throughout the spherical structure.

Further clues to the stability of buckminsterfullerene came from related studies published in October, 1986. Whilst Euler's formula allows a number of polyhedra with 20 hexagons and 12 pentagons, there was no way of telling their relative stabilities from the geometric structures alone. Tom Schmalz, W. A. Seitz, Douglas Klein, and G. E. Hite at Texas A&M University in Galveston examined the relative stabilities of five different C_{60} cages, of which the soccer ball structure was one, using Hückel theory and more quantitative (but still approximate) 'resonance' theories. Of all the possibilities, soccer ball C_{60} was

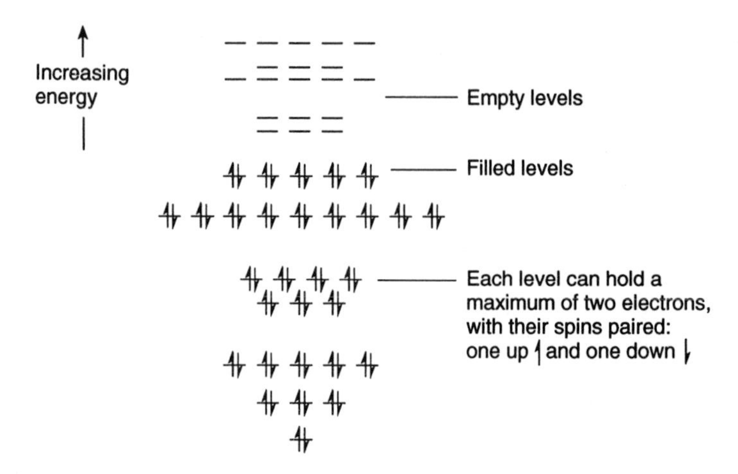

The Hückel method enables the pattern of pi-electron energy levels to be calculated. The results for soccer ball C_{60} are shown here. Each energy level can hold a maximum of two electrons with their spins aligned in opposite directions: one pointing up (↿) and one pointing down (⇂). Buckminsterfullerene has 60 pi electrons (one from each carbon atom) which must be fed into the energy levels, filling them from the bottom up. When this is done, it is found that all 60 electrons are paired up in 30 levels. As there are no unpaired electrons left over, the molecule is said to have a *closed electronic shell.*

the most stable. The chemists argued that this stability is derived from the fact that nowhere in the structure do two pentagons meet. By surrounding each of the 12 pentagons with a ring of hexagons, thereby isolating every pentagon from every other, certain destabilizing effects are avoided. In this respect, the soccer ball structure is *unique.*

This was all fine in theory. The soccer ball might be a unique structure for C_{60}, but it was still a long way from being proven and parts of the scientific community were beginning to get restless. To some, the 'bucky bandwagon' was already careering out of control. Here, supposedly, was a completely new form of carbon and a new kind of metal cluster compound. The formation of buckminsterfullerene was supposed to be intimately connected with the formation of soot, so it was being produced in every candle flame. It was also supposed to pervade the interstellar medium, the oldest of all molecules. And all this from apparently no more that a few mass spectrometer traces and some inspired guesswork. Some kind of backlash was inevitable.

The first note of dissent was sounded by Donald Cox, Andrew Kaldor, and their Exxon colleagues D. J. Trevor and K. C. Reichmann in an article provocatively entitled 'C_{60}La: A Deflated Soccer Ball?', submitted to the *Journal of the American Chemical Society* in January, 1986. If the Exxon scientists felt peeved that they had missed one of the most important discoveries in chemistry in recent memory, they did not show it. Instead, they pointed out that the mass spectra produced by a machine like AP2 requires very careful interpretation, and can sometimes be misleading. Their investigation of the C_{60} signal obtained using their own cluster beam apparatus suggested to them not some fabulously stable soccer ball structure, but rather a complex interplay between the efficiency of cluster ionization and ion fragmentation patterns.

They argued that there are two ways in which a large signal corresponding to C_{60} might be obtained. One way is through the direct ionization of neutral C_{60} to yield C_{60}^+, in which case the size of the signal would be related to the abundance of C_{60} in the cluster beam and would therefore depend on its stability, as the group at Rice had presumed. Another way is through the break-up of larger ionized clusters to produce C_{60}^+ fragments, in which case the size of the signal would have nothing at all to say about the relative abundance and stability of neutral C_{60}. Perhaps it wasn't the structure that was special, but certain aspects of the ionization laser frequency and intensity.

The Exxon scientists also had some problems with the interpretation of the mass spectra of the LaC_n complexes. They had no difficulty reproducing the kinds of effects that the Rice group had reported, but did not accept that the results were indicative of fullerenes with lanthanum atoms sitting in their centres. Under what they felt were more representative ionization conditions, the signals corresponding to the LaC_n complexes appeared much weaker than the Rice experiments had shown. In particular, the signal corresponding to LaC_{60} was only a few per cent of the size of the signal corresponding to bare C_{60}.

The group at Rice had made much of the fact that signals due to clusters with more than one lanthanum atom couldn't be seen in the mass spectra. The Exxon scientists argued that as the signal due to La_2C_{60} was expected to be much weaker than that for LaC_{60}, it was small wonder that it could not be observed. The fact that the clusters appeared to pick up only one lanthanum atom had nothing to do with closed-cage fullerenes, but had everything to do with the relatively low abundances of clusters with two or more metal atoms. To quote an old maxim: absence of evidence is not evidence of absence. Just because complexes of the type La_2C_n couldn't be seen didn't necessarily mean they weren't there.

At least the Exxon scientists ended on a conciliatory note. They warned how easy it is to be led astray in inferring structures from mass spectra, but urged that the search for the soccer ball should continue. However, the guesswork should be avoided in future by using techniques that could give more direct information about the structures.

There was a group at AT&T Bell Laboratories in New Jersey that had sounded much the same kind of warning in a paper that had appeared just before the *Nature* article. This paper described experiments on the laser vaporization of graphite that had been completed months *before* the experiments at Rice had even begun. L. A. Bloomfield, Mike Geusic (an ex-Smalley student), R. R. Freeman, and Walter Brown had sought to avoid the complications of ion fragmentation by dispensing with the ionization laser altogether. Instead, they monitored ions produced directly in the cluster source. By doing it this way, they found they could measure distributions of both positive and negative cluster ions.

Their positive ion distribution did show a slight overabundance of C_{60}^+, but this was on nothing like the scale of the signals that were subsequently seen at Rice and the Bell scientists did not pick it out as 'special' in any way.* However, their negative ion distribution appeared quite different. If these signals were all supposed to be related in some way to the distribution of *neutral* clusters then, they argued, the distributions of both positively and negatively charged ions formed from them should at least look similar.

The same argument was made again in September 1986 by a group from the Department of Chemistry and Biochemistry and Solid State Science Center at the University of California at Los Angeles (UCLA). M. Y. Hahn, E. C. Honea, A. J. Paguia, K. E. Schriver, A. M. Camarena, and Robert Whetten reported a positive ion distribution which showed the 'magic' natures of C_{50}, C_{60}, and C_{70}, but their negative ion distribution looked quite different. The conclusion once again was that the appearance of the distributions was strongly dependent on the fragmentation of larger clusters, with little to suggest that specific molecular structures were responsible for the magic numbers. Soccer ball C_{60} and the other fullerene structures were all very pretty to look at, but there was no compelling evidence to support the Rice group's assertion that these structures were being produced in the laser vaporization of graphite.

As the first anniversary of the discovery of buckminsterfullerene came and went, O'Brien, Heath, Kroto, Curl, and Smalley were hard at work defending their intellectual offspring. They were able to use AP2 to show that clustering conditions could be found under which both positive and negative ion distributions gave prominent signals for C_{50}, C_{60}, and C_{70}. They also reasoned that if the break-up of larger clusters were really responsible for the prominence of C_{60}^+, then as they systematically increased the intensity of the ionization laser, more and more clusters would fragment and the C_{60}^+ signal should grow larger and larger.

What they found was quite the reverse. With low laser intensities, they could fix the conditions to obtain the flagpole C_{60}^+ signal. As they then increased the laser intensity, the relative strength of the C_{60}^+ signal actually *diminished* and signals corresponding to neighbouring clusters began to appear. What was happening was that neutral C_{60} was indeed being formed and ionized. With too

*In fact, they did pick C_{60}^+ out for special treatment, using photons from a second laser specifically to break up the positive ions to see what kinds of fragments might be produced, but they drew no conclusions from their results regarding possible structures for C_{60}.

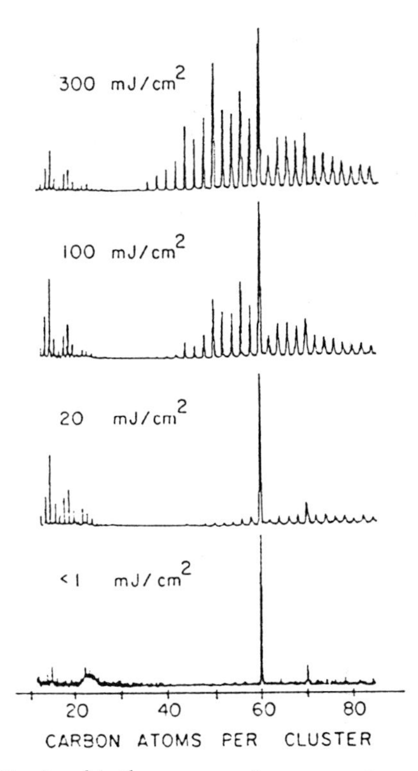

Instead of the C_{60} signal in the mass spectrum emerging as a result of ion fragmentation, it actually diminishes in size as the ionization laser power is increased. In the bottom panel, the ionization laser power is relatively low and the C_{60} signal is dominant (this is essentially the flagpole result). With increasing laser power, the relative size of the C_{60} signal decreases as a result of fragmentation and signals due to other even-numbered carbon clusters increase. These spectra have been plotted so that the C_{60} signal appears the same in each. Adapted, with permission, from O'Brien, S. C., Heath, J. R., Kroto, H. W., Curl, R. F., and Smalley, R. E. *Chemical Physics Letters*, **132**, 99.

high a laser intensity, the soccer ball structure was being broken up, producing smaller fullerenes and spitting out fragments that could combine to make a few larger fullerenes.

The scientists at Rice conceded the potential importance of ion fragmentation but concluded that these problems did not alter the essential correctness of their original assertions. C_{60} is magic because it is a soccer ball and there exists a wide range of related closed-cage molecules.

Between November 1985 and the end of 1986, 35 experimental and theoretical research papers were published on C_{60} and the other even-numbered carbon clusters. Arguments both for and against the idea of the soccer ball and other closed-cage fullerene structures had been presented. The theoretical evidence for the fullerenes was unarguable. These were predicted to be relatively stable structures with interesting relationships to polyhedra. The special nature of soccer ball C_{60} could be understood in terms of its geometrical shape (with its isolated pentagons) and its electronic structure—a closed shell with considerable stabilization achieved through electron delocalization.

But these were just predictions. Although the experimental evidence in favour of buckminsterfullerene now seemed reasonably secure, it was still all rather circumstantial. It relied too much on conjecture rather than hard fact. Kroto and Smalley continued to think about their pink or yellow vials, and Kroto dreamed of an NMR spectrum consisting of just one line. Over a year had gone by and they were no closer to making their dreams come true.

8

Pathological science

In the 18 months leading up to April, 1987, Kroto made a total of eight trips to Houston to work on fullerenes with the group at Rice. His eighth visit, which ended on April 29, was destined to be his last. Growing tensions between Kroto and Smalley finally spilled over into confrontation. Despite their outward desire to maintain that the discovery of buckminsterfullerene had been a team effort, inwardly they had come to disagree bitterly over specific details. Kroto and Smalley parted company with some considerable animosity, the collaboration that had given them the most exciting breakthrough of their scientific careers in tatters.

Kroto returned to Sussex, despondent but determined to defend his interests against what he saw to be an unjust attack, and to do what he could to continue with research in the new field of carbon chemistry that he had helped to establish. Meanwhile, Smalley, Curl, Heath, and O'Brien applied Smalley's elaborate machines in the search for further experimental proof that the fullerenes were real.

The Rice group was desperate to find some unambiguous tell-tale sign that C_{60} was really shaped like a soccer ball. By far the most compelling form of evidence would be a spectral signature. The perfect symmetry of buckminsterfullerene had implications for the appearance of its ultra-violet/visible and infra-red absorption spectra, and theoreticians had been busily utilizing some of their more sophisticated techniques to calculate what these spectra should look like. Generating enough C_{60} to measure its infra-red spectrum seemed out of the question, but the ultra-violet/visible spectrum was in principle more amenable to the sophisticated technology of machines like AP2. If the scientists could only measure this spectrum and show that it corresponded—even roughly—to the predictions of the theorists, this would be a strong indication that the soccer ball structure was right.

The earlier Hückel calculations had all shown that soccer ball C_{60} should be a stable molecule with a closed electronic shell. However, the Hückel method grossly simplifies the problem in order to make the calculations easier to carry out. Although the method provides estimates of the stabilization achieved through delocalization of the pi-bond electrons, the calculated energies are not only very approximate, they are also usually not absolute values, but values relative to an unknown 'resonance parameter'. The absolute values of the energies are much harder to come by, yet these (and more) are needed in order to infer the positions and intensities of features in the molecule's ultra-violet/visible absorption spectrum.

In the years since Erich Hückel devised his simplified approach, theoreticians have made many improvements on it, largely by including in the calculations many of the things that Hückel chose to neglect. All of these improved techniques are still 'semi-empirical', since their successful application relies on knowledge of the actual properties of molecules obtained from experiment. Experimental data on selected molecules are used to estimate the values of certain parameters of the theory, which are in turn used to calculate the unknown properties of other molecules. These methods differ from the non-empirical *ab initio* approaches that calculate everything from first principles and require no prior information from experiment. The semi-empirical methods represent a kind of half-way house.

Of the various semi-empirical methods available to theoreticians 'off the shelf', the so-called Complete Neglect of Differential Overlap (CNDO) method is among the most sophisticated. As its name suggests, its application still involves the neglect of certain contributions arising from the overlap of electron orbitals. There are also various modifications of the basic CNDO method which differ in the selection of empirical parameters and the nature of the experimental data from which the values of the parameters are estimated. Of these, the so-called CNDO/S method is particularly suited to the calculation of the energies of electronic transitions, including absorption transitions.

The results of CNDO/S calculations for soccer ball C_{60} were reported in a paper published in July 1987 by Sven Larsson, Andrey Volosov, and Arne Rosén at the Chalmers University of Technology and the University of Gothenburg in Sweden. From the series of predicted energy levels, they were able to determine the energies and intensities of a large number of absorption transitions. However, not all these transitions were expected to appear in the absorption spectrum. The strength (intensity) of a transition strongly depends on the symmetries of the initial and final electron orbitals involved, and certain combinations of symmetries render the transition 'forbidden', reducing its intensity to near zero.

The very high symmetry of buckminsterfullerene meant that many of the predicted transitions were expected to be forbidden on symmetry grounds, leaving only a handful of 'allowed' transitions that would appear as features in the absorption spectrum in the ultra-violet/visible region. The lowest in energy was a weak feature around 340 nanometres, with much stronger features around 260, 240, 230, and 220 nanometres.

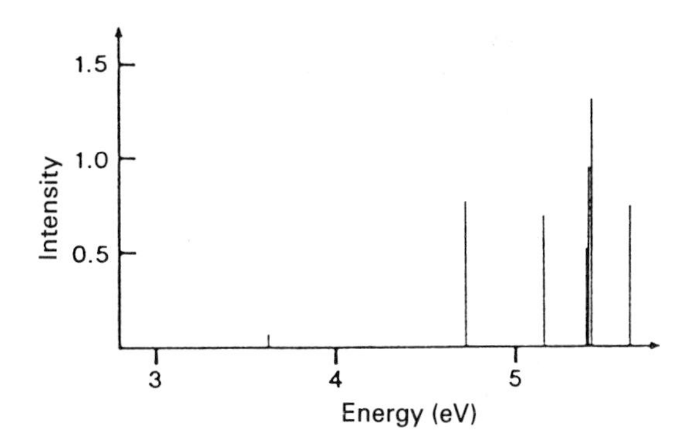

The ultra-violet/visible spectrum of soccer ball C_{60} predicted by Larsson, Volosov, and Rosén using the CNDO/S method. The theory predicts only the positions (energies or wavelengths) of the transitions and their total intensities, drawn as lines in this spectrum. In the 'real' molecule, the intensity of each transition would be distributed over a 'band', extending over a broad range of wavelengths. Each line in the predicted spectrum would therefore appear as a broad band in the measured spectrum, with the bands overlapping and merging. Apart from the six transitions shown here, a further 15 symmetry 'forbidden' transitions were also predicted but do not appear in the spectrum. Adapted, with permission, from Larsson, Sven, Volosov, Andrey, and Rosén, Arne (1987). *Chemical Physics Letters*, **137**, 501.

Smalley's cluster beam machines had been designed to do spectroscopy on exotic new molecular structures but, try as they might, the Rice scientists could not get AP2 to yield a spectrum of C_{60}. The technique that Curl had been keen to apply to the carbon clusters from the beginning of their collaborative research programme with Kroto—resonance-enhanced two-photon ionization—just did not seem to work with C_{60}. A tentative explanation was that the intermediate excited state of C_{60} produced by absorbing one photon was simply too short-lived. The probability of absorbing a second photon in the time available was just too low to give measureable ion signals.

In their desperation, they tried a radically different approach. As a postdoctoral associate with Leonard Wharton and Don Levy at the University of Chicago, Smalley had learned how to build and use big machines. He had also learned of the power of a supersonic expansion to cool molecules to extremely low temperatures, sufficient to produce and study very unstable molecules that would fall apart under normal, ambient conditions. With Wharton and Levy he had studied complexes formed by the loose association of two or more molecules, held together only by the slightest of interactions between the molecules' electron systems. These complexes—called van der Waals complexes after the Dutch physicist Johannes van der Waals—offered some fascinating new insights into chemical bonding and spectroscopy. They also offered a round-about route to the spectrum of C_{60}.

Heath, Curl, and Smalley studied van der Waals complexes of C_{60} with benzene (C_6H_6) and dichloromethane (CH_2Cl_2). The idea was to make these complexes in supersonic expansions, ionize them using a laser, and detect them in the usual way. The production of the complexes $C_{60} . C_6H_6^+$, and $C_{60} . CH_2Cl_2^+$ in separate experiments gave measured signal intensities that served as a baseline. They then irradiated the complexes with light from another laser which they scanned over a range of wavelengths in the visible and near ultra-violet. So delicate is the van der Waals bond holding C_{60} and C_6H_6 or CH_2Cl_2 together that even a small increase in the energy of the complex is enough to break it. So, if the C_{60} part of the complex absorbed light from the laser, the complex would break up and the measured ion signal would fall. By watching for a *depletion* of the signal as they scanned the wavelength of the laser, they could tell if C_{60} was absorbing light, and hence measure its absorption spectrum.

Scan after scan taken over several weeks produced just one tiny but reproducible absorption feature centred around 386 nanometres. That the feature was largely unchanged for both $C_{60} . C_6H_6$ and $C_{60} . CH_2Cl_2$ suggested that it was indeed an absorption due mainly to C_{60}, unaffected by the close proximity of the other molecule in the complex. It did not look like much, but it seemed they had measured at least one band in the spectrum of C_{60}.

Larsson, Volosov, and Rosén had predicted that soccer ball C_{60} should exhibit a weak feature around 340 nanometres. Given the limitiations of the theory the difference between 340 and 386 nanometres was probably bearable. The measured band was a lot weaker than predicted, but there were other mitigating factors that could explain this difference too.

In the short paper in which they described these experiments and gave their results, Heath, Curl, and Smalley emphasized a relatively straightforward observation. Ordinarily, a large unsymmetrical molecule with 60 atoms would be expected to have a broad absorption spectrum. That C_{60} produced a single, narrow band argued strongly for a rigid, highly symmetrical structure. That this band was close to a feature predicted for soccer ball C_{60} strengthened the scientists' case.

It was clear that as far as the ultra-violet/visible spectroscopy of C_{60} was concerned, the scientists at Rice could hope to do little better than the weak band they had measured at 386 nanometres. It was just as clearly not enough. They thought hard about other experiments they could do to prove that C_{60} had a soccer ball structure. They decided that they could add further to the case for buckminsterfullerene by watching how it fell to pieces under the sledgehammer blows of a high-power laser.

Together with Kroto, the Rice scientists had fought off the challenge presented the previous year by the groups at Exxon and UCLA regarding the interpretation of the mass spectra and the problem of fragmentation of the ions. Part of their response had been to show that C_{60}^+ itself could be fragmented using high laser intensities to produce a range of smaller and larger fullerene

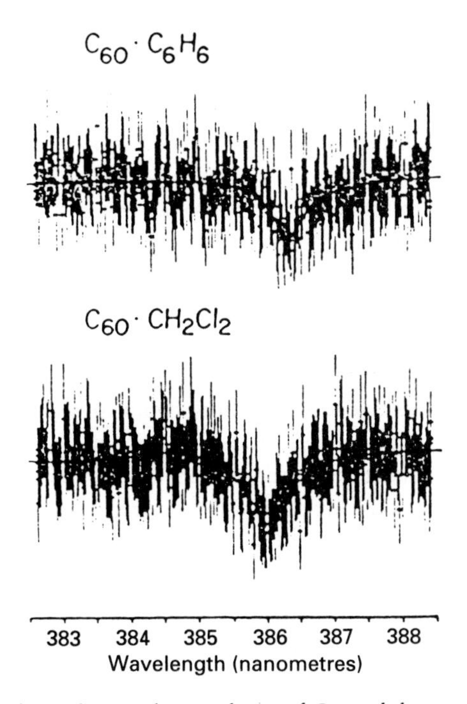

The spectra of van der Waals complexes of C_{60} with benzene (C_6H_6) and dichloromethane (CH_2Cl_2). The complexes were formed in a cluster beam by adding a little benzene and dichloromethane to the helium buffer gas inside AP2, and were ionized and detected using a time-of-flight mass spectrometer. By monitoring the signals of the $C_{60}.C_6H_6^+$ and $C_{60}.CH_2Cl_2^+$ ions whilst irradiating the cluster beam with light from a tunable dye laser, the Rice scientists could tell when the complexes absorbed light by a fall in the signals. The weak features around 386 nanometres observed for both complexes were ascribed to C_{60}. Adapted, with permission, from Heath, J. R., Curl, R. F., and Smalley, R. E. (1987). *Journal of Chemical Physics*, **87**, 4236.

ions. The Rice scientists believed there was much more to be learned from studying the ways in which these cluster ions broke apart under the influence of high energy radiation.

The group at Bell Labs had shown the previous year that the break-up of the smaller carbon cluster ions (C_n^+, $n<31$) occurs primarily by the loss of C_3 fragments. This was again entirely consistent with the idea that these small clusters have long-chain or ring structures. Theoretical calculations indicated that C_3 is a particularly stable fragment, making loss of C_3 the most thermodynamically favourable of the various possible ways of breaking up the C_n^+ cluster ions.

The larger cluster ions (C_n^+, $n>32$) didn't behave this way at all, as O'Brien demonstrated in a series of experiments on AP2 that was eventually to form the major part of his Ph.D. dissertation. Firstly, the larger cluster ions were much more difficult to fragment, requiring much higher laser intensities, with C_{60}^{+} the most difficult of all. Secondly, the fragmentation tended to produce other even-cluster ions, implying a loss of the less stable C_2 fragment rather than C_3. A mechanism based on closed cages seemed to be the only way of understanding this difference.

The Rice scientists reasoned that pumping energy into a fullerene cage causes its bonds to rearrange, eventually spitting out a C_2 fragment before closing up again to form the next smaller fullerene. This process could be repeated again and again, producing smaller and smaller fullerenes until C_{32} was reached. Further irradiation of C_{32} simply burst it apart to form linear chains. It was a logical mechanism, fully compatible with the experimental results and consistent with the fullerene idea. But how could they prove it?

Curl devised a particularly ingenious experiment. What would happen if they encapsulated a metal atom inside a fullerene cage which they then ionized and irradiated with intense laser pulses? If their fragmentation mechanism was right, then the same process should occur with the metal-containing fullerene. The cage of carbon atoms should shrink, two atoms at a time, until it eventually became too small to enclose the atom, at which point it would burst apart. This lower limit in cage size would depend on the size of the atom: the larger the central atom, the larger the fullerene needed to enclose it. They called it the 'shrink-wrap' mechanism. Demonstrating the process experimentally would support both the idea of fullerene cages and the possibility of encapsulation.

O'Brien began these experiments on AP2, before completing his Ph.D. and going on to further chemical physics research at Rice as a postdoctoral associate with Jim Kinsey. Heath completed his Ph.D. at around the same time and moved on to the University of California at Berkeley to work as a postdoctoral associate with Rick Saykally. Smalley and Curl continued the experiments with postdoctoral associate Jerry Elkind and graduate student Falk Weiss. They produced C_{60} cages containing a single potassium atom, ionized them, and trapped them in a device called a Fourier transform ion cyclotron resonance (FT–ICR) spectrometer. This apparatus enabled the scientists to store the trapped ions for relatively long periods of time (measured in minutes), allowing them to study their fragmentation patterns almost at leisure.

These experiments were a complete success. Like C_{60}^{+}, the potassium atom encapsulated KC_{60}^{+} also proved to be difficult to fragment except at very high laser intensities. But fragment it did, producing KC_{58}^{+}, KC_{56}^{+}, and so on all the way to KC_{44}^{+}, beyond which no further encapsulated cluster ions could be seen.

Cluster ions smaller than KC_{44}^{+} couldn't be made because the potassium atom sitting in the centre was too big. Shrink the carbon wrapping beyond 44 atoms and the cage could no longer stretch to cover the central atom: it would instead tear apart to give carbon chains.

Different-sized metal atoms should cause the fullerene wrapping to break up at different cluster sizes. The Rice group showed this to be the case with similar experiments on CsC_{60}^{+}, the ionized form of buckminsterfullerene containing a

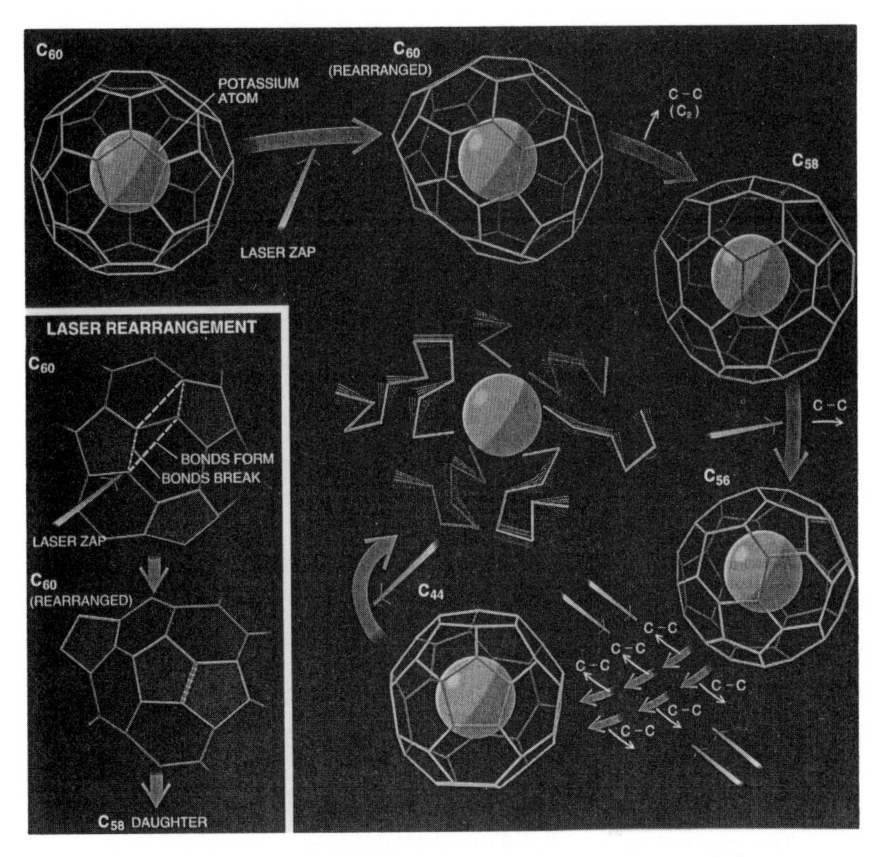

The shrink-wrap mechanism. Irradiation of KC_{60}, a buckminsterfullerene molecule with an encapsulated potassium atom, with intense light from a laser disrupts the carbon cage, causing it to spit out a C_2 fragment before closing up again. Subsequent irradiation of KC_{58} produces KC_{56}, and the process can be repeated again and again until KC_{44} is obtained. With each successive shrinking of the cage, it becomes more tightly wrapped around the central atom, and C_{44} is theoretically the smallest cage than can enclose it. Irradiation of KC_{44} shrinks the cage one step too far, and it bursts apart to form carbon chain molecules. Adapted from 'Fullerenes' by Robert F. Curl and Richard E. Smalley.

single caesium atom. A caesium atom is bigger than a potassium atom, and a simple estimate of the minimum size of fullerene needed to enclose a caesium atom suggested C_{48}, four more carbon atoms than the smallest cage that could fit around a potassium atom. When the scientists irradiated CsC_{60}^+ trapped in the FT-ICR spectrometer, they obtained signals corresponding to CsC_{58}^+, CsC_{56}^+, and so on all the way to CsC_{48}^+. Caesium-containing cluster ions smaller than this could not be produced.

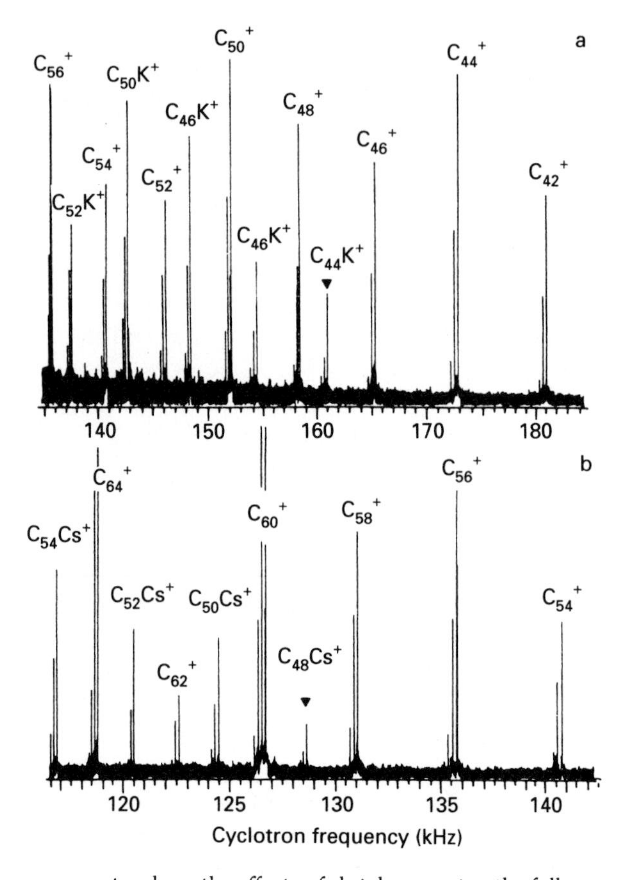

These mass spectra show the effects of shrink-wrapping the fullerene cage onto potassium and caesium atoms. The top spectrum shows signals due to KC_{52}^+, KC_{50}^+, KC_{48}^+, KC_{46}^+, and KC_{44}^+, as well as ions of 'bare' carbon clusters. The fullerene C_{44} is the smallest cage that can enclose a potassium atom, and, consequently, no KC_{42}^+ can be seen in the spectrum. For the larger caesium atom, C_{48} is the smallest cage it can fit inside. In the lower spectrum, no CsC_{46}^+ or CsC_{44}^+ can be seen. The results therefore support the idea of the shrink-wrap mechanism, the existence of a family of closed-cage molecules, and the possibility of incorporating metal atoms inside them. Adapted, with permission, from Curl, Robert F. and Smalley, Richard E. (1988). *Science*, **242**, 1017.

It all seemed to fit together neatly. For many scientists in the chemistry community, the experimental evidence in support of the fullerenes was now quite convincing, even though it was still all rather circumstantial. Not everyone was convinced, however. The group at Exxon continued to express doubts and had shown that it was possible to form clusters with as many as three potassium atoms attached to C_{60}. This, the Exxon scientists argued, was completely at odds with the soccer ball structure. The cage was too small to hold three potassium atoms inside it, and yet sticking atoms to the outside of the cage appeared implausible. With the bonding requirements of each of the molecule's 60 atoms met, there was nothing on the outside of the cage for the potassium atoms to hold onto. The only explanation was that C_{60} is not shaped like a soccer ball at all—it is some kind of graphite fragment with dangling bonds at its edges.

Now that he was out on his own, Kroto became acutely aware of the need to make a significant contribution to the unfolding fullerene saga; to establish his credentials independently of the work he had done in collaboration with the group at Rice. The only tried and tested method of making C_{60} and the other fullerenes was through the laser vaporization of graphite and subsequent expansion and cooling of the clusters. He therefore applied for (and eventually received) substantial funding from Britain's Science and Engineering Research Council to construct his own cluster beam apparatus in Sussex.

With one of his students, Ken McKay, he also carried out some preliminary experiments with an old arc evaporator. Kroto wondered if it might be possible to form C_{60} in the smoke and soot generated by evaporating graphite rods in an inert gas atmosphere. They drilled a hole in the base of the evaporator through which they admitted argon at pressures adjusted to match those in the clustering zone inside AP2. They then evaporated some graphite and studied the electron microscope images of the carbon soot they collected.

With high pressures of inert gas inside the evaporator, Kroto and McKay noted a change in the physical appearance of the soot deposit. With argon pressures between 40 and 100 torr the image became much grainier, leading Kroto to wonder if they were seeing the onset of the formation of spheroidal particles. The mechanism by which C_{60} was supposed to form inside AP2 suggested that even if it were possible, only minute quantities of C_{60} would be found in the soot. Kroto therefore reasoned that to stand any chance of detecting such small quantities of C_{60}, he would need a sensitive mass spectrometer. Over the next two years he applied to four different sources for funds to buy such a spectrometer, and was always led to believe that funding was just around the corner. But it never materialized.

As time had passed since the discovery of buckminsterfullerene, Kroto's initial elation had given way to doubt and uncertainty. He often wondered whether he and the scientists at Rice had stuck their necks out just a little too far. What he needed to calm his nerves was some additional piece of confirmatory evidence—something perhaps derived from the soccer ball concept that he could use to explain a different aspect of the experimental results obtained so far. Soccer ball C_{60} was the smallest fullerene structure containing 12 pentagons in which these pentagons could be isolated from one another, and this was clearly important to the structure's relative stability. So which was the next fullerene with isolated pentagons?

Kroto played around with some molecular models, but could not find a way to make closed cage molecules of C_{62}, C_{64}, C_{66}, or C_{68} without having two abutting pentagons somewhere in their structures. In fact, he already had the answer he sought. The paper he had published with the Rice scientists in which they had presented the first evidence for LaC_{60} complexes also contained a proposed closed cage structure for C_{70}. They had arrived at this structure by splitting C_{60} into two equal halves, introducing ten extra carbon atoms in a belt around the middle and stitching the lot together again. Of course, its pentagons were all isolated.

If C_{70} were the next fullerene after C_{60} to possess isolated pentagons, then perhaps this explained why the C_{70} signal always tended to appear faithfully alongside C_{60} in the mass spectra: Tonto beside the Lone Ranger. Isolating the pentagons gave C_{70} additional stability over neighbouring fullerenes and alternative C_{70} structures. Kroto had read the article on the stability of different C_{60} structures published by the group at Texas A&M in October 1986, and decided to contact Tom Schmalz to discuss this notion of the isolated pentagons. Schmalz explained that they had arrived at exactly the same principle. They had proven that structures for C_{62}, C_{64}, C_{66}, and C_{68} could not be constructed without abutting pentagons.

As a 'magic' number, 60 was readily explained in terms of the soccer ball, but until now there had been no obvious reason why 70 should be the next magic number. Isolated pentagons seemed to hold the key. Kroto wondered if it might be possible to pursue this line of reasoning to explain some of the other magic numbers that they had seen from time to time in the cluster distributions.

If this business about the pentagons was right, Kroto figured that there should be a hierarchy of stability amongst fullerene structures based on the number of pentagons that were obliged to be fused together. For a given fullerene (with a given number of carbon atoms) the most stable structures would be the ones with all the pentagons isolated, as in C_{60} and C_{70}. The next most stable structures would be those with two abutting pentagons, followed by three and then four. The argument for this hierarchy was simply one of strain energy: the more pentagons fused together in the structure, the more strained and less stable it becomes.

Looking back at the data obtained from AP2, Kroto noted that 50 appeared to be the next magic number below 60. Clearly, the structure of C_{50} had to contain some abutting pentagons, but could it be that it was the first in the series of fullerenes in which there were no fused triplets of pentagons? Schmalz, Seitz, Klein, and Hite later proved that this was indeed the case. So which was the first structure to avoid a fused quartet of pentagons? Playing around with molecular models, Kroto discovered that it was C_{28}. Now this was something. He recalled that in some of the cluster distributions they had obtained at Rice, the signal for C_{28} had sometimes appeared as large as that for C_{60}.

Kroto found that he could develop similar stability arguments to explain the 'magicness' of fullerenes with 24 and 32 atoms. Under the right conditions, C_{24} was the first of the large even clusters to show any kind of stability. After intense laser irradiation, C_{32} was the smallest cluster of this group that could be obtained by fragmenting and shrinking the larger clusters, as the Rice group had shown. The magic number sequence was therefore 24, 28, 32, 50, 60, and 70. At various times and under certain conditions, carbon clusters in this magic number sequence had all shown themselves to be 'special'.

The logic that had led to this interpretation of the stabilities of the magic number fullerenes was based on geometrical arguments combined with a little chemical intuition. However, the problem remained that for the larger fullerenes there appeared to be little to choose between the vast number of different polyhedral structures that were possible, and there was no easy way of telling which of these would have closed electronic shells, short of doing the Hückel calculations.

This particular problem was partly solved in 1987 at about the same time that Kroto was arriving at his isolated pentagon rule. Theoretical chemist Patrick Fowler and his colleagues at the University of Exeter had published a number of research papers on the relationships between polyhedral structures, their symmetries and relative stabilities. With J. I. Steer, Fowler developed some simple rules that could be used to predict which polyhedral structures of a given fullerene would be expected to have both isolated pentagons and closed electronic shells. This was Fowler's 'leapfrog' principle.

Start with any polyhedral structure composed of pentagons and hexagons, follow the rules of the 'leapfrog construction' and you can construct a polyhedron with three times as many vertexes as the original, with all its pentagons isolated from one another and which, in the context of a fullerene cage, is guaranteed to have a closed electronic shell.

The procedure involves first 'capping' the starting structure by drawing a point at the centre of each face and drawing a line from this point to each vertex of the face. Each pentagon then becomes a set of five triangles, each hexagon becomes six triangles. (This 'omni-triangulation' is reminiscent of Buckminster Fuller's energetic geometry.) Now take the 'face dual' of this capped structure by placing a point at the centre of each triangle and draw lines to connect adjacent points together. This series of connected points defines the final structure.

Through the leapfrog construction (so named because it 'leaps' over an intermediate deltahedron), a hexagon in the starting structure transforms to a hexagon in the final structure, rotated through 30°. A pentagon in the starting structure transforms to another pentagon, rotated through 36° and *surrounded by hexagons*. A vertex transforms to a hexagonal face and an edge transforms to another edge rotated through 90°. Applied to the smallest possible fullerene C_{20}—a cage of 12 pentagons unstable in geometical terms and with an open electronic shell—the leapfrog construction gives soccer ball C_{60}.

The value of Fowler's construction becomes apparent when it is used to deduce possible structures for some of the larger fullerenes. For C_{78}, there are 21 822 possible closed-cage structures, but there is only one possible structure for C_{26}. Applying the leapfrog construction to C_{26} yields a structure for C_{78}

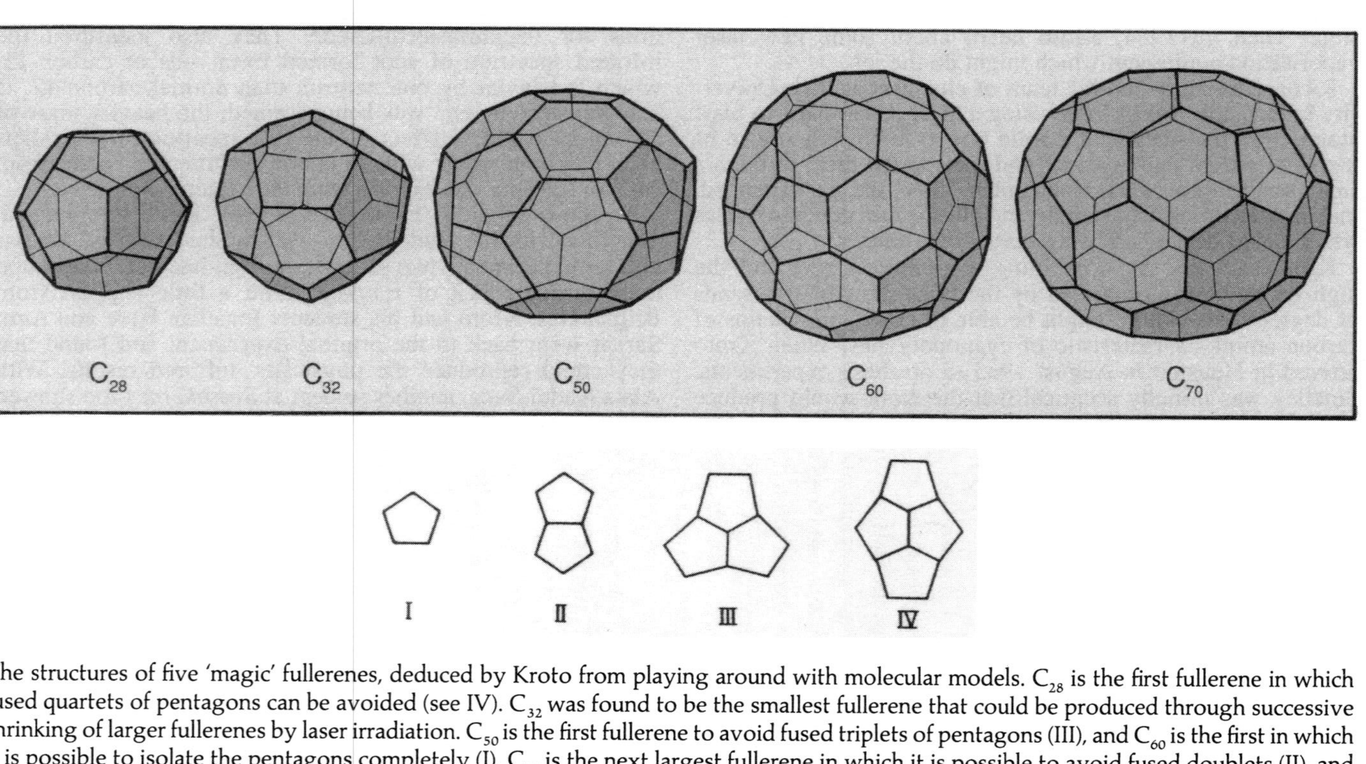

The structures of five 'magic' fullerenes, deduced by Kroto from playing around with molecular models. C_{28} is the first fullerene in which fused quartets of pentagons can be avoided (see IV). C_{32} was found to be the smallest fullerene that could be produced through successive shrinking of larger fullerenes by laser irradiation. C_{50} is the first fullerene to avoid fused triplets of pentagons (III), and C_{60} is the first in which it is possible to isolate the pentagons completely (I). C_{70} is the next largest fullerene in which it is possible to avoid fused doublets (II), and have all the pentagons isolated. Adapted, with permission, from Kroto, Harold (1988). *Science*, **242**, 1139; (1987), *Nature*, **329**, 529.

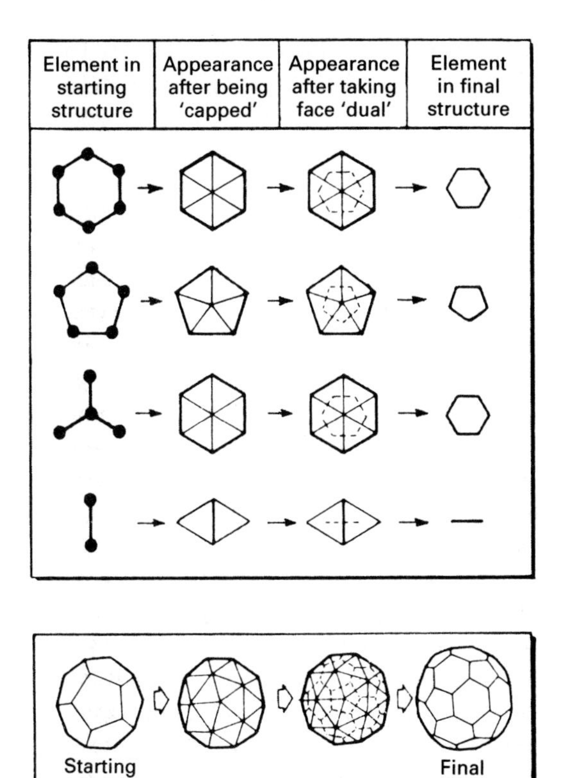

The leapfrog construction. Draw a point in the centre of each pentagonal and hexagonal face of some starting polyhedron, and join this point to each vertex of the face. The result is a 'capped' polyhedron, reminiscent of Buckminster Fuller's 'omni-triangulated' structures. Now take the 'face dual' of the capped structure by drawing a point in the centre of each triangle and connecting these points together (dashed lines). The resulting network is a new polyhedron with three times as many vertexes as the original, with all its pentagons isolated and (in the context of the fullerenes) with a closed electronic shell. Applying the construction to C_{20} produces soccer ball C_{60}. Adapted, with permission, from Baggott, Jim *New Scientist*, October 26, 1991.

which has both isolated pentagons and a closed electronic shell: one uniquely stable structure among 21 822 different possibilities. The leapfrog construction doesn't necessarily identify every stable structure from all the different possibilities, but it makes the process of searching for them much more straightforward.

Fowler found that he could generalize his principle to a simple formula. Fullerenes C_n have closed electronic shells when $n = 60 + 6k$, where k can be equal to zero or any integer number greater than one. Thus, C_{60}, C_{72}, C_{78}, C_{84},

C_{90}, C_{96}, etc. are all predicted to be closed-shell molecules. The rule applies only to structures that are leapfrog constructions of smaller fullerenes, and so is not exhaustive. Thus, C_{70} cannot be a leapfrog construction, and yet it does have a closed shell. Fowler and his colleagues later discovered relationships that could be applied to these other fullerenes, identifying two classes of carbon 'cylinder' with $n = 70 + 30k$ and $n = 84 + 36k$, where k is zero or a positive integer.

Every advance in understanding the factors that influenced the stabilities of (hypothetical) closed cage molecules lent support to the argument that they were real. But for the hard-nosed experimentalist, there was still no unambiguous proof. And the storm clouds were gathering.

Curl had warned of the dangers of dabbling in soot chemistry at the time he, Kroto, Smalley, Heath, and O'Brien had developed their icospiral nucleation mechanism. He was right to be concerned. As the bucky bandwagon continued to gain momentum, Curl, Smalley and Kroto did much to publicize and promote the virtues of the mechanism in review articles and lectures at international conferences. Here was an interesting new idea that might help to explain some of the long-standing mysteries of soot formation.

However, there were many in the close-knit community of soot specialists who didn't quite see it this way. What they saw was a bunch of chemical physicists relatively unfamiliar with the substantial body of accumulated knowledge in soot chemistry running around claiming to have solved all the problems of the field in one fell swoop. And with a mechanism that made no sense to those who knew better. They were outraged.

The broad consensus that grew within the more voluble section of the soot community was that, whilst the icospiral mechanism was no doubt an attractive notion, it had nothing whatsoever to do with soot. The most outspoken in their criticism of the mechanism were Michael Frenklach, at the Department of Materials Science and Engineering at Pennsylvania State University, and Lawrence Ebert, a combustion specialist at Exxon's Corporate Research Laboratory. There were many problems with the icospiral mechanism, they claimed, but the major ones were simply stated.

The mechanism couldn't be right because the speed with which the nautilus structure is likely to grow is just too slow to account for soot formation. Chemical analysis of soot samples reveals that it contains many elements apart from carbon (such as hydrogen, oxygen, sulfur, and nitrogen), whereas the icospiral mechanism requires soot particles made entirely from super-large carbon molecules. The results of X-ray diffraction and carbon-13 NMR spectroscopy studies of soot are all consistent with the traditional model of soot formation based on the coalescence of polycyclic aromatic hydrocarbons. Soot is chemically reactive, whereas the fullerenes were supposed to be relatively unreactive.

While Frenklach was prepared to concede that it may indeed be possible to form large closed-cage molecules of pure carbon, he was adamant that they could have nothing to do with soot. Ebert, on the other hand, was prepared to concede nothing. Like his Exxon colleagues Cox and Kaldor, soccer ball C_{60} and the other fullerenes were an unproven hypothesis.

Whilst there was much experimental evidence that could be brought to bear against the icospiral mechanism, much of it could be refuted. Smalley and Kroto repeatedly argued that their mechanism was intended to explain the very beginning of the process of soot formation, not necessarily the whole of it. It was this part of the process that the soot chemists themselves had to admit they didn't understand. Much of the experimental data—and particularly the presence of elements other than carbon—could be explained in terms of the later stages of the process after the nautilus-like structures had grown beyond a certain size.

There was even some experimental evidence that Smalley and Kroto interpreted to be in *favour* of the involvement of C_{60} and similar structures in soot and therefore a vindication of their mechanism. In 1987, the German soot chemist Klaus Homann and his colleagues at the Max Planck Institute for Physical Chemistry in Darmstadt reported that they had found evidence for the formation of C_{60} in 'sooting flames'. These were flames produced by the combustion of mixtures of oxygen and acetylene or benzene, in which the proportion of oxygen was insuffiсent to ensure complete combustion of the gases to carbon dioxide and water vapour. Without sufficient oxygen to burn all the carbon-containing material the result was a pall of smoke and soot.

According to the icospiral model, soccer ball C_{60} is the result of a chance closure of the structure, preventing further spiralling growth. Thus, C_{60} is a by-product of soot formation: any combustion system producing soot would therefore be expected to produce a little C_{60} also. Homann detected a molecule containing exactly 60 carbon atoms using a mass spectrometer and so was able to confirm that it is present in a sooting flame, but was unable to infer its structure from the data. Homann had no difficulty in accepting that this was buckminsterfullerene, but he did not believe it was being formed as a by-product of the icospiral mechanism.

Homann believed that C_{60} was being formed *after* the soot particles had been produced in the flame. His proposed mechanism was no less picturesque than the icospiral model, although it had the merit of not conflicting with the conventional wisdom of the soot community. He suggested that at high temperatures, the carbon bonding in an outside layer of a sandwich-like agglomeration of graphite fragments may rearrange—a pentagon appearing where before there had been a hexagon. The result of introducing this pentagonal defect is to kink the layer, forming a kind of 'bubble' on the surface. The bubble grows by further rearrangement and the appearance of more pentagons until the bubble is sealed off, releasing a soccer ball C_{60} molecule.

The experimental results were sufficiently open to interpretation for those on either side of the argument to conclude whatever they wished. For Smalley and Kroto, Homann's findings confirmed the role of the icospiral model in soot

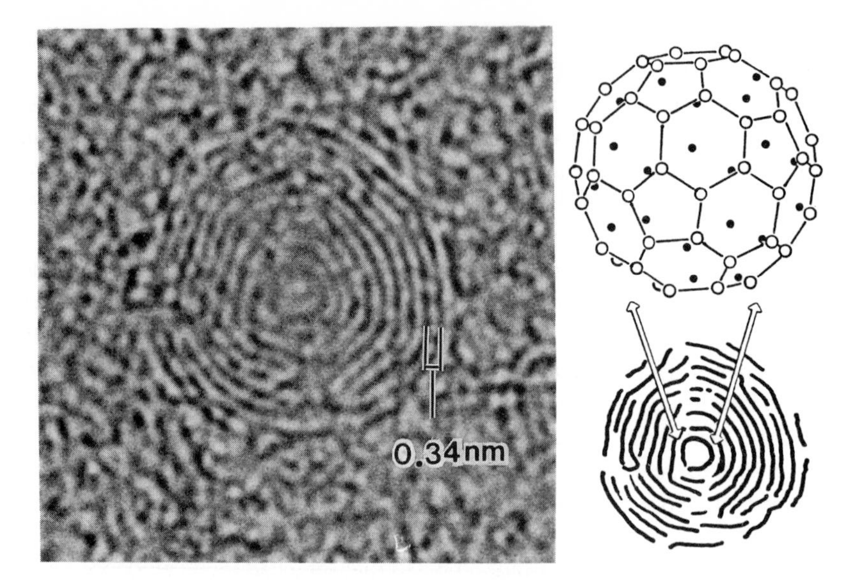

This electon microscope picture shows the random jumble associated with amorphous graphite, together with an intriguing structure of concentric spheres, with each layer separated by about 0.34 nanometres. The sphere right at the centre has a diameter of about 0.8 nanometres, corresponding to the diameter of soccer ball C_{60}. Reproduced with permission from Iijima, Sumio (1987). *Journal of Physical Chemistry*, **91**, 3466. Copyright (1987) American Chemical Society.

formation. For Homann himself and many others in the soot community, these data supported a radically different view: C_{60} molecules were produced from the surfaces of small graphitic particles.

A second piece of experimental evidence that was argued to be in support of the icospiral mechanism came from an unexpected but rather dramatic discovery. Whilst examining electron microscope pictures of films of carbon particles in 1980, Sumio Iijima at the NEC Corporation's Fundamental Research Laboratory in Japan had noticed some odd shapes. The pictures were of thin layers of amorphous carbon deposited on a substrate following evaporation of graphite rods in an electrical discharge. This was the same technique that Huffman, Krätschmer, and Sorg would use in their bell-jar experiments in Heidelberg two years later, and is a standard method for making carbon films for electron microscopy. Iijima noticed that amid the jumble of lines corresponding to the essentially random arrangement of fragments of graphite, there occasionally appeared a much larger, concentric arrangement of lines reminiscent of the layers of an onion.

Each dark line in these structures represents a nearly spherical shell of carbon, and a typical structure has about ten concentric layers. The spacing between each layer is about 0.34 nanometres, equivalent to the spacing between the flat layers of chicken wire in ordinary graphite. At the centre of the onion is a small sphere about 0.8–1.0 nanometres in diameter.

In 1980, Iijima had not quite known what to make of these spherical structures, but he had suggested that the layers might be a series of graphite shells, with the structure at the centre possessing the 12 pentagons necessary to ensure closure. After reading of the discovery of buckminsterfullerene in 1985, he quickly became convinced that the structures in his electron microscope pictures were fullerenes, one inside the other with soccer ball C_{60} at the centre. The method of making the films—arc discharge through two graphite rods—was after all not so different in terms of temperatures to the laser vaporization method used by the group at Rice. Iijima announced that the 60-carbon cluster had been 'revealed' in an article published in the *Journal of Physical Chemistry* in 1987.

For Smalley and Kroto, here was yet more experimental evidence supportive of the icospiral mechanism—electron microscope pictures of layers of graphite wound one around the other in a near-spherical structure. Of course, there was no telling the *molecular* structures from these pictures—still no proof of the soccer ball shape—but this was surely too much of a coincidence to be explained any other way.

In the end, the arguments between the radical icospiral enthusiasts (the 'bucky' pioneers) and the conservatives in the soot community became rather personalized. The protagonists in the debate clashed at a meeting held on Frenklach's home ground at Pennsylvania State University in the summer of 1989. Kroto presented an overview of fullerene research and described the icospiral model. Ebert then followed with a presentation directly aimed at dismantling these new ideas. Kroto did not take kindly to this treatment—he did not mind the scientific challenge but would have preferred an opportunity to prepare and present counter-arguments. Both Smalley and Kroto eventually dismissed the challenge, accusing the soot specialists of reactionary behaviour, with closed minds unprepared to consider any new idea in a field that had become a 'backwater' of modern chemistry.

Frenklach and Ebert could not be assuaged. Just because the icospiral mechanism was elegant, they argued, it didn't automatically follow that it must be right. This was true for the whole bucky bandwagon, still characterized in their opinion more by speculation than hard fact. They didn't like what was happening.

There was a recent precedent for this kind of abberant science. A similarly simple and elegant concept, backed up by nothing more than incomplete and irreproducible experimental data, had been dramatically announced at a hastily convened press conference at the University of Utah on March 23, 1989. It was inevitable that, sooner or later, somebody would compare buckminsterfullerene with 'cold fusion'.

In an article published in the American journal *Chemical and Engineering News* in February 1990, staff journalist Rudy Baum described how Frenklach and

Ebert had come to consider fullerene research to be like cold fusion, another example of 'pathological' science. This was science where the traditional values and methods of logic and reasoning had been abandoned and replaced by a pathological adherence to a self-justifying concept. Just as Martin Fleischmann and Stanley Pons insisted that cold fusion was right with apparently irrational conviction, so Smalley and Kroto saw fullerenes everywhere, including places where they had no business to be looking. Although Ebert later denied that he had ever used the term pathological in connection with fullerene research, it captured the mood of the debate. There were some scientists who felt that things had gone too far.

This kind of conflict is not unusual, but is relatively infrequent, in all branches of science (as it is in all aspects of human endeavour where personalities and interrelationships are important). At its heart was uncertainty: the uncertainty associated with building a new field of carbon chemistry on less than solid foundations. By the end of 1989, there were few scientists who disputed the evidence in favour of soccer ball C_{60} and the other fullerenes. But, such as it was, this evidence was not exactly unassailable.

What, the hardened sceptics would ask, did the bucky pioneers really have to go on? A series of blips in a mass spectrum, potentially explicable in a number of different ways none of which required closed cages of carbon atoms. Some measurements of chemical reactivity (or lack thereof) requiring too many assumptions to be conclusive. Some nice drawings of polyhedra and some theoretical constructions that were *supposed* to be stable (but then the theorists were notorious for getting things wrong when there were no solid experimental data). A 'shrink-wrap' mechanism that assumed what it was supposed to be proving. A weak absorption feature from some C_{60} van der Waals complexes that might (or might not) have something to do with absorption of light by C_{60} and which, if it did, proved nothing. A mechanism for soot formation completely at odds with virtually everything known about the properties and chemistry of soot.

In 1988, a further 30 research papers and review articles were published on buckminsterfullerene and its relatives. In 1989, this figure fell to 24. It seemed that the steam was beginning to run out. There was a limit to the ingenuity of the experimentalists: all the key experiments that could be devised to obtain circumstantial evidence had essentially been done. The theoreticians had calculated what was readily calculable. There were more games that could be played with polyhedra and magic numbers but these did not help solve the immediate problem. What everybody needed now was some *stuff*—a sample of buckminsterfullerene that could be tested, analysed, reacted, measured. Without such a sample there seemed little more that anybody could usefully achieve.

And yet, as far as the scientists actively involved in fullerene research could tell, they were as far away from that sample as they had ever been. What hope of making this stuff in large quantities when it formed at high temperatures only by chance? To build up soccer ball C_{60} literally atom by atom in a laboratory synthesis was a daunting task. The total synthesis of the much smaller cage molecule dodecahedrane ($C_{20}H_{20}$) had been reported by Leo Paquette and his colleagues at Ohio State University in 1983. It required 23 steps. A synthesis of buckminsterfullerene would probably require many more steps. Few organic chemists were sufficiently brave (or foolhardy) to attempt it. Orville Chapman was an exception.

The fullerene researchers had tried all the obvious things. They had battered the problem with logic and high technology. It seemed that a further breakthrough would take nothing short of a miracle.

9

A crazy idea

Ever since Lowell Lamb had told him about the discovery of buckminsterfullerene in December 1985, Huffman's mind had been working overtime. Lamb was a young, enthusiastic postgraduate student recently arrived in Huffman's group from Wall Street, where he had worked on computer systems. He continued to follow events in the financial world through the pages of the *New York Times*, and had read an article by Malcolm Browne on the discovery of soccer ball C_{60} published on December 3, 1985. He showed the article to Huffman.

Huffman's mind reeled. The experiments he had done on carbon soot with Krätschmer and Sorg during his sabbatical leave in Heidelberg in 1982–83 had turned up some puzzling results they couldn't explain. He had originally wondered if the camel humps might not be trying to tell him that there was something very unusual about some of the samples of soot they had made. But the humps had been so erratic and irreproducible. They were so typical of the kind of thing that can happen when something goes wrong with an experiment. He had come to be persuaded that Krätschmer was right and the humps were really due to some kind of junk.

Now there was buckminsterfullerene. This magical molecule had popped out of nowhere in an elaborate cluster-beam apparatus. It was so astounding it just had to be true. If Huffman was looking for something unusual, here was unusual writ large. Was it really possible that in that carbon evaporator in Krätschmer's lab in Heidelberg they had been making copious quantities of an all-carbon molecule shaped like a soccer ball? Huffman needed no persuading. As soon as he read about buckminsterfullerene he was convinced that this was what they had been making. But the more he thought about it, the more ludicrous it seemed.

The C_{60} molecule had formed in AP2 under favourable clustering conditions, adjusted to bring it into prominence. The interpretation of these experiments was that C_{60} was a survivor—its highly symmetrical and stable structure preventing its destruction or further growth in the cluster machine, in contrast to the other even clusters which were not so stable (the exception being, of course, the ever-faithful C_{70}). The end result was that, under the right circumstances, all the neighbouring even clusters could be stripped away, leaving C_{60} (and C_{70}), naked under the light. The presumption was that C_{60} was being formed almost by lucky chance: about one in ten thousand or even one in a million.

Nobody believed it was possible to form buckminsterfullerene *spontaneously* and in large quantities just by heating graphite rods electrically in an inert gas atmosphere. Such a proposition appeared simply unthinkable. And yet this was exactly what Huffman required if buckminsterfullerene was going to account for the presence of the camel humps in the ultra-violet spectra of the soot samples. When dealing with uncertainty in science, no scientist is averse to a little honest speculation. But this was not speculation, it was sheer madness.

Huffman contacted Krätschmer and asked him what he thought about the proposal. Krätschmer was forthright: he thought it was a crazy idea.

After some initial attempts to correlate features in the visible spectra of the small, matrix-isolated carbon clusters with features in the infra-red, the collaboration between Huffman and Krätschmer had faded. Krätschmer had become increasingly tied up with important work for the European Space Agency's Infra-red Space Observatory (ISO). This was to be an orbiting satellite with a liquid-helium-cooled infra-red detector designed to study infra-red objects in the universe, scheduled for launch in 1993. The funding for the ISO project had allowed Krätschmer to purchase a state-of-the-art Fourier transform infra-red (FTIR) spectrometer, an instrument far superior to the battered old spectrometer he had been using previously. The new spectrometer was needed to carry out careful calibration measurements of filters and optical components for the observatory.

Krätschmer had received news of the discovery of buckminsterfullerene from his colleague Alain Léger at the University of Paris. Léger was a specialist in the physical properties of polycyclic aromatic hydrocarbons, molecules believed to be responsible for strong infra-red emission from interstellar dust clouds heated by ultra-violet radiation. In November 1985, Léger had sent Krätschmer a preprint of the *Nature* paper (the manuscript version printed out from Smalley's office computer). Krätschmer had been absolutely thrilled by this discovery, and he had no difficulty accepting the proposal that C_{60} was shaped like a soccer ball. Having spent some years working in the field of interstellar dust and, particularly, interstellar carbon, he recognized the work as a revolution in thinking about this most familiar of all elements. Until Huffman contacted him, it had never occurred to him to connect buckminsterfullerene with the camel humps, which had long ago been banished from his thoughts as an irritating experimental artifact.

However, if Krätschmer was sceptical, Lamb was at least prepared to give Huffman the benefit of the doubt. He was intrigued by the idea and, as his postgraduate thesis topic hadn't been finalized at that stage, he asked Huffman to consider a project on C_{60}. Huffman was not encouraging. He had no funding for this kind of study, and figured that obtaining new funds for such a crazy proposal was going to be impossible. Instead, he put together a short proposal for a project on light scattering from small particles relevant to semiconductor manufacture, which was eventually supported by a consortium of local semiconductor companies. Lamb was a little disappointed, but such was Huffman's conviction that they had been making C_{60} in the Heidelberg evaporator that he was not averse to Lamb doing the occasional experiment in Tucson to check it out.

Nineteen eighty-six rolled on into 1987. The group at Rice continued to amass circumstantial evidence that the soccer ball structure of C_{60} was right. The Exxon group continued their attempts to deflate the proposal. More and more papers appeared by theoreticians containing predictions for the structures, bonding, energy states, and spectra of a 'hypothetical' soccer ball molecule.

Huffman was especially interested in the CNDO/S results published by Larsson, Volosov, and Rosén in July 1987. These calculations suggested that buckminsterfullerene should exhibit a very weak absorption feature in the ultra-violet corresponding to a wavelength around 340 nanometres. Further, much more intense features were predicted around 260, 240, 230, and 220 nanometres. Huffman recalled that the camel humps had appeared at around 340, 265, and 215 nanometres. The CNDO/S calculations could not be expected to reproduce the detailed appearance of the spectra, but the rough agreement between the predictions and the positions of the camel humps was rather suggestive.

Huffman sent the paper to Krätschmer and asked his opinion. Krätschmer agreed that this was very interesting, but it still all seemed to be too far-fetched. The camel humps were rather broad, they were not so prominent and they appeared in regions of the spectrum where many different kinds of molecules showed strong absorptions. In addition, these calculations were notoriously difficult to do accurately. Larsson, Volosov and Rosén had stated that their test calculations on naphthalene gave results within 0.5 electron volts of the experimental data, and this was *good agreement*. This level of accuracy placed the predictions for any of the strong ultra-violet bands of C_{60} anywhere between 275 and 200 nanometres. Krätschmer was not convinced that the apparent agreement between the CNDO/S predictions and the camel humps was anything more than coincidence. This was an accidental agreement between theory and experiment that Huffman had latched onto because it fed his crazy idea about buckminsterfullerene.

For Huffman, the sticking point finally came in a crowded bar in Tucson one Friday evening in September 1987. He had been discussing science with his colleagues over several pitchers of beer when the discussion turned to the protection of intellectual property rights through patents. Phil Krider, the patent counsellor of the University of Arizona's Technical Committee, argued that a scientist should attempt to patent every discovery, no matter how apparently trivial. His reasoning was simple. If the discovery turned out after all to be trivial, nothing was lost except a little time, money, and effort framing the patent disclosure. But if the discovery turned out to be really significant, well . . .

No matter how Huffman weighed the evidence, the proposal that buckminsterfullerene was being formed in the soot from a carbon evaporator did not appear to have much going for it. Now he was listening to Krider argue that scientists should patent even the most apparently insignificant discoveries—just in case. That clinched it for Huffman. He decided that if this absurd proposal had even the faintest chance of being right, he couldn't afford not to protect it.

Huffman began the process of drafting a patent disclosure the very next day. He explained his proposal regarding buckminsterfullerene to Krätschmer and described his experience in the bar. He said that he was planning to try to patent the process of making carbon soot containing the mysterious camel-hump-producing material which he believed could be buckminsterfullerene. Krätschmer thought this was just the kind of proposal one might make after several large pitchers of beer in a bar in Arizona, but in the cold light of day it was surely preposterous.

However, Huffman was determined. He was talking to a patent lawyer and drawing up a disclosure for the evaporation process—in his and Krätschmer's names—to be submitted through the University of Arizona. He planned some further experiments to try to improve the reproducibility of the camel samples. Krätschmer was too involved with the work on the ISO project to contemplate going back to the carbon evaporation experiments himself, so he wished Huffman well with his patent application and left him to it. Krätschmer could not help thinking it was all nonsense.

Krätschmer's scepticism was apparently justified when, at a conference on interstellar dust held at the Department of Astronomy at the University of Manchester in December 1987, he encountered Harry Kroto. Krätschmer raised with Kroto the question of the predicted ultra-violet spectrum of buckminsterfullerene, and Kroto pointed him in the direction of the depletion spectroscopy studies that Heath, Curl, and Smalley had reported that summer. Krätschmer was unaware of this work, and when he read the paper he discovered that this indirect method of determining the spectrum of C_{60} had revealed one weak absorption feature at 386 nanometres. This did not look much like the pattern of absorption bands predicted by the CNDO/S calculations, or the camel humps.

Lamb used the Tucson evaporator in early 1988 to repeat the experiments that Huffman, Krätschmer, and Sorg had done in Heidelberg six years before. The mysterious camel humps had appeared in the ultra-violet spectra of soot samples produced under relatively low pressures of inert gas, so Lamb used the same conditions. He found that he could sometimes observe the camel hump phenomenon, but this was still rather irreproducible. At least the phenomenon wasn't unique to the Heidelberg evaporator, which might have been anticipated if some experimental artifact was to blame. Huffman was bolstered by Lamb's results, but they got no closer to understanding what was going on.

By February 1988, Huffman and Lamb had failed to find a way to produce samples of carbon soot that would reproducibly show the camel humps, and Huffman had no choice but to withdraw his patent application. He was too busy with other, less frustrating, research projects and Krätschmer's apparent lack of interest did not encourage him to persevere. Inwardly resigned, Huffman's attention turned to other matters. But his conviction was unshaken: he was not prepared completely to let go his crazy idea.

Matters came to a head again in July 1988, when Huffman and Krätschmer got together at an International Astronomical Union (IAU) symposium on interstellar dust in Santa Clara, California. A presentation on the interpretation of the 217 nanometre interstellar band was followed by a panel discussion. Huffman was one of the panel speakers and Krätschmer sat in the audience. As the discussion turned to the possible role of graphite particles, Huffman gave arguments in favour but then changed track.

He described the camel humps that he and Krätschmer had seen in the ultra-violet spectra of some samples of the carbon soot produced by evaporation of graphite rods, shouting '. . . isn't that true, Wolfgang?'. Krätschmer was startled from his semi-wakeful state: he hadn't expected this. Huffman then went on to express publicly his crazy idea that buckminsterfullerene might be responsible, citing as evidence the recent theoretical predictions for the ultra-violet spectrum of soccer-ball C_{60}. 'After all,' Huffman maintained, 'what else can it be otherwise?'

Nobody really believed any of it. One of the features of the symposium was that the papers presented, as well as the panel discussion, were written up and published in a volume of proceedings. The proceedings of IAU Symposium No. 135, 'Interstellar dust', held in Santa Clara between July 26–30, 1988, make no mention of Huffman's proposal.

Huffman and Krätschmer met up afterwards to discuss what to do next. Huffman was still very keen to persuade his German colleague that they should both go back and do some more carbon evaporator experiments. Krätschmer had grown a little tired of Huffman's persistence, but bowed to the pressure. He agreed to try to do some more experiments using the bell-jar evaporator in Heidelberg, for no other reason than to lay this particular ghost to rest by showing Huffman that he was wrong. This was a burden Krätschmer could have well done without. He had no convenient student who could step in at short notice to do these experiments, and he was really far too busy writing non-conformance reports for the ISO project to do them himself.

Bernd Wagner arrived at the Cosmophysics Department at the Max Planck Institute for Nuclear Physics in the late summer of 1988, looking for something to do. He was an undergraduate student studying physics at the University of Cologne, and had decided that he wanted to spend two or three weeks breathing the rarefied atmosphere of frontier research and getting to know what a real research institute looks like from the inside. Krätschmer thought this unusual: most students seemed to prefer to be more familiar with the outside of the Institute. As nobody could find any gainful employment for this young idealist, Krätschmer asked if he would be interested in doing some experiments with the carbon evaporator.

Krätschmer explained how to use the evaporator, and Wagner set to work. The young student had no clear idea what he was doing, but the procedure was simple enough and he was soon producing samples of carbon soot for further analysis. It was Krätschmer's intention to repeat the experiments he had done with Huffman and Sorg in 1982–83 and try to reproduce the camel hump phenomenon. This time, however, he had a powerful Bruker FTIR spectrometer courtesy of the ISO project, and could look closely at both the infra-red and ultra-violet spectra of the soot samples.

Following Krätschmer's recommendations, Wagner kept the pressure of inert gas inside the bell-jar to 20 torr or less. The aim of these repeated experiments was still to generate graphitic particles that could serve as laboratory analogues of interstellar dust grains, and it was therefore important to avoid the clumping effects likely with higher gas pressures.

Wagner worked on the evaporator, and handed the soot-coated disks to Krätschmer, who then measured the spectra. They used disks of thin quartz for the ultra-violet spectroscopy and silicon or germanium for the infra-red. Immediately the mysterious camel humps reappeared in the ultra-violet spectra of some of the samples, but their irreproducibility was no less frustrating than before. The infra-red spectra of these new camel samples showed only the broad absorption characteristic of soot—due to stretching and bending vibrations of distorted graphite structures—together with some hints of sharp lines more characteristic of molecules. These lines were sufficiently faint to be dismissed as being due to contaminants.

It must be understood that at this stage, there was no good reason why Wagner should have done what he did next. A scientist who knew a bit more about what he or she was doing and why would almost certainly not have done it. Wagner had been preparing the samples according to Krätschmer's instructions, limiting the pressure of inert gas in the bell-jar to 20 torr. These experiments were supposed to be producing particles as small in size as physically possible, and it was most important to avoid the clumping of the growing particles by keeping to low pressures. It made no sense at all to try experiments using a high gas pressure, but in what he later described as an experiment done 'just for fun', Wagner produced a sample of soot by evaporating the graphite rods in 100 torr of inert gas.

He passed the sample to Krätschmer. Now the camel humps in the ultra-violet spectrum were much more prominent, suggesting that whatever was responsible for them had been formed in the soot in greater quantities. But it was the infra-red spectrum that was astounding. Superimposed on the broad background absorption of 'ordinary' soot were four strong lines. There were other lines but four were particularly prominent. Repeated experiments quickly showed that the four lines appeared only in samples that also gave the camel humps. Krätschmer could hardly believe it.

Undergraduate students of chemistry are taught that non-linear molecules consisting of N atoms will possess a total of $3N-6$ 'normal modes' of vibration. These modes arise through the subtle interplay of mechanical and electrical forces which govern the bonding and geometrical structures of molecules. Although the correct interpretation of these vibrations lies in the somewhat esoteric algebra of quantum mechanics, most chemists are usually more comfortable thinking about them in terms of old-fashioned 'ball-and-spring' models.

If we imagine a molecule to be composed of a series of spherical weights (the atoms) connected together by weak springs (the bonds) then the normal modes of vibration can be visualized as follows. Clamp one of the central weights in a vice and gradually increase the frequency with which it is forced to oscillate back and forth. This motion is communicated through the springs to all the other weights and the entire model jangles about chaotically. However, at certain characteristic frequencies of oscillation, patterns of motion emerge from the chaos in which all the atoms are moving in sync. These motions may involve the symmetric extension and compression of the springs (the atoms are moving 'in phase'), antisymmetric extension and compression (the atoms are moving 'out of phase') or bending vibrations. These are the normal modes of vibration.

A model of soccer ball C_{60} constructed from spherical weights and weak springs will wobble like a jelly. With $N=60$, the molecule is expected to have 174 normal modes of vibration. In any other molecule, such a large number of vibrations would make the infra-red absorption spectrum highly complicated (to say the least), difficult to analyse and interpret and therefore relatively uninformative. But there is a catch: only vibrations involving a change in the molecule's dipole moment can absorb radiation, and in C_{60} the full power of perfect symmetry is brought to bear. Of the 174 different vibrational motions in C_{60}, only *four* involve a change in the molecule's dipole moment and so its infra-red spectrum is expected to consist of just four lines.

Over the previous year, several groups of theoreticians had predicted the positions of these four lines, reporting their results in terms of the wavenumbers, the reciprocals of the wavelengths of the vibrational transitions expressed in units of reciprocal centimetres (or cm^{-1}).* Excited and somewhat agitated, Krätschmer looked back at the various predictions the theoreticians had made.

In June 1987, Z. C. Wu, Daniel Jelski, and Thomas George at the State University of New York at Buffalo had predicted four lines at 1655, 1374, 551, and 491 cm^{-1}. Further calculations by S. J. Cyvin, E. Brendsal, B. N. Cyvin, and J. Brunvoll at the University of Trondheim in Norway had produced 1434, 1119, 618, and 472 cm^{-1}. The following year Richard Stanton at Canisius College (also in Buffalo) and Marshall Newton at Brookhaven National Laboratory at Upton in New York reported the values 1628, 1353, 719, and 577 cm^{-1}, obtained from more sophisticated calculations. The four strong lines in Krätschmer's spectrum appeared at 1429, 1183, 577, and 527 cm^{-1}.

Given the likely accuracy of these calculations, the differences between the predicted positions and the lines in the experimental spectrum were not so great. It was enough that there were only four lines, so characteristic of the icosahedral symmetry of the soccer ball structure, and they formed the right kind of pattern in the spectrum. It looked to all the world as though he really had buckminsterfullerene in the soot.

But how much? For the four lines to be so clearly visible in the spectrum, there had to be quite a lot of buckminsterfullerene present. From the heights of the lines, Krätschmer estimated that buckminsterfullerene—if that indeed was what it was—must account for about one per cent or more of the soot. This was truly incredible! Other groups elsewhere in the world were struggling to make miniscule quantities of buckminsterfullerene by laser vaporization of graphite disks or rods using equipment costing hundreds of thousands of dollars. The amount of C_{60} produced in the cluster beam experiments was estimated at one part in ten thousand (or even one part in a million). And here were Wolfgang Krätschmer and Bernd Wagner in Heidelberg, making buckminsterfullerene *spontaneously* in quantities of the order of one per cent using an old bell-jar evaporator costing no more than a few thousand dollars!

Then Krätschmer's natural conservatism regained control, and he began to have doubts. If this was really true, if he really did have large quantities of buckminsterfullerene in his carbon soot, then this was, of course, absolutely wonderful. But what if it wasn't true? So many clever scientists were working on this problem with expensive apparatus—how could they all be wrong and Krätschmer right? If it was really so easy to make large quantities of buckminsterfullerene this way, how come nobody had discovered this before?

*This may seem somewhat archaic, but originates in the general desire of spectroscopists to avoid carrying around numbers associated with large powers of ten. When measured and reported in reciprocal centimetres, the wavenumbers of the infra-red transitions of molecules have values in the region of a few hundred to a few thousand.

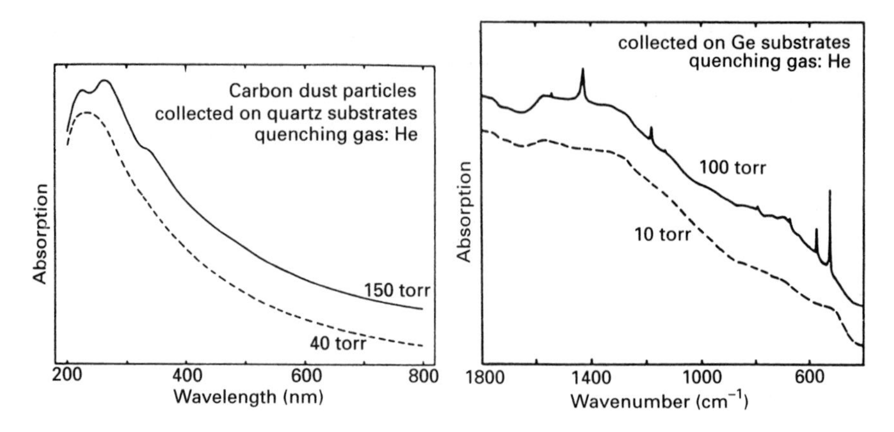

The diagram on the left is the ultra-violet absorption spectrum of carbon soot obtained with helium gas pressures of 40 and 150 torr. At the higher pressure, the three 'camel humps' are prominent. The diagram on the right is the infra-red spectrum of carbon soot deposited on a germanium substrate. The broad, featureless continuum characteristic of 'ordinary' soot arises due to stretching and bending vibrations of distorted graphite structures. At high helium gas pressures, four sharp lines at 1429, 1183, 577, and 527 cm^{-1} become apparent, superimposed on the broad continuum. Theoreticians had previously predicted that soccer ball C_{60} should have an infra-red spectrum consisting of just four lines, with positions roughly corresponding to the four sharp lines in this spectrum. Adapted, with permission, from Krätschmer, W., Fostiropoulos, K., and Huffman, D. R. (1990). *Dusty objects in the universe* (Bussoletti, E. and Vittone, A. A., eds.). Kluwer Academic Publishers, Amsterdam.

Krätschmer thought again about the possibility that it might still be some kind of junk, playing tricks on him. Contamination of the soot with oil from the vacuum pump could, perhaps, explain one of the infra-red lines, leaving only three others to be explained away. With only three lines, the case for buckminsterfullerene was not so strong. Whatever the explanation, it was clear that this result was not going to be enough. They had to get concrete proof that these lines were due to an all-carbon molecule.

Krätschmer described the results to Huffman over the telephone. As he had expected, Huffman was extremely excited: here was the first real evidence that his crazy idea might be right after all. In fact, Lowell Lamb had also been doing some further experiments using the Tucson evaporator and had independently discovered that the camel humps could be made more reproducible by raising the pressure of inert gas in the bell-jar. This was some relief, because it showed that whatever was going on was not just a function of the apparatus in Heidelberg. However, Huffman had no fancy infra-red spectrometer, so they could not confirm that the soot samples now also being made in Arizona gave the all-important four infra-red lines.

For Huffman, the infra-red spectrum was absolutely convincing, but Krätschmer urged caution. They agreed that they needed not only to repeat the experiments but also to extend them to obtain proof that an all-carbon molecule was responsible. What they needed to do was find a way to enhance the yield of buckminsterfullerene (or whatever) still further and extract enough of it from the soot so they could do some chemical analysis. This was not going to be easy.

One of Krätschmer's biggest problems was that he had no student to do the work. Bernd Wagner had now gone back to his undergraduate studies at the University of Cologne, conscious of the fact that he had been involved in something exciting but fairly oblivious to the implications of the work he had done.

Konstantinos Fostiropoulos had studied physics at Heidelberg University, completing a degree in chemical physics and obtaining his Diploma at the Institute for Physical Chemistry in the summer of 1988. Although he had grown up with his parents in nearby Mannheim, he had decided that he wanted to pursue the next stage of his research career in his native Greece. He went to Greece towards the end of that summer, touring various institutes trying to find a research adviser who could accept him as a Ph.D. student. But it was not a good time to be looking for a studentship in Greece and he was unsuccessful.

Disappointed, he returned to Heidelberg. Still casting around for a project to work on, he heard from a student friend that Krätschmer was looking for somebody to do experiments on carbon soot. He approached Krätschmer and explained that he was available.

Unable to see a clear way forward, Lamb had stopped work on the Tucson evaporator towards the end of 1988. Fostiropoulos restarted the experiments on the Heidelberg evaporator in February 1989. His first task was to strip down the evaporator, clean it out, and reassemble it to see if this made any difference to the ultra-violet and infra-red spectra. It didn't.

The next thing to do was to change the pumping system to see if that made any difference. They had so far been using an oil-vapour diffusion pump, and although the valve separating the bell-jar from the pump was always closed during the evaporation process, there was obviously still some opportunity for contaminating the soot samples with oil vapour or its degradation products. The physicists knew full well that such vapour consists of hydrocarbon compounds which would be expected to give rise to strong infra-red lines around 3000 cm^{-1}, characteristic of carbon–hydrogen bond vibrations. No such lines had been seen in the spectra of the soot samples but Krätschmer insisted they check every possibility. Fostiropoulos replaced the diffusion pump with a large turbo pump fitted with an oil absorber. It made no difference to the spectra.

Next they checked the possibility that there might be impurities in the inert gas they were using. They had now done experiments with both helium and argon, and found that the optimal pressures for producing the mysterious material in the soot were around 200 torr of helium or 100 torr of argon, admitted to the bell-jar from commercial gas cylinders. If impurities were present in these gases, then the use of higher pressures would increase the risk of contamination of the soot samples. They switched to an alternative supply of high-purity helium, with contaminants guaranteed to be less than one part per million by volume. It made no difference to the spectra.

With every check they made, the case for buckminsterfullerene grew a little stronger. However, Krätschmer was not going to rest easy until they had obtained definitive proof that an all-carbon molecule was responsible for the four infra-red lines. Fostiropoulos spent many frustrating months struggling to find a way to increase the yield of the material in the soot, but could not improve much on the yields they were already getting. Although they had 'large quantities' of the material compared to the amounts of buckminsterfullerene generated inside a cluster beam apparatus, these were still measureable only in micrograms. And Krätschmer was inclined to believe that extracting the material from the soot was going to be very difficult. He saw no reason to expect that buckminsterfullerene (if that was what it was) would have physico-chemical properties so very different from those of ordinary soot, making separation virtually impossible.

As the months of frustration had gone by, Fostiropoulos had begun to believe. The camel humps and the four infra-red lines were undoubtedly real and reproducible, and the argument in favour of buckminsterfullerene was supported by the theoretical predictions. But Krätschmer appeared to be forcing himself into a state of disbelief, or at least of suspended belief, convinced only of the fact that they could not afford to make a mistake. In March 1989, Pons and Fleischmann announced that they had obtained nuclear fusion at room temperature in a test tube. As the story unfolded, Krätschmer became increasingly dismayed at the damage that was being done to the credibility of science and scientists everywhere. The last thing he wanted to do was preside over another cold fusion débâcle.

What were they to do? They could not raise the yield of the material any higher and separation from the ordinary soot seemed out of the question, at least for the time being. What they really needed was an experiment that would confirm that the mysterious material consisted of all-carbon molecules without the need for greater yields or separation. Huffman, who needed no further convincing, recognized that his colleague had to have some additional proof. After some discussion, Huffman and Krätschmer agreed to try to generate soot from graphite made of carbon-13.

The positions of the absorption lines in the infra-red spectrum of a molecule are determined by the strengths of the bonds between its atoms and their masses. It is harder to move the nucleus of a heavy atom than a light atom, and this resistance to motion affects the period and frequency of molecular vibrations in much the same way that the period of vibration of a spring is affected by the mass placed on its end. As for the spring, the frequency of a molecular vibration is inversely related to the square root of the mass.

Replacing an atom of carbon-12 with an atom of the heavier isotope carbon-13 does not change the strengths of any chemical bonds in a molecule but it does change the frequency of vibrations involving that atom, and hence the position of any corresponding infra-red line. The change in line position relative to that produced by the same molecule containing carbon-12 is a constant factor, related to the ratio of the square roots of the atomic masses of carbon-12 and carbon-13. These masses can be looked up in a table. Huffman and Krätschmer estimated that if the four lines were really due to an all-carbon molecule like buckminsterfullerene, then substituting carbon-13 for carbon-12 would shift all four lines to lower energy (lower wavenumber) by a constant factor of 0.9625. Finding this would give them the confirmation they were looking for without the need to produce more material and without the need for an elaborate separation process.

This wasn't an easy thing to do either, but Krätschmer and Fostiropoulos agreed it was their only hope. Graphite rods made of carbon-13 were not commercially available, but it was possible to obtain samples of powder which contained 99 per cent carbon-13. Their task now was to figure out how to get from the powder to the soot.

Fostiropoulos purchased a quantity of carbon-13 powder in the summer of 1989. Initially, he tried laser vaporization of the powder, borrowing a ruby laser from his colleague Klaus Schneider and hoping to capture soot deposits which contained enough of the mysterious material to allow him to do the infra-red and ultra-violet spectroscopy. But the light powder tended to be simply blown away by the pulses of light from the laser and even at very high pressures of inert gas, he could not obtain soot deposits which showed the camel humps. Despite the fact that laser vaporization had been the technique which had revealed the existence and special nature of C_{60}, it did not appear to be a good way to make large quantities of it. Somehow Fostiropoulos had to be able to put the carbon-13 powder into the evaporator: he had to find a way to make graphite rods made of carbon-13.

As he and Krätschmer wrestled with this problem, the late summer conference season was fast approaching. Huffman and Krätschmer decided to gamble. They did not feel they had enough to argue a strong case that buckminsterfullerene was being produced in their soot samples, and they were certainly not ready to publish their findings in a peer-reviewed journal. But conference proceedings were a different matter. If they published their infra-red and ultra-violet spectra together with their speculations in the proceedings of some conference, then in the event that they were eventually proved right, nobody could accuse them of trying to hide their results from the outside world. If, on the other hand, they were eventually proved wrong, then the damage was not so serious. Nobody really paid too much attention to conference proceedings unless they were referred to in later publications, and if they were wrong their paper could quietly languish on the library shelves, unnoticed.

The conference they both wished to attend was the Fourth International Workshop of the Astronomical Observatory of Capodimonte, Capri, due to take place on September 8–13. The subject of the workshop was interstellar dust. They drafted a paper, entitled 'Search for the UV and IR Spectra of C_{60} in Laboratory-produced Carbon Dust', describing the camel hump spectra, the four infra-red lines, and the various checks they had made to ensure that these were not due to some obvious experimental artifact. They weighed up the supporting evidence from the theoretical predictions for both the ultra-violet and infra-red spectra of buckminsterfullerene. They referred to the apparent conflict with the depletion spectroscopy results of Heath, Curl, and Smalley, suggesting that possible distorting effects of the soot might be responsible for the differences. They mentioned that their results seemed to suggest buckminsterfullerene concentrations in the soot at the level of about one per cent.

Harry Kroto had been invited to attend the Capri workshop, but couldn't fit it into his schedule that summer. However, his friend and colleague, Mike Jura, did attend. Jura was an astronomer from the University of California in Los Angeles, and he had followed the buckminsterfullerene story with great interest. He picked up a shortened version of the Krätschmer–Fostiropoulos–Huffman paper that had been presented at the conference as an extended abstract, and over the top of the title he wrote 'Harry—Presented at Capri—Do you believe this?—Mike' He mailed the abstract to Kroto.

Along with most of the scientific community that had seen and read about the work that Krätschmer, Fostiropoulos, and Huffman had done, Kroto was initially sceptical. What strained his credulity was the inference that buckminsterfullerene was being formed in the Heidelberg evaporator in what amounted to microgram quantities. Nothing that he had learned about C_{60} and the way he believed it was formed inside AP2 had prepared him for this kind of proposal.

Kroto was also rather nonplussed: he had tried an almost identical graphite evaporation technique with Ken McKay two years before, and had since been strung along with the promise of funds to purchase a mass spectrometer, funds which had failed to materialize. He breathed a heavy sigh. The idea seemed crazy but the observations were undeniable. He was somehow going to have to resurrect his bell-jar evaporator and check the results out for himself.

Fortunately, Kroto did have another pair of hands that he could apply to such a project. With the help of Steve Wood from British Gas, he had secured a Collaborative Award in Science and Engineering (CASE) studentship for a combustion-related research project, funded in part by British Gas and by the UK Science and Engineering Research Council. With the prior agreement of his

Harry – Presented at Capri –
Do you believe this? – Mike

SEARCH FOR THE UV AND IR SPECTRA OF C_{60} IN LABORATORY-PRODUCED CARBON DUST

W.Krätschmer, K. Fostiropoulos *Max-Planck-Institut für Kernphysik, Heidelberg, W.-Germany*
and
D.R. Huffman *University of Arizona, Tucson, Arizona, USA*

Carbon dust samples were prepared by evaporating graphite in an atmosphere of an inert quenching gas (Ar or He). Changes of the spectral features of the carbon dust were observed when the pressure of the quenching gas was increased. At low pressures (order 10 torr), the spectra show the familiar broad continua. At high pressures (order 100 torr), narrow lines in the IR and two broad features in the UV emerge. The four strongest IR features are located in the vicinity of the lines predicted for the C_{60} molecule. One of the observed UV features may be related to the known 368 nm transition of C_{60}. It thus appears that at high quenching gas pressures C_{60} is produced along with the carbon dust.

Mike Jura sent Kroto a copy of the abstract of the Krätschmer–Fostiropoulos–Huffman paper, annotated with the comment: 'Harry—Presented at Capri—Do you believe this?—Mike.'

industrial sponsor, the terms of the award were sufficiently flexible to allow Kroto the freedom to choose the specific nature of the project. Although it was becoming increasingly difficult to find students to take up these awards, he had managed to sign up Jonathan Hare a few months before.

Hare had graduated in physics from the University of Surrey in Guildford earlier that summer. Part of his degree course had involved an 'industrial' year at the National Physical Laboratory in Teddington, where his dormant interests in astronomy had been rekindled. As his final examinations had drawn closer, he had written to the University of Sussex about the possiblity of pursuing postgraduate studies in experimental astronomy. However, there were no openings in the Astronomy Department and his interviewer had recommended that he try another university. Then, a few weeks later, a further letter from the Astronomy Department arrived suggesting that he contact Kroto. From Kroto he learned for the first time about this magical new molecule called buckminsterfullerene, and was instantly captivated, by both the concept and Kroto's evident enthusiasm for it.

Hare arrived in Sussex to start his postgraduate studies in mid-October. Kroto told him about the paper by Krätschmer, Fostiropoulos and Huffman and explained that he wanted to try to repeat these experiments to see if the four infra-red lines were really there. Together with Simon Balm, another Kroto student, Hare reassembled the old carbon evaporator that Kroto and McKay had been forced to abandon in 1987. The equipment was old but it was simple and reasonably serviceable. Within a matter of days, Hare and Balm were obtaining their first samples of soot. Balm moved on to work on Kroto and Stace's new cluster beam apparatus and Hare was joined by Amit Sarkar, a chemical physics undergraduate planning to work on the evaporator as part of his final-year research project.

Hare and Sarkar learned for themselves that the pressure of inert gas inside the bell jar was critical. On October 22, they confirmed that under certain conditions the carbon soot did show the four infra-red lines that Krätschmer and Wagner had seen a year before, although the lines were very weak and not very reproducible. Things had moved quickly for Hare: he had been working on his Ph.D. for only 12 days.

Then it all started to go wrong. Hare and Sarkar struggled to reproduce the infra-red spectrum and then the old, tired evaporation equipment finally gave up the ghost. The insulation inside the large electrical transformer they were using burned out and the experiments crashed to a halt. They had been offered a tantalizing glimpse of their goal, only to have it cruelly snatched away by the vagaries of their decrepit equipment. Hare decided that this was an opportunity to rebuild the evaporator more or less from scratch.

Steve Wood was acting as the industrial supervisor for Hare's CASE studentship and through his actions British Gas chipped in with £80 to buy a commercial power supply of the kind normally used with an arc-welding kit. This replaced the burnt-out transformer. Hare rebuilt the vacuum system and the electronics, and by early December was ready to start the experiments all over again. He and Sarkar glimpsed the four infra-red lines once more, although they were again very weak and irreproducible. This was enough to give Sarkar something positive to write about in his project report, but was not enough to provide the certainty that Kroto was looking for. Hare pressed on as Christmas approached, changing every experimental variable he could think of in attempts to enhance the production of the mysterious material in the soot.

In Heidelberg, Fostiropoulos conducted a search of the available literature and found that the general consensus of opinion was that amorphous carbon could not be used to make graphitic solids. Powder into rods was supposed to be impossible. Undaunted, he talked to colleagues at the Max Planck Institute for Chemistry in Mainz. These scientists had developed an apparatus to compress materials at high pressures (up to 80 times atmospheric pressure) with simultaneous heating to temperatures approaching 2000 kelvin. He took some of the expensive carbon-13 powder to Mainz to see if they could compress it at high temperatures into a solid. To his surprise and great delight, they succeeded.

As Hare was slowly rebuilding the Sussex evaporator, Fostiropoulos was building his own high-pressure rig modelled on the Mainz apparatus and beginning the laborious process of making carbon-13 rods. On New Year's Eve, with the Berlin Wall crumbling under the pressure of social forces released by the collapse of the communist system in East Germany, Fostiropoulos worked on in the deserted laboratory, quiet and undisturbed.

The resulting carbon-13 rods were not very stable, but they were good enough for the evaporation experiments. Fostiropoulos mounted the rods in the bell-jar early in February 1990, one year from the start of his Ph.D. He opened the valve to the pumps and evacuated the bell-jar before sealing it off and admitting 100 torr of helium. He slowly increased the current flowing through the contacted rods until the inside of the bell-jar was bathed in the brilliant light emitted by the arc. He repeated this procedure at 20-second intervals until the contact could be maintained no longer.

After waiting impatiently for the apparatus to cool down, he carefully lifted out the substrate with its precious coating of carbon-13 soot, and carried it to the FTIR spectrometer. He ran the spectrum and there they were: four infra-red lines at 1375, 1138, 556, and 508 cm^{-1}. The ratios of the positions of these lines to their counterparts in the spectra of the carbon-12 samples were 0.9622, 0.9620, 0.9636, and 0.9639, an average ratio of 0.9629. The predicted ratio was 0.9625. There could be no mistake. The pattern of lines and the shifts due to isotopic substitution all pointed to one conclusion: they had buckminsterfullerene in the soot. Fostiropoulos let out a cheering cry.

In the midst of his reflections on the significance of the spectrum drawn on the computer screen before his eyes, Fostiropoulos was interrupted by Hugo Fechtig, who was showing a visitor around the Department. For the benefit of the visitor, Fechtig asked what it was they could see on the monitor screen. With a distinct gleam in his eye, Fostiropoulos explained that it was an infra-red spectrum of carbon soot containing C_{60}, a completely new, all-carbon molecule shaped like a soccer ball. Both Fechtig and the visitor seemed somewhat taken aback by this answer.

Krätschmer had heard Fostiropoulos's cry, and entered the lab just as Fechtig asked his question again, adding '. . . is this *really* C_{60}?' Krätschmer could not tell from the general appearance of the patterns on the monitor screen that this was the spectrum of carbon-13 soot, and assumed instead that it was an old spectrum of carbon-12 soot that Fostiropoulos had loaded from the computer files. 'No, no . . . we can't be sure': Krätschmer sounded apologetic. Then Fostiropoulos announced that this was actually the first spectrum he had obtained for the carbon-13 soot. Krätschmer had been prepared for some news, and this was news indeed. Fechtig and his visitor quietly stepped out of the lab to leave them to their energetic discussion.

A few weeks later Krätschmer received a letter from England. It was from Jonathan Hare.

Kroto had proposed to Hare that they present their preliminary results on the infra-red spectra as a poster at a forthcoming conference on astrochemistry. He

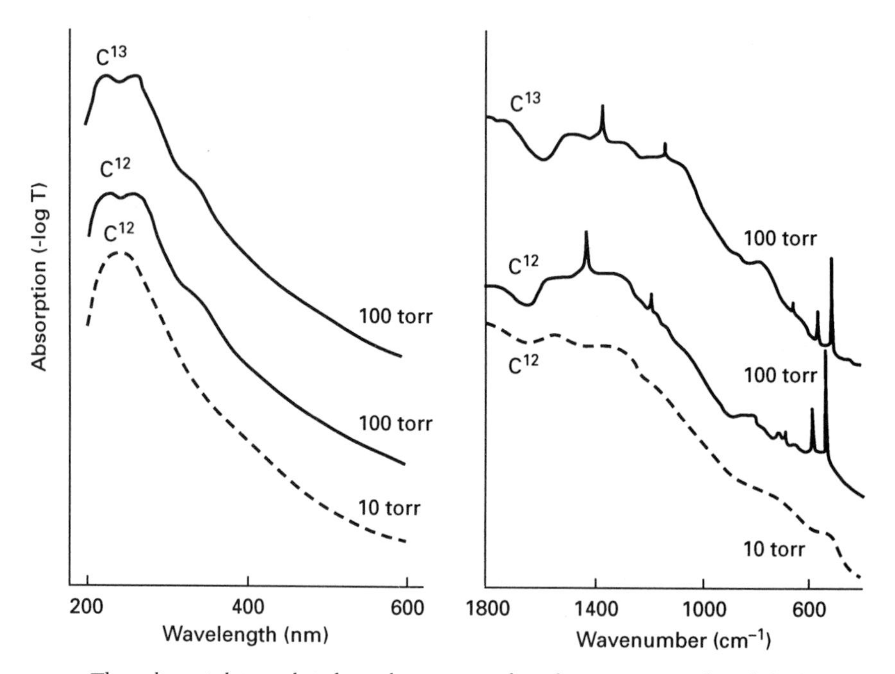

The ultra-violet and infra-red spectra of carbon soot produced by evaporating graphite rods made of carbon-13, compared with the equivalent carbon-12 spectra. The ultra-violet spectrum of the carbon-13 soot demonstrates that the camel humps are unaffected by isotopic substitution, as expected. In contrast, the four infra-red lines all shift their positions in the spectrum of the carbon-13 soot by an amount consistent with isotopic substitution in an all-carbon molecule. Adapted, with permission, from Krätschmer, W., Fostiropoulos, K., and Huffman, Donald R. (1990). *Chemical Physics Letters*, **170**, 167.

wanted to acknowledge the earlier work of the Heidelberg/Tucson group, but was not sure what reference to give for the paper the group had presented at Capri. Kroto had therefore urged Hare to write to Krätschmer to ask if the paper had been published, and if so what the correct reference was. Hare wrote on February 28, and took the opportunity to ask Krätschmer for some more details of the experimental conditions that the Heidelberg group was employing to make the 'special' soot. He explained that the Sussex group had managed to confirm the presence of the four lines in the infra-red spectrum of the soot, but that these were rather weak.

Krätschmer was shocked. He had been fairly confident that nobody would take the Capri paper seriously, and was very surprised to discover that Kroto and his group were catching up so quickly. Of course, Krätschmer, Fostiropoulos, and Huffman were still quite far ahead. They had long ago mastered the evaporation conditions and now had the wonderful results with the carbon-13 soot. But if they were going to hold on to their lead, they would have to move faster.

Krätschmer thought long and hard about how he should reply to Hare's letter. He sent the full text of the Capri paper (it didn't contain much more than Kroto and Hare already knew) and in his friendly letter he gave a detailed description of their experimental arrangement and conditions. He did not mention the results they had just obtained with soot made from carbon-13.

Whether they liked it or not, the race for buckminsterfullerene was now well and truly on.

10

Fullerite

Drawing encouragement from Krätschmer's letter, Hare was able to make samples of carbon soot which showed the four infra-red lines with considerably improved reproducibility. He methodically varied everything that could be varied, and found that with pressures of helium above 10 torr, the distance between the tips of the contacted graphite rods and the substrate became a crucial factor in determining the nature of the soot deposited. If buckminsterfullerene really was being formed, then maybe putting the substrate too close to the rods caused it to get too hot, driving off any C_{60} that had already been collected.

Hare presented the now fully reproducible results at a conference on interstellar clouds held in Manchester on March 25–29, 1990. Now the reception was much more favourable than the one Krätschmer and Huffman had received for their paper presented in Capri, or for the original crazy idea that Huffman had espoused in Santa Clara. The difference was that Hare was presenting independent *confirmation* that the results produced by the Heidelberg evaporator were correct.

Of course, any talk about buckminsterfullerene in the soot was still all very speculative, but there were these four lines and they did form the kind of pattern in the spectrum that had been predicted by the theoreticians for soccer-ball C_{60}. The astronomer Pat Thaddeus remarked to Hare that he thought the Sussex group 'really had something'.

At Kroto's instigation, Hare and Simon Balm spent the next few months analysing some spectroscopic data that had been obtained by the space probe Giotto as it had passed through the tail of Halley's comet in March 1986. The data had been analysed and assigned to a specific molecule by other astrochemists, but Kroto did not believe this assignment was correct.

This was not an entirely unwelcome diversion. The next step for the Sussex group was to accumulate as much of the 'special' soot as possible—to produce enough to carry out various analytical studies. This meant cranking the handle on the evaporator and, whenever the infra-red lines appeared in strength, scraping the soot off the substrate and adding it to a growing stockpile in a small glass vial. Hare was able to do this every few days or so while working on the Giotto data and during these few months collected a small stock of the special soot. According to Krätschmer's estimates, one per cent or more of this soot was supposed to be buckminsterfullerene.

During a break from his university teaching and adminstrative duties over the Easter period, Kroto travelled to California to spend about one month working with Mike Jura in the Astronomy Department at UCLA. While he was there, he was introduced to François Diederich, Orville Chapman's one-time postdoctoral associate who had now returned to UCLA's Department of Chemistry and Biochemistry from Heidelberg. Diederich appeared very excited and asked that Kroto accompany him to his office, as he had something important he felt that Kroto should see. Kroto asked somewhat facetiously: 'You've not made C_{60}?'. Diederich was not amused. He froze, with a serious, quizzical expression on his face. 'How do you know?', he asked.

Diederich had moved back to UCLA in 1985, just when the great rush of interest and enthusiasm for C_{60} had begun with the publication of the *Nature* paper. Several of his UCLA colleagues in Chapman's and Whetten's research groups subsequently became sceptical that the mass spectral signatures observed in the experiments at Rice had anything to do with soccer ball C_{60}. In particular, Whetten had become interested in exploring alternative structures for C_{60}, and talked to Diederich of the possibility of forming a flat 60-atom structure from six C_{10} rings.

Diederich hadn't seen an easy way to make C_{10} rings, but he believed that larger C_{18} and C_{30} rings containing alternating single and triple carbon–carbon bonds were feasible targets for synthetic chemistry. These are the cyclic equivalents of the long-chain polyynes in which the dangling bonds at either end connect up with each other like a snake biting its own tail. Diederich had set to work and, with colleagues Yves Rubin, Carolyn Knobler, Whetten, Kenneth Schriver, Kendall Houk, and Yi Li, reported the first successful synthesis and characterization of the cyclic molecule C_{18} in September 1989.

With Rubin and Knobler, Diederich had continued to work on ways to extend this method to larger ring molecules, and had succeeded in making $C_{18}(CO)_6$, $C_{24}(CO)_8$, and $C_{30}(CO)_{10}$. These molecules all possess the same basic structural framework: a large ring of carbon atoms with alternating single and triple bonds. The carbon monoxide (CO) groups sit at the apexes of a rather rounded triangle, square or pentagon, respectively, with each side of the geometric figure consisting of a $-C{\equiv}C-C{\equiv}C-$ unit. These were the largest carbon oxides that had ever been prepared and, as the carbon monoxide groups were relatively easily removed, they were also very good precursors for the all-carbon ring molecules C_{18}, C_{24}, and C_{30}.

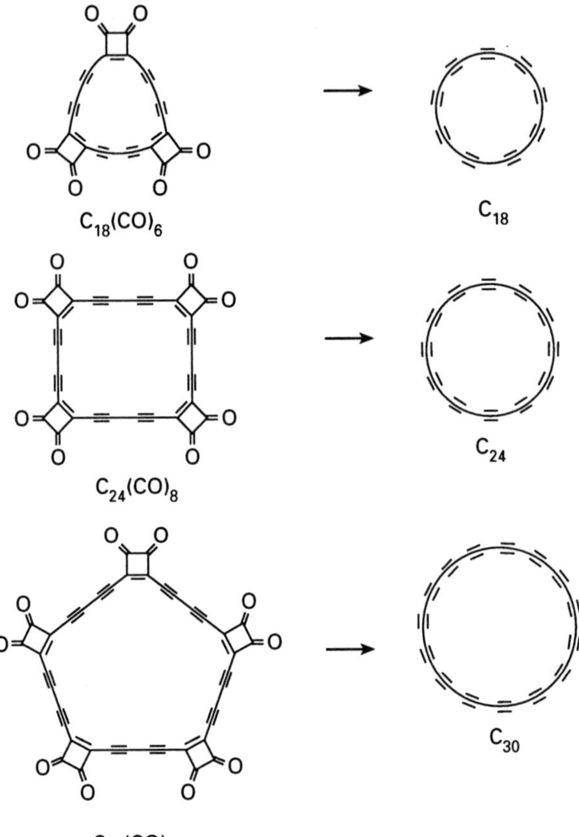

The large carbon oxides synthesized by Diederich, Rubin, and Knobler were believed to be suitable precursors for the all-carbon ring molecules C_{18}, C_{24}, and C_{30}. The oxides decompose through gentle heating, giving the rings and carbon monoxide (CO) molecules. These rings are the cyclic equivalents of the long-chain polyynes, the dangling bonds being eliminated by joining one end of the chain to the other like a snake biting its own tail.

The UCLA group had studied the oxides using a variety of techniques, including laser-desorption mass spectrometry carried out in collaboration with Michael Kahr and Charles Wilkins at the University of California at Riverside. This technique involved the use of infra-red radiation from a pulsed carbon dioxide laser to unstick molecules from a surface, getting them into the gas phase where they could be detected by mass spectometry. Unlike the laser vaporization process used in AP2, the laser desorption process required much gentler conditions. The purpose was not to disrupt the surface and create exotic new molecules in the resulting plasma, but simply to unstick molecules that had already been synthesized and deposited as a solid film on a substrate.

They found that the mass spectra of positive ions produced from desorbed $C_{18}(CO)_6$ showed a strong signal corresponding to C_{18}, indicating that the carbon monoxide groups had been removed as a result of heating the molecule during desorption. Presumably, this was the positive ion of the C_{18} ring molecule, although there was actually no way of telling the structure from the mass spectral data. However, also rather prominent in this spectrum were peaks due to C_{36}, C_{50}, C_{60}, and C_{70}. There was a similar story for $C_{24}(CO)_8$, except that now the signal corresponding to C_{24} was not very strong. The spectrum was instead dominated by C_{48}, C_{50}, C_{60}, and, strongest of all, C_{70}. For $C_{30}(CO)_{10}$, the story was quite different. There was no discernible signal from C_{30}, but the C_{60} signal was simply huge. It was the flagpole result all over again. Of course, there was also a small signal from the ever-faithful C_{70}.

The UCLA/Riverside group had come to the conclusion that this last spectacular result indicated that C_{30}^+ was not very stable and rapidly reacted with an uncharged C_{30} molecule to form C_{60}^+ with high efficiency. There was also the possibility that C_{60}^+ was being formed spontaneously by building up from smaller fragments produced by laser desorption. However, the fact that the C_{60} signal was such a conspicuously large product of the desorption of $C_{30}(CO)_{10}$ seemed to suggest that spontaneous 'dimerization'—two molecules of C_{30} combining to give C_{60}—was primarily responsible.

To Kroto, this result was almost as startling as the original AP2 results from September 1985. It offered a possible explanation for many of the puzzles surrounding carbon chains and the formation of C_{60} both in AP2 and (who could tell?) in samples of carbon soot. Was chains to rings and rings to spheres a plausible mechanism?

But at the same time Kroto was also extremely apprehensive. Could the UCLA group use this technique to make large quantities of buckminsterfullerene? Could the group start with $C_{30}(CO)_{10}$ and simply heat this to make C_{30}, which would spontaneously dimerize with high efficiency to give C_{60} on a large scale? Kroto had been dreaming of that one-line NMR spectrum for far too long. It was his line and he wanted to be the first to measure and report it. He certainly did not want to pipped at the post after all this time.*

When he returned to Sussex, Kroto discussed the UCLA/Riverside results with David Walton. Walton was able to draw on his wealth of experience with aspects of the synthesis of long-chain polyynes and reassure Kroto that it would be extremely difficult, if not impossible, for the UCLA chemists to scale up their process to make buckminsterfullerene in the kinds of quantities needed for comprehensive chemical analysis. Their results were fascinating, and perhaps they provided some key explanations for how the closed-cage molecules could be formed. But this was not an easy route to the large-scale production of

*Kroto was right to be apprehensive. In the spring of 1990, Diederich and his colleages had received funding from the US National Science Foundation to purchase a dedicated laser desorption apparatus. They aimed to produce C_{60} on a preparative scale from the coalescence reactions of C_{30}.

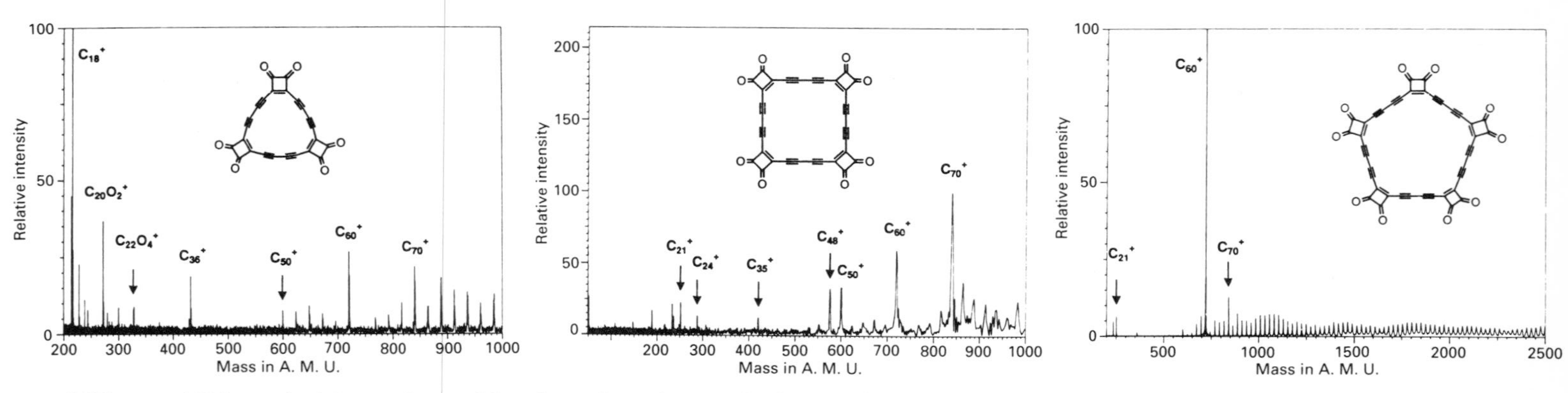

Wilkins and Kahr studied the products of desorbing the carbon oxides from surfaces using infra-red light from a carbon dioxide laser. The mass spectra showed they got more than they bargained for. The strongest signal obtained from the desorption of $C_{18}(CO)_6$ (left) corresponds to C_{18}^+, as expected, but C_{36}^+, C_{50}^+, C_{60}^+, and C_{70}^+ are also prominent. For $C_{24}(CO)_8$ (middle spectrum), the signal due to C_{24}^+ is actually quite small, and C_{50}^+, C_{60}^+, and (particularly) C_{70}^+ dominate the spectrum. But most incredible of all is the result for $C_{30}(CO)_{10}$ (right), in which no C_{30}^+ can be seen at all and the spectrum is instead overwhelmed by a C_{60}^+ flagpole. Adapted, with permission, from Rubin, Yves, Kahr, Michael, Knobler, Carolyn B., Diederich, François, and Wilkins, Charles L. (1991). *Journal of the American Chemical Society*, **113**, 495. Copyright (1991) American Chemical Society.

buckminsterfullerene. Nevertheless, Kroto became very conscious of the need to move faster in the race to be first to isolate the new molecule.

In the last few days of May, Hare decided that he had collected enough of the carbon soot to measure a solid-state NMR spectrum. This was something that Kroto's Sussex colleague Ken Seddon had suggested some time before, but Kroto had been sceptical that the black soot, which looked so unpromising, could really contain quantities of C_{60} large enough to produce a measureable NMR signal. Now Kroto was beginning to wonder if this crazy idea could be true after all. His encounter with Diederich had served to remind him that he couldn't afford to wait for the answer to drop in his lap.

Hare submitted a sample of the carefully collected soot for analysis using the laboratory NMR facility, but the machine was faulty and no results could be obtained. As he finalized his plans for a hill-walking holiday in Scotland, he passed a small sample of the soot to Ala'a Abdul-Sada, an Iraqi postdoctoral researcher with considerable expertise in mass spectrometry. When he returned from his holiday a week later, an excited Abdul-Sada explained that the mass spectrum showed strong peaks corresponding to C_{60} and C_{70}. Hare was thrilled. Kroto wanted the result confirmed. He was wary of the possibility that the C_{60} could be forming during the sampling process, thereby misleading them into thinking that it was present in the soot. But the mass spectrometer had broken down and the experiment couldn't be repeated.

Hare was nevertheless convinced that the mass spectrum confirmed the presence of C_{60}. He had the four infra-red lines, so characteristic of the soccer-ball structure, and a mass spectrum which gave a strong signal for C_{60}. The next step was to find a way to separate the C_{60} from the soot.

I telephoned Kroto at his office in Sussex sometime early in June. I had read the article by Rudy Baum in the February 5, 1990 issue of *Chemical and Engineering News*, in which he described how Ebert and Frenklach had compared the work on carbon clusters to cold fusion, and had dismissed both as examples of 'pathological science'. Ebert and Frenklach had published a letter in the May 14 issue of the journal, claiming they had made no such criticism. I judged the story sufficiently interesting and controversial to warrant a further article for the UK popular science magazine, *New Scientist*, and decided to contact Kroto to get some more background details. Having once been an academic scientist with research interests in chemical spectroscopy, I knew the community quite well. I had known Kroto personally for more than seven years.

We discussed the soot controversy at some length over the telephone. However, Kroto was of the opinion that this was old news, and felt that there was something in the air that was likely to make the buckminsterfullerene story really take off all over again. He described the results that Diederich had shown him in Los Angeles, and explained the business about the four infra-red lines that the Heidelberg/Tucson group had found in samples of carbon soot and which had since been confirmed by his student Jonathan Hare in Sussex. He suggested that I wait a little while to see what was going to happen next. Kroto's premonitions were completely justified: we were all blissfully unaware that the dam had already burst in Heidelberg.

Krätschmer, Fostiropoulos, and Huffman worked on the paper describing their breakthrough with carbon-13 through March and April, corresponding by mail, fax and telephone. In it they summarized the evidence from the ultra-violet and infra-red spectra of both the carbon-12 and carbon-13 samples, and listed the various experimental checks they had made to ensure the results were both real and reproducible.

They speculated on possible explanations for some of the weaker lines that were barely visible in the infra-red spectra. For the first time, they noted the connection between C_{60} and C_{70} that was apparent in the time-of-flight mass spectra reported in the original *Nature* paper in 1985, and speculated that C_{70} might also be present in the soot in smaller quantities, explaining the extra infra-red lines. They repeated their estimate that about one per cent of the soot was buckminsterfullerene. They submitted the paper to *Chemical Physics Letters* at the end of April. It was published two months later, in the July 6 issue.

Meanwhile, Fostiropoulos cranked the handle on the Heidelberg evaporator to collect as much of the special soot as possible. They now addressed themselves to the difficult task of isolating the buckminsterfullerene from the soot. Krätschmer was adamant that this was something they just had to do.

Krätschmer was experiencing a kind of schizophrenia. One half of him believed that they really could make large quantities of buckminsterfullerene in their evaporator. The other half of him refused to believe anything until they had established it beyond the shadow of a single doubt. His greatest fear was that, despite all the evidence they had accumulated so far, they could still be horribly wrong. He was afraid that this whole buckminsterfullerene business could blow up in their faces. The analogy with cold fusion was constantly nagging.

By March 1990, the cold fusion story had degenerated into a sordid, complex mess of unjustified, unverifiable, and spurious claims, American institutional and state politics, and patent nonsense. For scientists the object lesson was that for any such singular discovery, you made damned sure you knew the truth and could prove it before you called a press conference. So far, the papers that Krätschmer, Fostiropoulos, and Huffman had published had merely given interesting new results along with some speculation. Now they had to know for sure. Somehow, they had to get the buckminsterfullerene out of the soot.

Had they been chemists working in a chemical laboratory, then this task would have been much more straightforward. But they were astrophysicists working an in institute for nuclear physics, and they did not have the first idea how they could get this stuff separated from the soot. They were the wrong people in the wrong place, but it was their discovery all the same.

Kosta Fostiropoulos (pictured here on the right) and Wolfgang Krätschmer had to tread warily in their efforts to prove they had formed buckminsterfullerene by the simple evaporation of graphite. By March 1990, the cold fusion story was still making headlines all around the world, for all the wrong reasons.

Then came a stroke of good fortune. Having completed the carbon-13 paper, Krätschmer sent a copy of the manuscript to his friend Léger in Paris. Léger responded with a fax containing a simple message—'Bravo!'—and contacted his colleague Werner Schmidt at the Institute for Polycyclic Aromatic Hydrocarbon Research. Schmidt routinely supplied Léger with samples of exotic polycyclic aromatic hydrocarbons that this private institution synthesized on a small scale.

Prompted by Léger, Schmidt wrote to Krätschmer and suggested they try to sublime the buckminsterfullerene out of the soot by heating it to about 800 or 900 kelvin under vacuum or in an inert gas. Alternatively, they could try to extract the buckminsterfullerene by shaking the soot with an aromatic solvent. The ordinary soot particles themselves would be insoluble, but Schmidt believed that soccer-ball C_{60} ought to behave like a large aromatic molecule, and should therefore quite happily dissolve in a suitable solvent. Conscious of the carcinogenic nature of benzene, Schmidt suggested they try a less hazardous benzene derivative.

Not being chemists, Krätschmer and Fostiropoulos found it difficult to believe that the separation of C_{60} could be so easy, but they decided to try Schmidt's suggestions anyway. What did they have to lose? As they were working in a physics lab, they had none of the solvent that Schmidt had recommended, so they first tried sublimation.

Very late one night in early May, Fostiropoulos placed a little of the soot and a thin quartz substrate in a glass tube. He then filled the open tube with argon, which forced out the air above the soot. He heated the bottom of the tube with the naked flame of a Bunsen burner. At first, the substrate did not appear to have changed: he could see no sign of a coating. But as he looked more closely, he noticed that the reflected light from the surface of the substrate did appear different: *something had been deposited.*

He was extremely tired, but nothing was going to keep him from measuring the spectrum. He placed the substrate in the ultra-violet/visible spectrometer and set the machine to scan the wavelength. He watched the recording pen intently as it moved over the chart paper and, for the second time in his life he felt the electric thrill of scientific discovery. There they were, three of the strongest, most beautiful camel humps he could ever wish to see. Gone, or at least significantly reduced, was the background absorption due to ordinary carbon soot. The sublimation process had worked: it really was that easy. He was the first person in the world to see the ultra-violet spectrum of almost pure buckminsterfullerene.

Fostiropoulos left the spectrum on Krätschmer's desk and headed home. It was time for sleep.

Two days later, Fostiropoulos and Krätschmer embarked on a period of intense activity. Fostiropoulos used the same sublimation technique to deposit the material on both quartz and silicon substrates and repeatedly measured its ultra-violet and infra-red spectra. He confirmed that the four infra-red lines were there, stronger than ever. Krätschmer alerted Huffman to these latest developments and Huffman re-activated Lowell Lamb, who started to repeat some of the experiments that Fostiropoulos had done. Lamb quickly confirmed the results of the ultra-violet spectroscopy.

Lamb was just finishing off his thesis work, and Huffman was concerned that this business with the carbon soot would prove to be a major distraction at a crucial time. Not that Huffman had much choice; he had heard the news about the separation from Krätschmer on May 18, just two days before he was due to leave for Paris. He needed somebody to work on the problem while he was away.

In Heidelberg, Krätschmer and Fostiropoulos discovered that the solid material sublimed onto their coated substrates was readily washed off and dissolved in benzene. This prompted them to try the other approach that Schmidt had suggested. They still had none of the recommended solvent, so they continued with benzene.

Fostiropoulos dispersed some of the soot in a glass tube partially filled with benzene, and placed the tube in an old centrifuge. He switched the instrument on and leaned heavily on it (as prescribed by lab folklore) so that it wouldn't jump about too much. When the procedure was complete, he found that he had

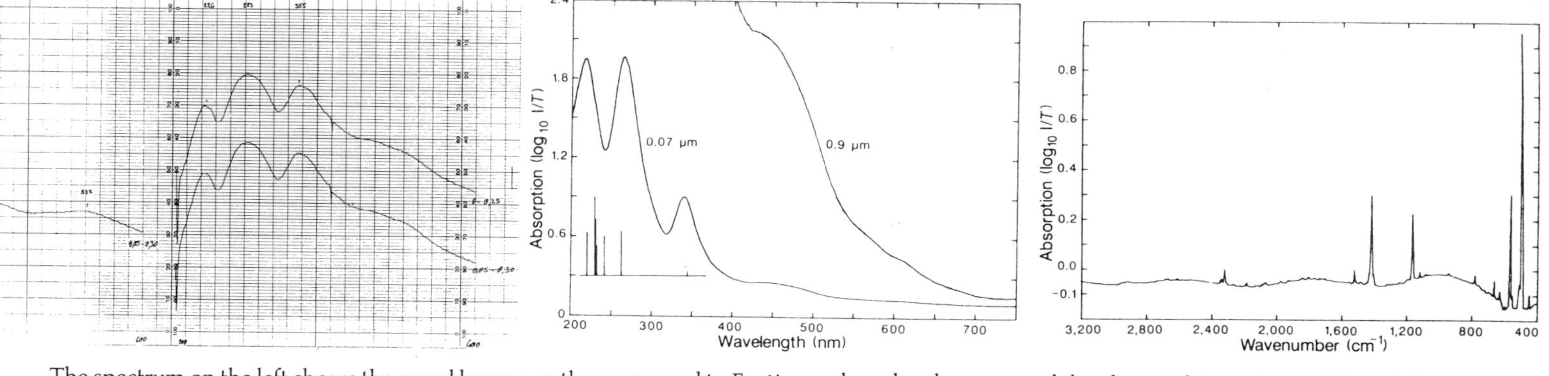

The spectrum on the left shows the camel humps, as they appeared to Fostiropoulos when he measured the ultra-violet spectrum of the solid material sublimed from the carbon soot. With further purification, the spectrum appeared as shown in the middle, which also compares the experimental features with the theoretical predictions of Larson, Volosov, and Rosén for soccer ball C_{60}. The infra-red spectrum on the right shows the four distinct lines so characteristic of the soccer ball structure. The smaller lines that can be seen are due to impurities (including some C_{70}). The middle and right most spectra are adapted, with permission, from Krätschmer, W., Lamb, Lowell D., Fostiropoulos, K, and Huffman, D. (1990). *Nature*, **347**, 354. Copyright (1990) Macmillan Magazines Limited.

obtained a deep red solution the colour of a Bordeaux wine, with the insoluble soot compressed as a deposit at the bottom of the tube. Fostiropoulos decanted off the red solution, and then evaporated the benzene with mild heating to produce a dark powder. Subsequent sublimation of the powder onto quartz and silicon substrates showed the same ultra-violet and infra-red spectra as before.

Gentle evaporation of the solvent from the red solution produced tiny crystals of the new substance which Krätschmer and Fostiropoulos studied under a microscope. These were beautiful orange–brown crystal shapes: hexagonal rods, platelets, and star-shaped flakes. By placing a drop of the red solution under a microscope, Huffman and Lamb subsequently found they could actually watch the crystals form as the solvent evaporated. This was to be one of the most compelling images in the whole buckminstefullerene story: crystals of a totally new form of carbon taking shape right before their eyes. Nobody on earth had ever seen this before.

As the buckminsterfullerene story had unfolded, many scientists had made reference to C_{60} and the other fullerenes as a new form, or 'allotrope' of carbon. Huffman disagreed with the use of this term. What had been discovered was a new series of all-carbon molecules. Huffman argued that you didn't have a new allotrope until you held the new crystalline form in your hand and could prove that it was different. For this reason, he and Krätschmer decided to give a new name to the solid material they had produced. They called it 'fullerite'.

Fostiropoulos in Heidelberg and Lamb in Tucson now scrambled to get the analytical work done. Their problem was again that they were the wrong people to have made this discovery, and a physics laboratory was not the best place to have made it in. With the help of his colleagues Harry Zscheeg and Ghaleb Natour in the Cosmophysics Department, Fostiropoulos managed to measure the mass spectrum of the new substance and confirm the presence of strong signals due to C_{60} (and a little C_{70}, as always). However, the resulting spectra, measured using a mass spectrometer purchased for the CRAF satellite mission, were not as clean as they had hoped. The first measurements on this instrument had been simply awful, showing many other signals in addition to the expected peaks corresponding to C_{60} and C_{70}.

The small crystals they could now watch growing on a microscope slide were ideal for electron diffraction studies, and the sublimed or solvent-extracted powder was suitable for X-ray powder diffraction. Neither of these techniques could provide detailed information on the *molecular* structure of the new substance, but they could be used to determine if it really was composed of tiny spheres of carbon. Like a large pile of real soccer balls, the C_{60} molecules were expected to pack together in a hexagonal or cubic close-packed arrangement, and both electron and X-ray powder diffraction studies would give them the packing order and the average spacing between the balls.

Fostiropoulos took a sample of the crystals he had obtained to the nearby European Molecular Biology Laboratory where, with the help of Werner Kuhlbrand, he measured the electron diffraction pattern of the fullerite. The

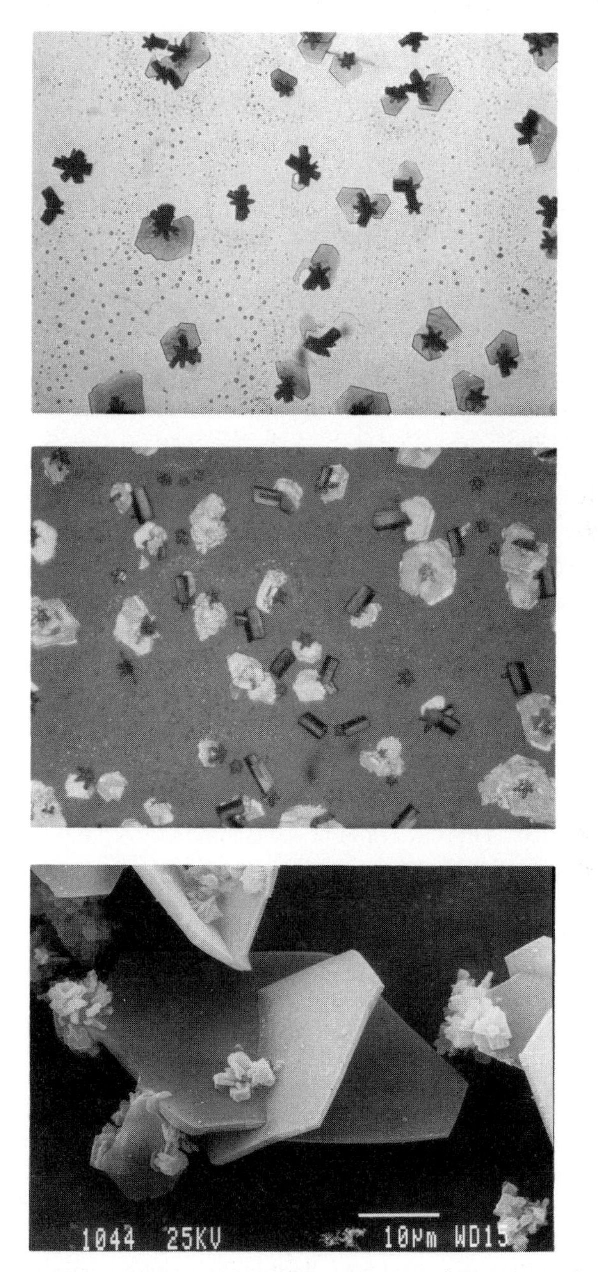

Photomicrographs (upper and middle picture) and scanning electron microscope image (lower picture) of crystals of fullerite. Nobody on earth had ever seen these crystals before.

The diffraction of a beam of electrons through a crystal produces a series of spots in a pattern which is related to the spacings between the crystal planes. Analysis of this pattern, obtained by Fostiropoulos and Werner Kuhlbrand from a large crystal of fullerite, showed that the molecules that make up the solid are spherical, with an average spacing between them of about one nanometre.

close-packed arrangement has crystal planes that are expected to give certain characteristically strong signals in the diffraction patterns. Some of these signals were not as strong from the fullerite as they should have been, indicating that the packing of the spheres was somewhat disordered.

However, the patterns did yield an average spacing between the centres of the C_{60} soccer balls of about one nanometre. This seemed entirely reasonable, as the diameter of an individual soccer ball molecule was expected to be about 0.7 nanometres, leaving about 0.3 nanometres for the clouds of electrons around the outside of each sphere. The spacing between the layers in graphite, determined by similar electron clouds, is 0.335 nanometres.

What they wanted next was unambiguous proof that the C_{60} molecules had the proposed soccer ball structure, for which they needed an X-ray diffraction pattern from a single crystal of the fullerite. The pattern of spots formed by the diffracted X-rays, whose measurement and interpretation had increasingly become the preserve of specialist crystallographers, would reveal the locations of the carbon atoms, allow the lengths and angles of the carbon–carbon bonds to be deduced and prove beyond doubt that C_{60} was a soccer ball.

The Research Division of International Business Machines Corporation has two large research centres in the United States, at Yorktown Heights in New York State and at San Jose in California. IBM's Almaden Research Center lies on the outskirts of San Jose, about 60 miles south of San Francisco down Highway 101. Many of the staff scientists at Almaden are engaged in research and development projects that have a direct bearing on IBM's products, but there are a few who are given free rein to conduct whatever research appeals to them, irrespective of its potential to yield a commercial return.

There are many in the scientific community who regard this as a rather enlightened approach for a commercial research and development organization, but through it IBM gain three important benefits that can have potentially commercial spin-offs. Firstly, IBM retains a pool of top-class scientific expertise on which the company has a direct claim. Although the work of these scientists might be esoteric, 'blue-skies' stuff, there is inevitably scope for drawing on their expertise to feed projects which have a more commercial orientation. Secondly, these top-class scientists act as magnets, pulling fresh talent into the system, if only for relatively short periods through IBM's postdoctoral research associate programmes. Finally, there is always the possibility of finding the kind of unlooked-for breakthrough that can often be missed in more directed research projects which, of necessity, must be strongly disciplined. In 1986, the high-temperature superconducting properties of a lanthanum–barium–copper oxide ceramic were discovered by IBM scientists Alex Müller and Georg Bednorz working at the Division's European research centre in Zürich.

After completing his undergraduate studies at Stanford University and a Ph.D. in physics at the University of California at Berkeley, near San Francisco, Don Bethune's IBM research career had begun at Yorktown Heights. He moved to the Almaden Research Center to work on aspects of non-linear optics and the interactions between gases and solid surfaces, and first heard about the fullerenes through a lecture given by Smalley at a conference in New Hampshire in 1987. He had subsequently kept a semi-watchful eye on the growing body of fullerene literature.

Towards the end of 1989, he decided to translate his growing interest in clusters, and particularly the fullerenes, into a new research project. It had by then become clear to him that some kind of new approach to the study of carbon clusters was needed if the proposed soccer ball structure of C_{60} was ever going to be conclusively demonstrated.

The weak 386-nanometre band obtained in the depletion spectroscopy studies of Heath, Curl, and Smalley was still the only published experimental signal extracted from a C_{60} apparently reluctant to yield its secrets. By itself this feature was not very much to go on. Bethune was not an astrophysicist and was unaware of the Krätschmer, Fostiropoulos, and Huffman paper presented at Capri the previous year.

Bethune thought carefully about the problem and devised a new approach involving an ion cyclotron resonance spectrometer similar to the one that

Smalley and his colleagues had used to trap ions. He discussed his proposals with Heinrich Hunziker, one of the research managers at Almaden, who was, in turn, also surprised by the lack of hard experimental data on C_{60}. However, Hunziker thought that Bethune's approach was perhaps unnecessarily complicated. He wondered if a simpler technique might not yield results more quickly.

With his colleagues Mattanjah de Vries and Dutch postdoctoral research associate Gerard Meijer, Hunziker had developed a laser desorption apparatus similar to the one that Charles Wilkins and Michael Kahr at Riverside had used to study the mass spectra of the products obtained from the large carbon oxides. In fact, the IBM machine used supersonic expansion, laser ionization, and time-of-flight mass spectrometry. In terms of the technology it was just like AP2, except that the purpose of the first laser was to unstick molecules that had already been produced and deposited an a surface.

Bethune and Meijer agreed to embark on fullerene research using this apparatus and carried out their first preliminary experiments in May 1990. They began by increasing the power of the desorption laser to the point where it was vaporizing the surface of a graphite target, effectively repeating the cluster experiments first conducted by the Exxon team in 1984. Although they succeeded in obtaining the now familiar cluster distributions, they were fully aware that five years of experimentation with this technique had not produced a definitive structure for C_{60}. The IBM scientists still needed a different approach.

Meijer and de Vries stressed to Bethune that their apparatus had been designed not for vaporization, but for the desorption of molecules already formed and deposited on a surface. If they could somehow make buckminsterfullerene and stick it to the surface of a suitable substrate, then theirs was a very powerful technique for analysing the resulting solid film. The problem was that, as far as they knew, nobody had been able to trap even the minute quantities of C_{60} required for this kind of experiment.

The community of scientists working on C_{60} appeared to have become resigned to the fact that it couldn't be made in large quantities. To the small group of fullerene physicists now forming at IBM, this seemed at odds with the claimed structural stability and chemical inertness of the molecule. If buckminsterfullerene was really so stable, there was in principle no reason why the IBM scientists couldn't accumulate quantities of it on a substrate surface.

The only sure-fire way of making C_{60} in detectable amounts was in a cluster-beam apparatus like AP2. Bethune, Meijer, and de Vries wondered if they could actually collect the material formed in the cluster beam, rather than ionize and detect the individual clusters. If they could collect enough C_{60} this way, then maybe they could carry out the gentler laser desorption experiment. Perhaps they could even do some spectroscopy on the desorbed C_{60} molecules.

On impulse, Bethune contacted Eric Rohlfing, who had by now moved from Exxon to Sandia National Laboratories at Livermore, just 30 miles north-east of San Jose. He asked Rohlfing what he thought about their idea. Rohlfing laughed and commented that whoever could measure a spectrum of C_{60} would probably get a Nobel Prize. But he did suggest they forget about cluster beams and just try to deposit some carbon soot on a surface. Who could tell? Maybe they would find some C_{60} in it.

This meant giving up the sure-fire way of producing C_{60} and trying something highly speculative. But somehow Rohlfing's suggestion appealed to the IBM physicists. After all, everybody was claiming that C_{60} formed spontaneously in the plasma created by the vaporization laser and was such a stable molecule. Why shouldn't it also form in more conventional systems, along with soot?

Persuaded by the logic of their own arguments, the physicists elected to plunge themselves into the murky world of combustion chemistry. This was a field about which they knew virtually nothing, but they did know that in order to get soot you have to burn something. Surrounded on all sides by high-tech equipment in one of the most well staffed, equipped, and funded laboratories in the world, they resorted to backyard science.

With a sudden sense of urgency, they found a discarded peanut can and filled it with alcohol. They cut a hole in the plastic lid through which they stuffed a rag to act as a wick. They placed the can in a fume cupboard, lit the rag and stood back to watch what happened next. The alcohol burned cleanly, without producing any smoke or soot, but as the plastic lid caught fire a pall of thick, black smoke was formed. Bethune jabbed a copper rod into the smoke and within seconds it was covered with a deposit of soot. The experiment may not have been very elegant, but the physicists had just taken their first dabble in soot chemistry.

They placed the soot-coated rod in the laser desorption apparatus, and found—to their surprise and delight—clusters of carbon atoms running to very high masses. The mass spectrum revealed that atoms other than carbon were present (particularly sodium, left over from the salt on the peanuts), but the main thing was that the soot deposits *did* contain carbon clusters. Experiments with soot produced by an acetylene torch gave similar results. By May 25, they had found that soot produced by burning plastic or acetylene contained nearly pure carbon clusters which they could desorb from the surface and detect. It was that simple.

These spectra showed that the clusters contained small amounts of elements other than carbon (most likely hydrogen). To produce the fullerenes, they decided they would need a cleaner source of soot. They therefore went back to the laser vaporization of grapite but this time they evaporated the graphite under a static pressure of about 500 torr of argon and simply collected the soot that was formed on either copper, gold, or quartz substrates placed about one centimetre from the target. The coated substrates were then removed, carried across the lab, and placed in the laser desorption apparatus. The results from the first of these experiments, carried out on June 18, were as clear as they were striking: large signals due to C_{60} and C_{70} showed that these molecules were

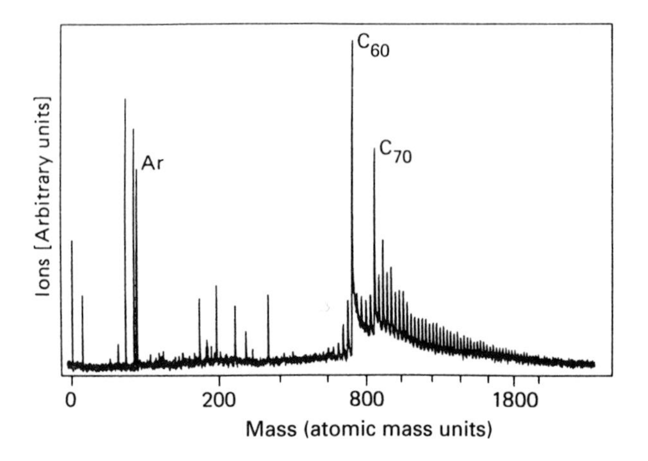

The IBM scientists deposited carbon soot on a substrate surface using a laser to vaporize graphite. They then analysed the soot using laser desorption and mass spectrometry. The results showed that the soot contained fullerenes, with C_{60} and C_{70} predominant. They estimated that each square centimetre of deposit contained about 100 billionths of a gram of C_{60}. Adapted, with permission, from Meijer, Gerard and Bethune, Donald S. (1990). *Journal of Chemical Physics*, **93**, 7800.

indeed present in the soot deposits, as Rohlfing had suggested. There were also signals for large even clusters up to C_{200}. The IBM scientists felt as thought they had stumbled on Aladdin's cave.

Over the next month, Bethune and Meijer assembled an apparatus to produce C_{60} by the laser vaporization of graphite. They experimented with different pressures of inert gas and looked at the effects on the C_{60} yields of changing the focus of the vaporization laser. They looked at the fragmentation of the fullerenes in the mass spectrometer. They even tried to find C_{60} in drops of benzene, alcohol, and many other solvents that they trickled over the soot deposits, but without success.

On July 19, Bethune described their results to other IBM scientists gathered for one of the regular in-house research seminars. He explained what he and Meijer had done so far and estimated that about one square centimetre of soot-coated substrate contained on average about 100 nanograms (100 billionths of a gram) of C_{60}. Admittedly, this did not sound like much, but it was considerably more than they had seen reported by anybody else and it opened up the possibility of accumulating much larger quantities.

The discussion that followed Bethune's short presentation quickly became focused on the analysis of the samples, and particularly the importance of carbon-13 NMR. They were all agreed that the one-line NMR spectrum would provide a unique confirmation of the soccer ball structure. But how much C_{60} would they actually need for this kind of measurement? Their colleague Nino

Yannoni suggested they would need something between one-tenth of a gram and one gram, about a million times more than Bethune and Meijer were collecting in the soot.

Subsequent discussions with Robert Johnson, a specialist in high-sensitivity NMR in Almaden's Polymer Science and Technology Department, revealed that they might be able to manage with 300 micrograms of C_{60}. This was still several thousand times more than they were making, but Bethune and Meijer felt that with time and effort, accumulating this amount was just about possible.

They would stand a better chance if they could find a way to scale up the production process. Bethune knew of several very high-power excimer lasers around the lab and thought they could get access to one of these. Perhaps they could blast large chunks of graphite to atoms and so form the C_{60}-containing soot in greater quantities.

On the other hand, there were other analytical techniques available that were not quite so demanding in the amount of material required for study. And there were certainly many IBM scientists only too willing to join in the fun. Robert Wilson indicated that the films they were currently making would be suitable for generating a Scanning Tunneling Microscope (STM) image of the soot layer on the surface. This was a new technique in which an electrically charged, extremely fine tip was moved slowly over a surface, only a few atomic diameters above the topmost layer. Tunneling of electrons from the tip to the layer (a purely quantum phenomenon) gives a measureable electric current. By adjusting the height of the tip to keep this 'tunneling current' constant, the contours of the layer could be traced out. Scanning the tip over the layer allowed a three-dimensional image of the surface layer to be built up. A surface layer of soccer ball C_{60} molecules was expected to appear as row upon row of tiny spheres.

In addition, Hal Rosen thought they might be already making enough C_{60} for him to measure the Raman spectrum. However, although these techniques required less material, the material itself had to be reasonably pure if they were to be able to interpret the data. Bethune and Meijer reasoned that it might be possible to purify the C_{60} by subliming it out of the soot, depositing it as a solid film on another substrate.

There were all kinds of ways they could tackle the problems, but first Bethune and Meijer had to conclude their laser vaporization experiments. As they started to write up their results, they realized they had no way of knowing whether the C_{60} molecules they had been detecting were formed by the vaporization laser and deposited as stable molecules (as they had assumed up to now) or whether they were instead formed during the laser desorption process.

To find out, they hit on the idea of making a soot deposit consisting of alternate layers of pure carbon-12 and carbon-13. If the C_{60} molecules were being produced during the desorption process, then the carbon isotopes would necessarily become scrambled. The resulting mass spectrum would then show several peaks, corresponding to C_{60} predominantly composed of carbon-12 but with one or more atoms of carbon-13, and C_{60} composed predominantly of carbon-13 but with one or more atoms of carbon-12. On the other hand, if the C_{60} molecules desorbed from the deposit intact (as they had assumed), then the mass spectrum would show only two peaks: one corresponding to C_{60} composed only of carbon-12 and the other corresponding to C_{60} composed only of carbon-13.

There was an agonizing delay of a week as they waited nervously for the isotopically pure sample of carbon-13 powder to arrive. When at last they were able to perform these crucial experiments, Bethune and Meijer were relieved to observe only two peaks in the mass spectra. As a further check, they mixed together some samples of the carbon-12 and carbon-13 powders and produced soot by vaporizing the result. Subsequent desorption of this isotopically mixed material gave spectra showing a single mixed-isotope C_{60} signal. There could now be no doubt about what they had produced.

They completed a short paper describing these experiments which they submitted to the *Journal of Chemical Physics* on August 7, and turned their attention fully to the problem of scaling up their method of soot production. Hunziker commented that if C_{60} could be made so readily in the plasma created by laser vaporization, perhaps it could also be made in an arc discharge. Bethune and Meijer were sceptical, but agreed that this was something they could try . . . later.

On July 26, Hare read the Krätschmer, Fostiropoulos, and Huffman carbon-13 paper in *Chemical Physics Letters*. Here was yet more convincing evidence that buckminsterfullerene was present in the soot at the level of about one per cent. The paper was also a stark reminder that Kroto and Hare were up against some stiff competition in the race to be first to separate and properly characterize the new substance.

Once again, Hare thought hard about the problem of isolating the C_{60}. Like the members of the Heidelberg/Tucson group, Hare had trained as a physicist, and these problems of chemical separation and analysis did not come particularly easy. However, unlike Krätschmer, Fostiropoulos, Huffman, and Lamb, Hare was a physicist working among chemists in a chemistry laboratory. Maybe something had rubbed off on him, because he *was* able to make the intuitive leap and consider that dissolving the C_{60} out in a solvent might be worth a try.

Benzene seemed to him to be an obvious choice of solvent, and on Friday, August 3, he placed about half his precious stock of the soot in a small glass vial. To this he added about 25 millilitres of benzene. Nothing happened, which was no great surprise as everybody knew that carbon soot is insoluble. He shrugged his shoulders and put the vial up on a high shelf, went home for the weekend and promptly forgot about it.

When he returned to the lab on the morning of Monday, August 6, the vial still contained a substantial amount of insoluble soot, but the solvent above it had turned a faint red colour. Now this was something. Getting a coloured solution from soot was simply impossible. Hare filtered out the undissolved solid and gently heated the solution to concentrate it by evaporating some of the benzene. The faint red colour intensified to a deep wine-red. Hare went around the lab with his prized solution, pointing it out to his colleagues and telling them: 'C_{60}'s in here!'. Although his colleagues humoured him, they really didn't believe it. Hare wasn't sure he believed it either.

Kroto didn't know what to believe. Could this be the miracle he had dreamt of for five years? Or was it something else? Kroto knew that colloidal suspensions scatter light and can appear red under certain conditions. They needed to measure the mass spectrum of the stuff in the solution, to make sure that C_{60} and, probably C_{70} too, were indeed present. Hare tried to measure the mass spectrum of the solution three days later, but the lab spectrometer was still giving problems and he could obtain no signal.

The next day, Friday, August 10, Kroto received a telephone call. It was Philip Ball at the London office of *Nature*. He had received a paper on C_{60} and wanted to know if Kroto was prepared to review it. It appeared to be quite an important paper. Kroto agreed, and the manuscript was faxed to Sussex a little while later. Kroto asked if Hare would collect the fax from the store room, where the fax machine was situated. On his way back up the stairs, Hare flicked through the pages of the manuscript intently. The concerned expression on Kroto's face belied another of his premonitions, and as he looked through the pages his expression became more grave.

'Oh my God,' he said, 'they've done it'.

11

The one-line proof

Time was moving on. Huffman arrived in Heidelberg for a short visit on June 4 and resolved to sit down with Krätschmer to make a start on a draft paper describing the results they had obtained so far. From previous experience, Huffman knew very well that writing papers with Krätschmer was not so straightforward. Although Krätschmer's English was very good, he would often use words or phrases that didn't read quite correctly to a native English speaker. Krätschmer was happy for Huffman to correct what he had written, but would then agonize over the changes, wondering if the sense of what he had wanted to communicate had been affected by the revisions. Huffman decided that this time he would pre-empt this agonizing by drafting the paper on Krätschmer's own word processor in Krätschmer's own office. It made no difference. Krätschmer still agonized over the English, and they got no more than half the paper done.

They decided that, as solid-state physicists first and foremost, they would make the discovery of *fullerite* the key feature of their paper. They would announce the discovery of a new crystalline form of carbon, and present evidence that it was composed of soccer-ball shaped C_{60} molecules without getting too caught up in the molecular aspects of the substance, hopefully avoiding attention that could otherwise be drawn to the inevitable loose ends. The chemistry they could happily leave to the chemists, while they concentrated their efforts on the solid state physics.

As June progressed, they faxed the manuscript back and forth between Heidelberg and Tucson. They also continued to amass more experimental data and to refine the data they already had. Further mass spectrometry studies in Tucson confirmed that the main component of the fullerite was C_{60}, contaminated with some ten per cent C_{70}. Huffman and Lamb took a sample of the fullerite powder to the neighbouring Geology Department, and watched as a technician measured its X-ray diffraction pattern. The first of a series of peaks indicative of reflections from different crystal planes emerged, and the

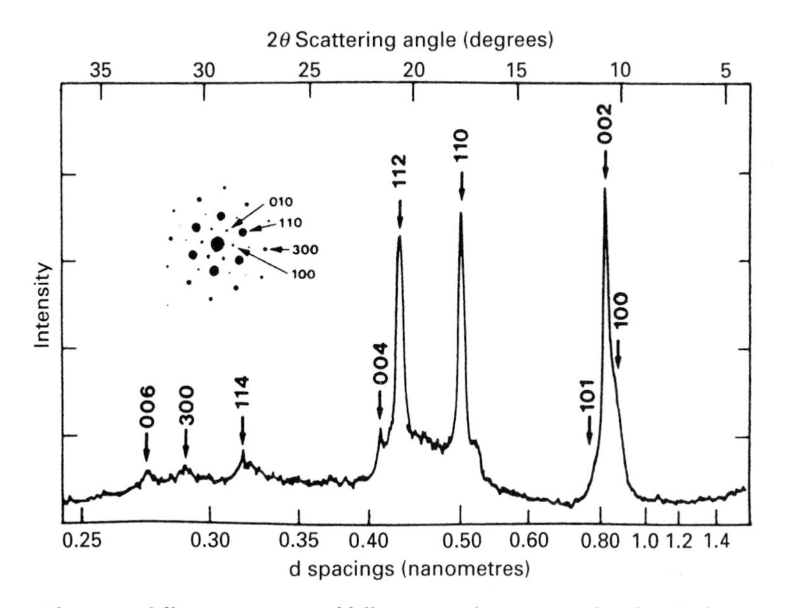

The X-ray diffraction pattern of fullerite powder, measured in the Geology Department in Tucson. The peaks represent reflections from different crystal planes, confirming that the material *is* crystalline. The fact that this pattern of peaks and spots had never been seen before in a sample of pure carbon also confirmed that Huffman and Lamb were dealing with a totally new form. Like the electron diffraction pattern, this result suggested that the substance was composed of tiny molecular spheres about one nanometre in diameter, but problems of interpretation remained. The absence of certain peaks characteristic of the close packing of spheres indicated that there might be some long-range disorder in the crystal structure. Adapted, with permission, from Krätschmer, W. Lamb, Lowell D., Fostiropoulos, K., and Huffman, D. (1990). *Nature*, **347**, 354. Copyright (1990) Macmillan Magazines Limited.

technician remarked that he had never—but never—seen anything like this before. The peaks showed that what they had *was* a crystalline form of carbon, and the fact that it had never been seen before proved that it was totally new.

But the definitive analysis was still X-ray diffraction from a single crystal. Growing a large single crystal of the fullerite turned out to be much more difficult than growing smaller ones under the microscope, and both Fostiropoulos and Lamb spent long hours in their respective labs before obtaining suitable crystals, grown to about 500 micrometres in diameter and with clearly discernible faces. In the end, it was all a bit of a disappointment. The large single crystals simply refused to yield the expected diffraction pattern. There were analogies with the crystal structure of cobalt. Perhaps the soccer ball molecules were not packing together in the orderly way the physicists had anticipated. Problems with solvent molecules or possibly even some C_{70}, which was supposed to be an elongated or squeezed sphere, could be enough to cause long-range disorder in the crystal structure leading to a loss of definition of the crystal planes. They were not going to get the unambiguous result they had hoped for.

Fostiropoulos spent a few days at the Max Planck Institute for Medical Research in Heidelberg in the laboratory of Heinz Staab, then President of the prestigious Max Planck Society. It was here that Fostiropoulos discovered that it was possible to separate the red solution, using techniques taught in school science classes all around the world. He found that the fullerite was soluble in hexane as well as benzene, and that the hexane solution could be separated into three distinct bands by thin layer chromatography. However, he did not attempt chromatography on a preparative scale, and did not pursue any further analysis of the individual components.

During this brief period in Staab's lab, one of the scientists from the Institute's NMR group asked Fostiropoulos if he had thought of measuring the carbon-13 NMR spectrum of his sample. Fostiropoulos wasn't familiar with this technique and didn't know what it could do. The scientist explained that if, as he had said, C_{60} was supposed to have a soccer ball structure, then its NMR spectrum should consist of a single line. Okay, said Fostiropoulos, but what would he learn from this? Not convinced that the NMR spectrum would be worth the effort, he decided not to measure it. Krätschmer later discovered that to the chemist, such is the importance of NMR as an analytical tool that 'when the NMR machine breaks down, the chemists go home'.

They now had to decide what to do with their results. Apparent disorder in the crystal structure had produced less-than-clear results from the single crystal diffraction studies. However, the mass spectra, and the electron and X-ray powder diffraction data all seemed pretty conclusive. They also had the excellent ultra-violet and infra-red spectra, the latter clearly supporting the buckminsterfullerene proposal. Was this enough?

Krätschmer and Fostiropoulos were beginning to feel the effects of this extremely intense period. Each night they worked on in the lab until the early hours of the next morning—sometimes two, three, or four a.m.—when Krätschmer would drive Fostiropoulos home through the dark shadows of the Odenwald. Krätschmer's wife was often on the telephone, wanting to know when her husband was coming home. The scientists were very excited by what they were doing, but they were also exhausted.

The time came to cut their losses. Krätschmer wasn't sure what to do next, but Huffman was ready to go with the data they had. They completed the paper by the end of June, but now Huffman had another problem. It was clear that the evaporation process could be used to produce a new form of carbon after all, and he wanted to redraft and resubmit the patent disclosure he had had to withdraw in February 1988. His patent advisers at the University of Arizona were screaming at him not to publish any details of either the process or the fullerite itself until the patent application had been lodged with the relevant authority. Placing such details in the public domain too soon would render the patent application invalid. His advisers recommended that the fullerite be included because patented processes are notoriously easy to circumvent unless they are absolutely watertight, and they figured that, no matter how it was being made, fullerite was fullerite. Once protected by the patent, any process that produced fullerite would need to be licensed.

There followed a whole month of nervous anticipation as the patent was drafted and processed. Although Krätschmer stood to benefit—his name was on the application alongside Huffman's—he regarded the business as a rather unpleasant distraction from his first duty, which was to report on the breakthrough in the name of scientific progress. He could not be persuaded that scientists have a duty to themselves to protect their intellectual property, particularly where this property may lead to commercial applications.

Their greatest fear was that this delay would endanger their claim to priority—that they would not be the first to report the discovery of fullerite. They knew that Kroto and Hare were on the trail but had no idea how far behind they might be. They were completely unaware that the well-resourced scientists at IBM's Almaden Research Center were also close behind.

Finally, at the end of July, they got the green light from Huffman's patent advisers. The application was made and they could now publish their results. Huffman agreed to submit the paper to *Nature* and handle any further correspondence with the editors on Krätschmer's behalf. This gave Krätschmer the opportunity to get away with his wife on their annual holiday and forget all about buckminsterfullerene. They left Heidelberg on August 1, bound for Hungary.

Philip Ball received the manuscript of the paper on Tuesday, August 7. At IBM's Almaden Research Center, Bethune and Meijer were just submitting their first paper on the detection of C_{60} and C_{70} desorbed from carbon soot to the *Journal of Chemical Physics*. In Sussex, Kroto and Hare were wondering what to do about the wine-red solution they had obtained from the soot just a day earlier, trying to bite back on their frustration at the sorry state of the departmental mass spectrometer.

Ball faxed the manuscript to Kroto on August 10. When Kroto read it through, he finally became aware that he and his Sussex colleagues had been involved in a race for buckminsterfullerene without their knowing it, and they had lost. The Heidelberg/Tucson group had put together a totally convincing case. The fact that Hare had obtained the red solution quite independently just days before deepened Kroto's conviction but was also very disappointing. They had come so close.

After a rather gloomy lunch, Kroto tried to pull himself together. He telephoned Ball and announced that the Krätschmer, Lamb, Fostiropoulos, and Huffman paper was simply great and should be published as it stood without delay. He also wished to waive the anonymity of the refereeing process and asked Ball to congratulate the authors on his behalf. However, it was standard practice at *Nature* to secure the opinions of at least two referees for the longer articles, and so Ball asked Kroto if he could name a suitably knowledgeable scientist to act as a second referee. Without hesitation, Kroto suggested Bob Curl.

Kroto faxed his formal referee's report to *Nature*, and then began to assess the damage. He read through the manuscript once more, and this time noticed that there were a couple of relatively important pieces missing from the jigsaw puzzle of evidence. These were not crucial to the paper—which was convincing enough—but finding the missing pieces offered Kroto and his colleagues the chance to come a good second.

Kroto noticed that the Krätschmer, Lamb, Fostiropoulos, and Huffman paper did not contain a mass spectrum, although mass spectral data were referred to in the text. Abdul-Sada had already used mass spectrometry to show that the soot samples obtained by carbon-arc evaporation contained substantial quantities of C_{60} and C_{70}, confirming the arguments made in the *Nature* paper.

Most importantly, the paper did not contain a carbon-13 NMR spectrum. All the other analytical evidence the Heidelberg/Tucson group had presented appeared to be solid, but the one-line NMR spectrum was undoubtedly the jewel in the crown and it was still up for grabs. The Sussex group had been trying to get an NMR spectrum of the carbon soot for some time, without success. Now they had a solution containing C_{60} and a little C_{70}. If they could get these separated somehow, NMR spectra of solutions of C_{60} and C_{70} should be readily measureable. It became clear that all was not lost for the Sussex group.

There was still a good chance that Kroto and his colleagues could be first to that coveted NMR line. But if they were going to be first, they had to move very, very quickly. Kroto knew that in this age of rapid telecommunications, scientists active in a particular field of research no longer had to wait for the often slow mechanics of publication. For all he knew, the Heidelberg/Tucson manuscript could already be on its way to a dozen or more laboratories as a 'preprint'. These laboratories would be filled with scientists only too eager to repeat the experiments and measure the NMR spectrum of buckminsterfullerene. After all, it was such an obvious thing to do, at least to a chemist.

Curl received the manuscript by fax at noon the same day, *Nature* benefiting from the five-hour time difference between Sussex and Houston. He took the manuscript home and worked on it over the weekend. It was clearly a paper of the first importance, heralding a new era in carbon chemistry and materials science. Curl was in no doubt that, on reading this paper, literally thousands of scientists all over the world would rush to repeat the carbon evaporator experiments and reproduce and extend the results that Krätschmer, Lamb, Fostiropoulos, and Huffman were reporting. Under normal circumstances, he felt that the manuscript easily met the standards required. But these were not normal circumstances. A huge volume of resources—scientists and equipment—would likely be thrown at carbon soot following publication of this paper. If all this effort was not to be mis-directed, there was no room for error whatsoever.

A single-crystal X-ray diffraction pattern would have clinched it, but obtaining such a pattern appeared to be out of the question. Curl was prepared to accept the argument that over long ranges the packing of the spherical molecules in the fullerite must be somewhat disordered. However, without a single-crystal diffraction pattern, and with electron and X-ray powder diffraction patterns that were not completely unambiguous, Curl concluded that it was important to have as much supporting evidence as possible from other analytical techniques, particularly mass spectrometry and NMR.

The mass spectrum of the fullerite was conspicuous by its absence. The evidence for the existence and special structure of C_{60} had come from mass spectrometric studies of carbon clusters. It was such a key piece in the jigsaw puzzle—a strong, clear signal corresponding to a molecule with a mass of exactly 60 carbon atoms—that Curl felt it just had to be included. Unlike Kroto, he did not have the benefit of the work that Hare and Abdul-Sada had done at Sussex. He did not have his own mass spectrum to serve as independent confirmation that the Heidelberg/Tucson group had got it exactly right.

On Sunday, August 12, Curl faxed a detailed, five-page report back to Ball at *Nature*. He had examined every assertion in the manuscript in minute detail, making suggestions as to how the arguments might be better developed. He also recommended that the mass and NMR spectra of the fullerite be included. Dealing with these detailed comments would inevitably cause delay, which Curl was at pains to emphasize should be kept to a minimum. He too believed it most important that the Heidelberg/Tucson group should receive due credit for being first.

When Huffman and Lamb received Curl's report, they had to admit it was honest and painstaking. Every point that Curl had raised would undoubtedly strengthen their arguments and make the paper better, although there was a limit to how much they could keep adding and stay within the bounds placed on the manuscript by the rules laid down by *Nature*.

Philip Ball, following Curl's recommendations, was asking for a figure showing the mass spectrum. They had a perfectly good mass spectrum measured by a colleague at Tucson which they had chosen not to include in the original manuscript because the colleague had asked for his name to be added to the list of authors. They had judged that the credit was really theirs alone, and so had preferred simply to record the results of the mass spectrometry in the text.

Krätschmer was still in Hungary, soaking up the sunshine. It was therefore up to Huffman and Lamb to find a way to meet at least some of Curl's recommendations.

For a few heady weeks Bethune and Meijer believed they had the field all to themselves. As far as they knew, they were the only scientists with access to a method capable of generating quantities of buckminsterfullerene sufficient for spectroscopic analysis, and they fully intended to make the most of it.

Their research programme almost suffered a major setback when Bethune and his wife were offered an all-expenses-paid one-week holiday in Hawaii, where their three sons were due to compete in a swimming contest. The opportunity had arisen because of a mix-up over the number of parents attending the contest with their children, and as a result too many flight tickets and hotel rooms had been booked. It was too good an opportuniy to miss, and Bethune would have accepted had he been able to find somebody to look after his two younger daughters. Instead, he decided that he would stay behind but his wife should go. It was, with hindsight, a fortuitous decision. Life was about to become very hectic.

On Monday, August 20, Chris Moylan, one of Bethune's IBM colleagues, sent him a note asking if he had seen a paper by Krätschmer, Fostiropoulos, and Huffman just published in the July 6 issue of *Chemical Physics Letters*. Bethune had not. He immediately read the paper and cursed himself. Of course, he and Meijer had been on the right track, but scaling up the production process required not the high-tech approach utilizing an even bigger laser, but a low-tech approach with a bit of graphite and a power supply.

Bethune told Meijer about the paper and they scrambled to get hold of the largest power supply they could find. They fed the terminals from the supply through into a vacuum chamber and connected them to either end of a thin graphite rod. They did not have a bell-jar apparatus, so they were just going to fry the hell out of a single graphite rod and see what they got. They admitted 100 torr of helium into the chamber and increased the electric current passing through the rod until the light emitted from the glowing material became too intense for the naked eye to stand. In the light projected onto the ceiling through a window on top of the chamber they saw the shadows of thin wisps of carbon smoke literally boiling off the graphite surface. After only a minute of this torture, the rod snapped, breaking the electrical circuit it completed.

On the viewing window at the top of the chamber, a yellowish-brown layer of soot had been deposited. Bethune and Meijer removed this window and placed it in their laser desorption apparatus. A single shot from the desorption laser was enough for them to see the characteristic mass spectrum of the fullerenes on the oscilloscope screen, dominated by the signal due to C_{60}.

The next day Bethune took his wife and sons to San Francisco airport to catch their flight to Hawaii. He still had his daughters to look after, but had secured the child-minding services of several of his colleagues' wives and daughters, so he was free to return to his experiments during the day. On August 22, he telephoned Huffman in Arizona to tell him they could now confirm that C_{60} was present in the soot formed by electrically vaporizing graphite. Huffman explained that since the publication of their *Chemical Physics Letters* paper on the carbon-13 soot experiments he, Krätschmer, Lamb, and Fostiropoulos had confirmed the presence of C_{60} for themselves. They had extracted C_{60} (fullerite) from the soot and had analysed it using mass spectrometry among other techniques. They had submitted a paper describing their results to *Nature* and Huffman was now wrestling with a problem raised by the referee. He wasn't sure that they would be including any mass spectral data in the final form of their paper.

Huffman was reluctant to say more, but he recommended that Bethune go ahead and publish the mass spectral results he had obtained with Meijer at IBM. The IBM physicists duly wrote up their results in a paper which they submitted to *Chemical Physics Letters* on Friday, August 24. As Rick Smalley was an editor of this journal, they sent their manuscript to him.

Bethune and Meijer continued their experiments over the weekend. By vaporizing numerous small graphite rods, they collected about 20 milligrams of the C_{60}-containing soot. It was clear from Bethune's telephone conversation with Huffman that the Heidelberg/Tucson group had somehow managed to extract the C_{60} from the raw soot, but Huffman had not said how they had done it. Bethune had earlier thought of trying to sublime the material, and had had a small electrical heating device constructed to fit inside the vacuum chamber.

He and Meijer tried this out on the following Monday. They placed the 20 milligrams of soot in a metal sample container fixed to the heating device. In front of a small hole drilled in the container they mounted a quartz substrate. They raised the temperature of the container to 700 kelvin and watched as a thin yellow film of solid material formed on the surface of the quartz. Subsequent laser desorption mass spectrometry showed that this material was fullerite—nearly pure C_{60} with some C_{70} contamination. They were now ready to try for the Raman spectrum, carbon-13 NMR spectrum, and STM image.

Back in Sussex, Kroto and Hare decided that they needed help—fast. They discussed their options over coffee with Roger Taylor, a physical organic chemist at the School of Chemistry and Molecular Sciences at Sussex, and Taylor immediately offered his all-important expertise in the preparation and purification of organic compounds. Taylor quickly discovered that the fullerite was soluble in other solvents, notably hexane.

The solubility of the fullerite in hexane suggested to Taylor that it might be possible to separate the carbon clusters chromatographically on a large scale. With assistance from another Sussex chemist, Jim Hanson, he chose to try a chromatographic column packed with alumina. As the red solution was gradually washed down the column, it separated into three distinct bands. The first was a delicate magenta. The third was a deep wine-red.

The departmental mass spectrometer was still out of action so Kroto sent Hare off with samples of the fractions separated from the soot to the Manchester laboratories of VG Analytical, a commercial company specializing in state-of-the-art technology for chemical analysis. At the VG labs, Hare handed over the sample to Paul Scullion for analysis using Fast Atom

Bombardment (FAB) mass spectrometry. In this technique, the solid sample is bombarded with neutral atoms (typically argon) and the desorbed molecules or fragments produced are ionized and detected. This is particularly suitable for the analysis of samples of unknown composition, since it produces large quantities of the molecular ions—the singly ionized forms of the desorbed molecules—allowing rapid identification of the molecules present.

Kroto, Taylor, and Walton sat waiting for Hare's call to say that everything checked out, their patience wearing thin. Hare was having problems getting the samples into a matrix form suitable for analysis. After some frantic telephone calls, and some crucial advice from Taylor, Hare and Scullion finally managed to get the sample dissolved into a matrix of *meta*-nitrobenzyl alcohol. The mass spectrometry showed that the magenta fraction was C_{60}. The more strongly coloured red fraction was C_{70}, its intense colour masking the magenta in solutions of fullerite even though C_{70} accounted for only ten per cent of the total material dissolved in solution. The middle fraction was a mixture of C_{60} and C_{70}.

So there it was. After all these years of joy, frustration, doubt, argument, and counter-argument, the Sussex chemists were the first to hold in their hands a solution of pure buckminsterfullerene. It was hardly credible. This was the fantastic molecule that had been foretold in the 1960s. By September 1985 it had become a large blip in a time-of-flight mass spectrum and a purely speculative three-dimensional construction wrought from some highly active imaginations. Now it was real. Buckminsterfullerene was a real molecule with real properties. A thin film of the solid was mustard-coloured, appearing a darker brown or black with increasing film thickness. Mix it with benzene and other solvents and it gave solutions coloured a delicate magenta. Now, what about that NMR spectrum?

By the end of August, Curl still didn't know what had become of the Krätschmer, Lamb, Fostiropoulos, and Huffman paper, and he was getting worried. Looking back through his five-page referee's report, he felt increasingly guilty about the possibility that he had erected somewhat artificial barriers to the rapid publication of the paper. He decided to set aside formalities, and telephoned Huffman on August 31.

He explained to Huffman that he hadn't wanted his recommendations to be taken so seriously that meeting them would unduly delay the paper's publication. He would settle for a mass spectrum showing that the fullerite was composed predominantly of C_{60}. To Huffman, this news was a tremendous relief. He explained that they had a mass spectrum showing exactly what Curl wanted to see, but they had been unable to include it in their manuscript because of the claims on authorship that had been made. With this, Curl's major objection to the paper evaporated.

To make up for lost time, Curl urged Huffman to send the manuscript as a preprint to a number of distinguished chemists, drawing their attention to the discovery. He gave Huffman a long list of names and addresses. This was about the last thing Huffman wanted to hear. The previous few weeks had been among the most difficult of his life and he felt he was close to collapse. There was just so much conflict between his and Krätschmer's concerns to be extremely careful—to avoid another cold fusion—and the now desperate need for urgency.

Krätschmer arrived back from his holiday, and Huffman bought him up to date with everything that had happened. They decided to go with the mass spectrum of the fullerite that had been obtained for them by Zscheeg and Natour in Heidelberg, and re-worked the paper to accommodate some of Curl's recommendations. They did not include an NMR spectrum.

Curl contacted Philip Ball at *Nature* and explained that the Heidelberg/Tucson group had had a perfectly acceptable mass spectrum all along. This resolved his major concern over the paper and he urged Ball to publish it as quickly as possible. Huffman submitted the revised paper to *Nature* on September 7.

Having successfully separated the C_{60} and C_{70} from the carbon soot by sublimation, Bethune and Meijer decided to try for the Raman spectrum first. In collaboration with Hal Rosen and visiting scientist Wade Tang, they measured the spectrum of the thin films of fullerite they had obtained. As they had by now come to expect, the spectra were entirely consistent with the theoretical predictions for the soccer ball structure.

In particular, they observed strong lines at 273, 497, and 1469 cm^{-1} which they assigned to C_{60}. Of these, the line at 273 cm^{-1} was the most conclusive. All the earlier theoretical studies had predicted that buckminsterfullerene should possess a 'squashing' or 'bouncing' vibration, in which the sphere is distorted into an ellipsoid. While there were large variations in the predicted positions of the higher-energy vibrations, all the studies placed this bouncing vibration within 20 cm^{-1} of the measured position. Here was another key piece in the jigsaw puzzle.

The IBM scientists submitted their results in a paper to *Chemical Physics Letters* on Friday, September 7. The following day Bethune was due to travel to the German town of Konstanz on the shores of the Bodensee, to attend the fifth ISSPIC conference which was to start on Monday, September 10. He had tried several times to persuade the conference co-organizers, Olof Echt and E. Rechnagel, to allow him time to present the results the IBM team had obtained on buckminsterfullerene, but had been told that the programme was full. The last response to his request had, however, contained a merest glimmer of hope. The programme was full but if somebody should withdraw their paper at the last minute, then a presentation from Bethune could be considered, along with all the other hopefuls. This was not much of an encouragement, but it was enough for Bethune. He prepared a short lecture—just in case.

Meijer drove Bethune to San Francisco airport on Saturday morning. On the way they discussed the experiments they should attempt next. Bethune had

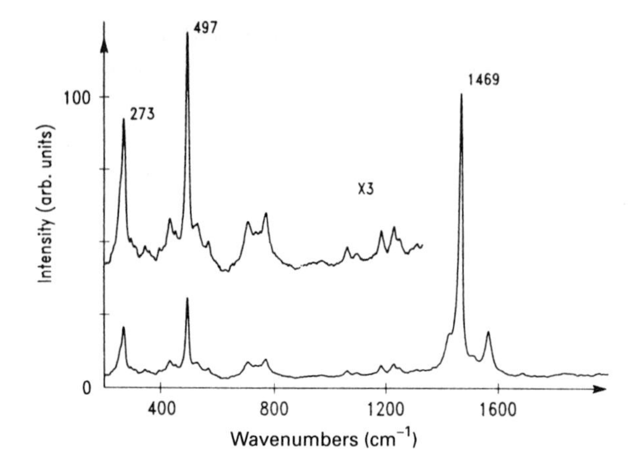

The Raman spectrum of a purified film of C_{60} reported by IBM scientists Bethune, Meijer, Tang, and Rosen. Three strong lines appear at 273, 497 and 1469 cm^{-1} which can be assigned to C_{60}. Of these, the 273 cm^{-1} line is the most compelling, corresponding closely to theoretical predictions for the 'bouncing' or 'squashing' vibration of soccer ball C_{60}. Adapted, with permission, from Bethune, Donald S., Meijer, Gerard, Tang, Wade C., and Rosen, Hal J. (1990). *Chemical Physics Letters*, **174**, 219.

passed some of the thin film samples to Robert Johnson, who was trying to get the solid to dissolve in various solvents and begin the carbon-13 NMR measurements. Bethune wasn't sure if the Heidelberg/Tucson group hadn't already made these measurements (Huffman had not talked about NMR on the telephone), but Meijer was convinced that it was too spectacular a prize not to pursue vigorously. They agreed that Meijer would give Johnson whatever assistance was necessary to get that one-line proof.

But the IBM scientists were not destined to be the first to that one-line NMR spectrum. Just as Bethune and Meijer were working with Rosen and Tang on the Raman spectrum of buckminsterfullerene towards the end of August, and as Curl was making enquiries about the fate of the *Nature* paper, so the chemists in Sussex had been busy with their solutions of C_{60} and C_{70}.

Hare had passed a sample of the magenta solution to Gerry Lawless and Tony Avent, who ran the departmental NMR facility at Sussex. They measured the first NMR spectrum of buckminsterfullerene, but when Kroto got sight of it he could not fight off an overwhelming sense of anti-climax. The spectrum consisted of a large, single line due to the solvent (benzene) with a chemical shift of 128 parts per million (ppm)* and a tiny, almost insignificant line at 143 ppm. Avent assured him that this was the signal due to buckminsterfullerene and the fact that there was only one line confirmed it had the soccer ball structure. Kroto overcame his sense of anti-climax. The one-line proof was his at last.

They then measured the NMR spectrum of the red solution. If the 'squeezed-sphere' fullerene structure proposed for C_{70} was correct, then the NMR spectrum was predicted to show five lines, consistent with the five different types of environment that the carbon atoms could have in the structure. The relative intensities of the lines should reflect the different numbers of carbon atoms that had these different environments—10:20:10:20:10.

Avent measured the spectrum on Sunday, September 2, and he and Kroto studied it intently. It actually showed five lines in the region where C_{60} had shown one, but one of these lines coincided exactly with the C_{60} signal. This could just be a coincidence, but it could also mean that C_{60} was present as a contaminant in the red solution, leaving only four lines that could be assigned to C_{70}. Kroto felt that it was important to eliminate any trace of doubt by ensuring that all the C_{60} was removed from the red solution. Taylor repeated the chromatography, and the C_{60} NMR line duly disappeared.

This left only four lines, but then they noticed that there was a fifth, lying so close to the huge benzene solvent line that it had almost been swamped. That was it. The ten carbon atoms that run around the 'waist' in the C_{70} fullerene structure have an environment not so different from the carbon atoms in benzene, which was why the signal due to these atoms appeared so close to the solvent line. Any closer and they would have missed it completely.

It was Walton who convinced Kroto that, in many respects, the five lines in the NMR spectrum of C_{70} were more conclusive than the single-line they had at last obtained for C_{60}. The mass spectrometric data and the single line NMR spectrum of C_{60} did not preclude a large ring structure of 60 carbon atoms, with alternating single and triple bonds (although explaining the four infra-red lines would have been very difficult with such a structure). However, the NMR spectrum of C_{70} not only confirmed that the closed-cage fullerene structure was right, it also supported the idea that there could be a whole family of fullerenes of varying stability.

Buckminsterfullerene was the smallest closed-cage structure with no adjacent pentagons, and its spherical symmetry made it by far the most stable of the fullerene family. C_{70} was the next largest fullerene with no adjacent pentagons, and it should therefore come as no surprise to find that the carbon evaporation

*This is the standard unit for reporting lines in NMR spectra. The exact positions of NMR signals depend on the size and strength of the magnet used. The usual way of getting around this instrument dependence is to record simultaneously the NMR signal of a reference standard (usually tetramethylsilane). NMR lines are then reported in terms of the differences between the measured positions and the position of the standard expressed as a fraction of the resonance frequency of the standard. From the numbers that typically come out, units of parts per million are most convenient.

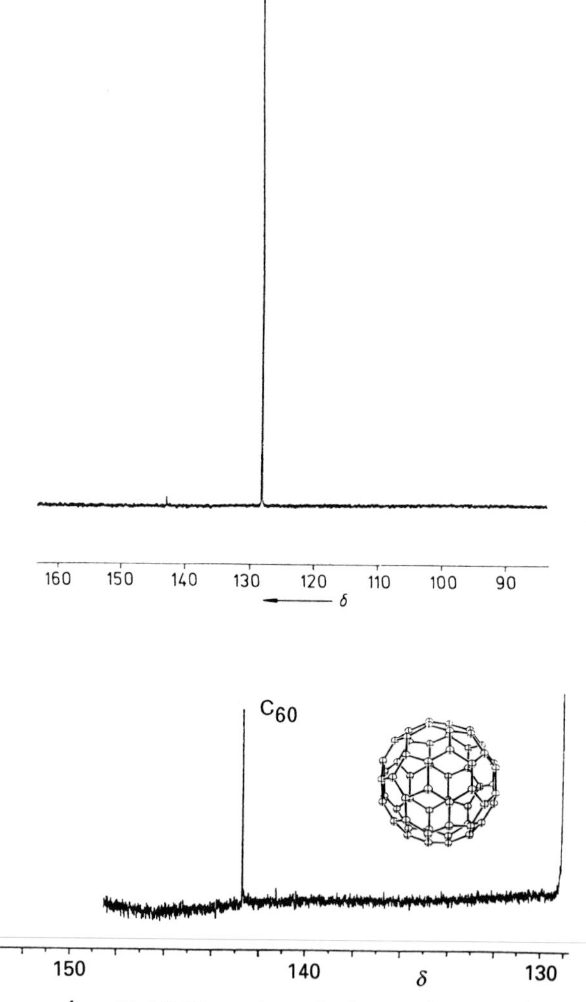

The upper carbon-13 NMR spectrum is the one-line proof as it first appeared to Kroto at the end of August, 1990. The one line in question is *not* the striking signal at 128 ppm, which corresponds to carbon-13 nuclei in natural abundance in the benzene solvent. It is, rather, the almost insignificant blip at 142 ppm. Despite the fact that Kroto had been dreaming of this line for five years, he could not fight off an overwhelming sense of anti-climax. The one line is seen more clearly in the lower spectrum. Adapted, with permission, from Kroto, Harold W. (1992). *Angewante Chemie (international edition)*, **31**, 111 and Taylor, Roger, Hare, Jonathan P., Abdul-Sada, Ala'a, and Kroto, Harold W. (1990). *Chemical Communications*, 1423.

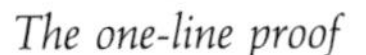

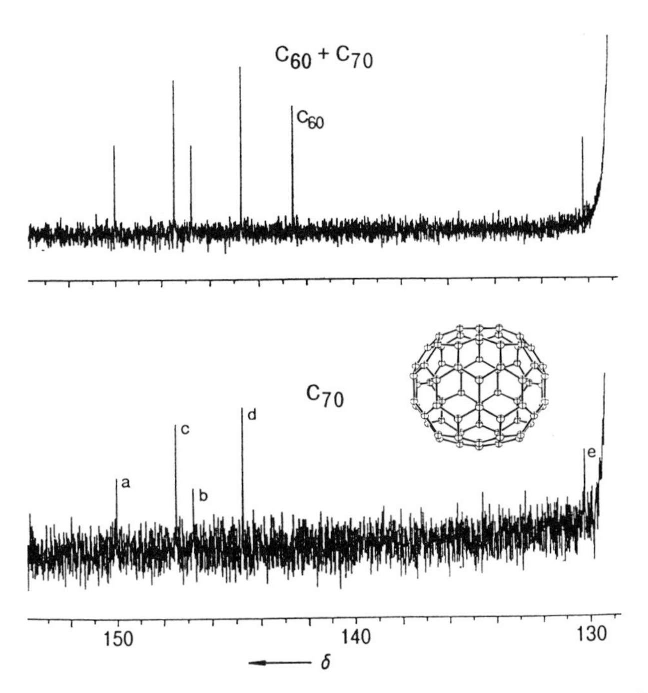

The carbon-13 NMR spectrum of the red solution as it appeared on Sunday, September 2 (upper spectrum) and after further purification to remove the last traces of C_{60} (lower spectrum). The proposed 'squeezed sphere' structure of C_{70} has carbon atoms in five different types of chemical environment, implying an NMR spectrum consisting of five lines. The spectrum of the purified sample did indeed show five lines, but with one dangerously close to the large signal from the benzene solvent. Adapted, with permission, from Taylor, Roger, Hare, Jonathan P., Abdul-Sada, Ala'a, and Kroto, Harold W. (1990). *Chemical Communications,* 1423.

process produced a little C_{70} along with the C_{60}. All the various threads were now being pulled together into a coherent story. It was not madness, nor even a mere flight of fancy. The fullerenes represented a whole new form of carbon in addition to graphite and diamond, and Kroto and his colleagues now had the evidence to prove it.

12

There's lots of it to go around

When Smalley heard the news of the Heidelberg/Tucson breakthrough from Curl, his immediate reaction was to celebrate with champagne. It didn't really matter all that much that a couple of physicists had done what the chemists had struggled so hard to do, and failed. It was enough that buckminsterfullerene was real at last.

Smalley was to give a lecture at the forthcoming ISSPIC conference in Konstanz, which was due to begin on September 10. Neither Huffman nor Krätschmer was intending to be present, and yet Heidelberg was not too far from the conference venue. Smalley reached a decision. If there was a chance that Krätschmer could get to the conference in time for his lecture, then he was prepared to give up the first five minutes of his allotted time to allow Krätschmer the opportunity to announce their breakthrough in person. He contacted Krätschmer directly and made the offer.

Krätschmer accepted, although he also noted with his typically dry humour that it was to Konstanz that the Czech reformer Jan Hus, leader of the 'Hussite' revolution in medieval Bohemia, had been invited to present his theses by the Reich emperor, with full assurances for his safety. Hus had presented his theses and had been promptly burned as a heretic. Admittedly, these were more enlightened times, but Krätschmer had no wish to suffer ostracism because of his possibly heretical views concerning a third form of carbon.

Smalley contacted Olof Echt and made the necessary arrangements. Krätschmer would announce the discovery on the morning of Wednesday, September 12. At the same time Smalley initiated a major research programme involving Rice scientists from the departments of chemistry and physics and the Rice Institute for Biochemical and Genetic Engineering. Their collective aim was to adapt and improve the Heidelberg/Tucson evaporation technique so they could go all out for buckminsterfullerene.

Kroto wasn't entirely satisfied with the first carbon-13 NMR spectrum of the red C_{70} solution, fearing some problems from contamination with a small amount of C_{60}. However, he was prepared to begin the task—with Hare, Taylor, and Abdul-Sada—of drafting a short paper describing the results they had obtained thus far. They began this task on Monday, September 3, just as Taylor was beginning a further chromatographic separation of the red solution to remove the last traces of C_{60}.

In the paper they described the FAB mass spectrum of the fullerite, the mass and NMR spectra of the chromatographed magenta and red solutions, and reported the band maxima and line positions from the ultra-violet/visible and infra-red spectra of both C_{60} and C_{70}. If Taylor was ultimately successful in eliminating the last traces of C_{60} from the red solution, then they could add a new C_{70} NMR spectrum to the manuscript before submitting it for publication. For the time being, it was important that they had something ready to submit to a journal. Psychologically, Kroto was now more than satisfied with the one-line C_{60} spectrum, but an unambiguous five-line C_{70} spectrum would be more than icing on the cake.

As the Sussex team worked on the chromatographic separation and measured the first NMR spectra, Kroto and Walton talked frequently about the need to inform the Heidelberg/Tucson group of the progress they were making. After all, Krätschmer's letter of the previous March had been so helpful to Hare, describing the conditions under which he and Fostiropoulos were generating carbon soot in the Heidelberg evaporator. Consequently, Kroto and Walton agreed that Krätschmer should be kept up to date. Although his letter hadn't told them everything, Krätschmer's advice had been crucial to the early success of the Sussex research programme on carbon soot, for which Kroto was very grateful. In truth, Kroto was also wary of the sensitivities that scientists can sometimes exhibit when working in such intense competition. He wanted to be sure that Krätschmer was completely comfortable with what the Sussex team had done quite independently. In particular, he wanted to make it clear in the text of their paper that they had discovered the extraction process—and obtained the red solution—*prior* to their subsequent knowledge of the Heidelberg/Tucson group's great achievement.

Kroto was due to attend an interdisciplinary science conference organized between the Universities of Sussex and Zagreb, to be held in Brioni, Yugoslavia, beginning in the second week in September. He had already decided to combine the trip with a short holiday, taking his wife and choosing to journey to Yugoslavia by car. By setting off a day earlier, they could stop off in Heidelberg on their way to Brioni, giving Kroto the opportunity to inform Krätschmer of the results they had obtained in Sussex and to congratulate him in person for his achievements in isolating and partially characterizing the fullerite.

Having confirmed by telephone that Krätschmer was going to be in Heidelberg during the next few days, and was happy to receive visitors, Kroto and his wife set off for Germany on Thursday, September 6, arriving in Heidelberg the following evening. Over dinner at the *Bierhelderhof*, nearby the Max Planck Institute for Nuclear Physics, Kroto, Krätschmer, and Fostiropoulos exchanged their stories, telling of the trials and tribulations of pioneering the fullerene field through this testing early period. Kroto explained the significance of the NMR measurements and passed a copy of his draft manuscript to Krätschmer. He was pleased to discover there were no sensitivities. It was then that Kroto learned that publication of the *Nature* paper had been delayed over the issue of the mass spectrum.

Kroto also had with him samples of the magenta and red solutions, which he proudly showed to Krätschmer and Fostiropoulos. As they later parted company, Kroto presented Krätschmer with a gift. It was a bottle of *Chateauneuf-du-Pape*, a deep red wine from the Rhône vineyards, overlooked by the ruined summer palace of the Avignon popes. Kroto had not intended the gesture to be in any way symbolic, but the wine was the colour of fullerite in solution, and the label showed that it was a 1985 vintage. This had undoubtedly been a good year, a year in which the signals from AP2's time-of-flight mass spectrometer had alerted Kroto, Heath, and O'Brien to the fact that C_{60} might be special in some way. It was almost exactly five years since those signals had first appeared on the monitor screen in Smalley's lab at Rice.

As Kroto was journeying across Eastern Europe, Taylor was succeeding with the further purification of the red C_{70} solution by chromatography. The NMR spectrum of the resulting purified sample showed the expected five lines and the draft NMR paper was rewritten to accommodate the new result. On Monday, September 10, the Sussex chemists sent the paper to *Chemical Communications*, a journal published by the Royal Society of Chemistry specifically for the rapid communication of new results. Hare brought the revised paper with him as he joined Kroto in Brioni.

During the week beginning September 3, Whetten was in Italy at a conference on molecular and ionic clusters. There was very little on carbon clusters presented at this conference, but it was in any case what Whetten heard during the lively social intercourse amongst its participants that interested him the most.

Amongst the attendees was Kroto's Sussex colleague Tony Stace. In an exchange over dinner one evening with German chemical physicist Ed Schlag, Stace had intimated that there had been some dramatic developments in Heidelberg and Sussex concerning buckminsterfullerene, but that he was not at liberty to say more at this stage. Whetten witnessed the exchange, and telephoned Diederich back in Los Angeles.

Just like the Sussex and IBM groups, Whetten and Diederich had also seen the Krätschmer, Fostiropoulos, and Huffman carbon-13 paper in *Chemical Physics Letters* earlier that summer. On the telephone from Italy, Whetten asked Diederich if he had heard anything more, wondering if the dramatic developments that Stace had referred to had originated within the community of organic chemists in Heidelberg. Diederich had heard nothing, and was very surprised to hear Whetten's news.

Whetten had no choice but to press Stace for more details. Under pressure, Stace told Whetten what he knew, which was in any case rather limited. He had heard that there had been a breakthrough in generating and isolating C_{60} and C_{70} in large quantities from soot produced by carbon arc evaporation. Kroto and his colleagues had managed to purify samples of both C_{60} and C_{70} and had measured their carbon-13 NMR spectra. The spectrum of C_{60} showed the expected one line. The spectrum of C_{70} showed five lines.

To Whetten, this was extraordinary news. He gradually digested this new information and thought about its implications over the weekend of September 8–9, as he made his way from Italy to Germany. The following week he was due to give a presentation—unconnected with carbon clusters—at the ISSPIC conference in Konstanz.

Bethune arrived in Zürich on the morning of Sunday, September 9, completing his journey to Konstanz by train. Going through the registration formalities in the early evening, he was informed by Echt that his slim hopes had in fact been fulfilled. One of the scientists due to give a presentation at the conference had pulled out at short notice, so Bethune could have a 10-minute slot just before lunch on Tuesday. Bethune was well pleased, and relieved that he had taken the trouble to prepare a short lecture.

The first day of the conference proceedings was devoted largely to metal clusters and little was said about carbon or the fullerenes. Bethune spent that evening marshalling his thoughts and going through his lecture material. He had only ten minutes to describe what the IBM group had done and wanted to make sure he got the most out of every one of them. He also sought as much drama as possible: as far as he knew, his was to be the first public announcement of the large-scale production and partial characterization of buckminsterfullerene, and he fully intended to enjoy the moment.

Bethune stood to give his short lecture shortly after noon on Tuesday. He described the IBM group's early studies on carbon-containing soot using laser desorption mass spectrometry, and their astonishment at the infra-red results in the Krätschmer, Fostiropoulos, and Huffman paper published in *Chemical Physics Letters* in July. He explained how they had immediately set up an apparatus—'. . . essentially a pencil lead and a battery'—to produce large quantities of carbon soot and showed that this too contained C_{60} in abundance. He then described the sublimation process and displayed the mass spectra of the purified C_{60} films. After presenting the Raman spectra of the films and explaining how the strongest peaks could be assigned to vibrational modes of the soccer ball C_{60} structure, he concluded by speculating that chemists and materials scientists would soon be able to obtain buckminsterfullerene from commercial chemicals suppliers.

To most of the assembled audience, this was surprising and fantastic news. An electron microscopist, who had been using graphite evaporation for many years as a means of coating samples with thin films of carbon, asked Bethune if he had been making buckminsterfullerene all this time, without knowing it. Bethune explained that the pressure of inert gas inside the vacuum chamber was critical in determining the yield of buckminsterfullerene in the soot. He doubted that an electron microscopist would employ the kinds of pressures needed for the efficient production of C_{60}.

It was a high point of the conference, and Bethune was proud and exhilarated. Later that afternoon, Smalley approached him in the hallway outside the lecture theatre and in a jocular fashion accused him of having 'let the cat out of the bag'. At the time, Bethune wasn't sure what Smalley meant by this, but in fact he had pre-empted the formal announcement of the breakthrough achieved by the Heidelberg/Tucson group that Krätschmer was planning to give during the first five minutes of Smalley's lecture the following morning.

Krätschmer travelled to Konstanz with Fostiropoulos that same day, arriving at the conference in the early evening. There they met up with Bethune, who gave them preprints of the IBM group's forthcoming papers, all beautifully confirming the very things that Krätschmer had spent the last four months agonizing over. Krätschmer urged Bethune to publish the Raman spectra as quickly as possible. From his meeting with Kroto, and the draft manuscript that Kroto had given him, it was clear that the Sussex group were also producing and purifying C_{60} and C_{70} in large quantities. It was surely only a matter of time before they turned their attention to the Raman spectra.

Bethune was surprised to learn that Krätschmer and Huffman had been working on carbon soot since 1982, and had first obtained an infra-red spectrum of soot consisting of just four lines no less than two years before. For the IBM group, progress in the short time since May had been fast and systematic, aided by the carbon-13 paper published in *Chemical Physics Letters* and the details of the evaporation technique it contained. That the IBM group had been able to catch up so quickly was due in part to the strengths of the interdisciplinary collaboration fostered by the IBM research environment.

But Bethune and Meijer had come an awfully long way entirely on their own. If they had had just another few months they could have probably been mass-producing fullerenes entirely independently of the Heidelberg/Tucson or Sussex groups. Bethune was resigned: such is life.

Was there still a chance to come a good second? It appeared that neither Krätschmer nor Huffman had thought to measure the carbon-13 NMR spectrum of either the solid fullerite or its solution in benzene. The importance of the one-line proof had been reiterated during the conference in Bethune's conversations with Smalley and Whetten, and Whetten had said that he thought Kroto already had a one-line NMR spectrum of C_{60} in solution. It wasn't clear to Bethune just how complete a result this was, or whether the

Sussex team had yet submitted it for publication. He couldn't help wondering how far Johnson and Meijer had got with the spectrum back at Almaden.

On Wednesday, September 12 at 9 a.m., Krätschmer rose to explain how he, Fostiropoulos, Huffman, and Lamb had stumbled on their major discovery: 'It all started in the fall of 1982 . . .'. He described their method for producing fullerite in bulk quantities in a bell-jar evaporator. He showed colour photomicrographs of fullerite crystals and presented the X-ray powder and electron diffraction results and the ultra-violet and infra-red spectra. He showed the title page of the forthcoming *Nature* paper.

His short presentation was extremely well received, with Smalley leading sustained, 'table-rapping' applause. The conference participants recognized that they were witness to an event unique in the science of carbon. Seemingly out of the blue, here was this mild, rather self-deprecating, German physicist showing them crystals of a totally new form of carbon. For many in the audience this was a highly memorable experience.

It was surely a hard act to follow. Smalley yielded to the buzz of excitement that had swept through the audience and, after briefly mentioning the material he had been due to present (on ICR probes of cluster surface chemistry), he abandoned his scheduled lecture and instead gave an *ad hoc* talk on C_{60} and the fullerenes. He went back over the origins of his collaboration with Curl and Kroto and the spectacular mass spectrometry results they had obtained with Heath and O'Brien at Rice in September 1985. He then went on to discuss some of the properties of buckminsterfullerene, the higher fullerenes, metal-encapsulating fullerenes, and the icospiral nucleation mechanism.

At the end of Smalley's lecture, Whetten again remarked that Kroto and his colleagues in Sussex had successfully isolated C_{60} and C_{70}, and had obtained a one-line NMR spectrum for C_{60} and a five-line spectrum for C_{70}. It was striking that buckminsterfullerene was finally being introduced to the world together with an almost complete set of analytical data, all beautifully confirming its soccer ball structure.

There were two further lectures that morning, including one on gold clusters from Cox and Kaldor and their Exxon colleagues, before a scheduled break for coffee. As the audience broke up, Krätschmer was congratulated by a number of scientists. Among them was Arne Rosén, who had to confess that he was quite impressed with the accuracy of his own predictions for the ultra-violet absorption spectrum of soccer-ball C_{60}, predictions that had given Huffman the necessary encouragement to pursue his crazy idea.

Krätschmer was also approached by Whetten, who asked if he knew anything of Kroto's NMR results. Krätschmer admitted that Kroto had visited him in Heidelberg the previous week and had given him a manuscript of a paper which contained the carbon-13 NMR spectra of C_{60} and C_{70}. As they talked, Whetten mentioned that his travel plans after the Konstanz meeting included a visit to Berlin, and Krätschmer asked if he would have the opportunity to come to Heidelberg. He and Fostiropoulos were still struggling with the problem of the single-crystal X-ray diffraction pattern of the fullerite and was interested to hear if Whetten had any advice. They agreed that Whetten would delay his trip to Berlin and visit Krätschmer in Heidelberg the following week.

Krätschmer did not stay for the remainder of the conference. It was during this period that his telephone began to ring almost constantly, and there were many things to be done. The conference proceedings closed that day at 12:30 p.m. After lunch the participants were given the opportunity to detach themselves from their science and enjoy the beauty of their surroundings in an excursion by boat across the Bodensee. They soaked up the scenery in the September sunshine and sipped a glass or two of wine before taking their seats for a lavish conference dinner. There are no marks for guessing the main topic of conversation.

The conference ended shortly after midday on Friday. Bethune travelled back to Zürich by train, accompanied by Smalley and his wife, Lynn Chapieski. Fullerenes were again the main subject of their conversation. Bethune described how he and his IBM colleagues had become embroiled in fullerene research back in May.

Smalley was intrigued by the relative ease with which the IBM scientists had been able to observe C_{60} and C_{70} in soot formed from laser vaporization. In what Smalley referred to as the 'search for the yellow vial', Jim Heath had spent nearly a year at Rice trying to extract buckminsterfullerene from soot produced by the same method. In the end they had abandoned hope of getting enough buckminsterfullerene this way, concluding that there was just too little of it in the soot to give a measureable yield by solvent extraction. It was not clear why the Rice group had failed where the IBM group had succeeded. In fact, Krätschmer and Fostiropoulos had also tried to produce their mysterious camel samples using laser vaporization of graphite, and they had failed too. Smalley was rather upset to discover that he and his Rice colleagues had apparently been so close to making bulk quantities of buckminsterfullerene for themselves.

On the evening of Saturday, September 15, a weary Bethune returned to San Francisco airport after his long flight home from Zürich. At the airport he was greeted by Meijer, who was clearly excited about something—he was jumping up and down and emitting strange Dutch whooping noises. Meijer drove the 50 miles from the airport to the Almaden Research Center, insisting that Bethune shake himself free of his jet-lag and go with him directly to the lab.

They arrived at about 11 p.m. Waiting for them in Robert Wilson's darkened laboratory was a picture the like of which Bethune had never seen before. Wilson had succeeded in using scanning tunneling microscopy to generate an image of the surface layer of a thin film of fullerite deposited on a gold surface. It was an image of row upon row of molecule-sized spheres. If seeing is believing, then this was truly a revelation. Dotted here and there were slightly elongated spheres sticking up above the rest. These were presumably C_{70} molecules, present as contaminants in the fullerite.

Seeing is believing. The scanning tunneling microscope (STM) image of a thin layer of fullerite on a gold surface, obtained by Robert Wilson at IBM. The image is of row upon row of tiny molecular spheres (C_{60}) with a scatter of slightly larger, elongated molecules (presumed to be C_{70}). From this picture there can be no doubt about the spherical nature of the fullerenes, but the detailed positions of the carbon atoms in the molecules cannot be seen, despite the fact that the technique has the resolution to image individual atoms. The reason for this emerged from further studies by the IBM group.

Bethune then accompanied Meijer to another wing of the research centre, and another darkened lab. Robert Johnson indicated a monitor screen displaying a trace with a single sharp peak near the centre. He had just completed his double checks and was proud to inform his late-night visitors that they were looking at the single-line NMR spectrum of C_{60}. It took Bethune, Johnson, and Meijer just two days to draft a short paper describing this result, which they faxed to the *Journal of the American Chemical Society* on Monday, September 17.

Bethune told his colleagues of his experiences in Konstanz, and the comments that Whetten had made about the Sussex group having obtained the one-line C_{60} spectrum some weeks before. He decided to telephone Kroto to tell him of their result and to ask for a copy of the Sussex group's paper. Bethune received a copy by fax from Kroto, together with the message: 'Congratulations on the one line. It is nice that C_{60} can make a lot of people happy. There's lots of it to go around.'

By a stroke of good fortune, Diederich had some months before booked a flight from Los Angeles to Frankfurt scheduled to depart on Tuesday, September 18. He planned to attend the BASF 125th Anniversary Symposium, to be held between September 24–26 in Ludwigshafen, and had intended to spend a few days beforehand with his family in Luxembourg. When Whetten called again to inform him what he had found out during the Konstanz meeting, Diederich quickly revised his plans.

The two UCLA scientists met up at Frankfurt airport on the morning of September 19. Diederich picked up a hire car and they drove straight to the Max Planck Institute for Nuclear Physics in Heidelberg. This was all familiar territory to Diederich: he had obtained his Ph.D. at the nearby Max Planck Institute for Medical Research just five years before.

Krätschmer told them everything about the carbon evaporator, the soot, the sublimation, and the solvent extraction. They suggested a short collaboration. With a sample of the fullerite obtained from soot produced by the Heidelberg evaporator, Whetten and Diederich and their colleagues in Los Angeles would repeat and extend the analytical work and attempt to isolate C_{60} and C_{70}. Their primary aim was to try to clarify the situation with regard to C_{60} and the diffuse interstellar bands.

Krätschmer also showed them Kroto's manuscript. Apart from mentioning that the Sussex group had used a chromatographic separation step, the draft paper that Kroto had passed to Krätschmer gave no further details. It was clear, however, that the Sussex group had already measured the one-line NMR spectrum.

Whetten flew back to Los Angeles with a sample of the Heidelberg fullerite on Thursday, September 20. With the help of Diederich's graduate student, Yves Rubin, he immediately set up a carbon arc evaporator to generate more of the soot, initiated a series of analytical measurements on the sample of fullerite he had obtained from Krätschmer and started to search for a chromatographic method for separating C_{60} and C_{70} from the fullerite.

In the meantime, Diederich moved on to Ludwigshafen and the BASF Symposium. Returning to his hotel room at the end of the first day's proceedings, he found a fax from Los Angeles containing the first FAB mass spectrum of the fullerite obtained by his colleagues at UCLA. This spectrum showed not only large signals due to C_{60} and C_{70}, but also prominent signals due to larger fullerenes between C_{76} and C_{96}.

Subsequent evenings afforded Diederich little time for sleep, as he quickly built up a large fax and telephone bill. Being so far away from all the action was obviously irritating, but had the advantage that he was still within easy reach of Krätschmer. He was able to go back to Heidelberg on the evening of September 25 to clarify some aspects of the experiments.

By the time Diederich returned to Los Angeles on September 28, Rubin had already drafted a paper describing the UCLA group's results. The long list of authors included Fostiropoulos, Krätschmer, and Huffman. They submitted the paper to the *Journal of Physical Chemistry* on October 3.

Huffman called a press conference in Tucson on Friday, September 27. He announced the discovery of a new form of carbon, comparing it to the discovery of a new planet. The Krätschmer, Lamb, Fostiropoulos, and Huffman paper was finally published in the September 27 issue of *Nature*. Like the original announcement of the discovery of buckminsterfullerene in November 1985, the further discovery of a whole new form of crystalline carbon was featured on *Nature*'s front cover, illustrated with a photomicrograph of the fullerite crystals.

It was a tremendous, personal triumph for Krätschmer and Huffman, who had come through an awful lot together since they had first observed the camel hump phenomenon in Heidelberg in 1982. What they had dismissed as 'some kind of junk' had turned into one of the most exciting developments in chemistry and materials science in decades. It was no less a triumph for Fostiropoulos and Lamb. Their dogged persistence and experimental skills had brought forth a remarkable new form of carbon from some distinctly unpromising starting material. All they had had to go on was their conviction that somewhere in there they would find buckminsterfullerene.

As the summer months had gone by, there had been many nervous moments for the Heidelberg/Tucson scientists. To a certain extent they were aware that they were involved in a race (although they didn't know just how many scientists were competing), but their personal and scientific integrity would not allow them to cut corners: to make claims they felt they couldn't justify. It is gratifying—and desperately important for the future of the scientific enterprise—that they succeeded.

Ever since March 1989, there had always been a tendency to compare work on buckminsterfullerene with the 'pathological science' of cold fusion. So many things could have gone wrong as the Heidelberg/Tucson group struggled to turn Huffman's crazy idea into reality. But Huffman's long shot had paid off. Whatever this was, it sure wasn't another cold fusion.

I was in contact with Kroto shortly before he left for Germany and Yugoslavia, and received breathless details of the separation and NMR spectra of C_{60} and C_{70} that were literally 'hot off the press'. On September 19, I received more details in a fax from Kroto (who was by then in Zagreb), and described the work in a short article which was eventually published in the Science section of *New Scientist* on October 13. This article announced the birth of a new field of carbon chemistry.

The field expanded rapidly. Within a few months, a large number of research papers appeared confirming and extending the results obtained by the Heidelberg/Tucson group. There were the papers from the IBM group confirming that C_{60} and C_{70} formed in carbon dust by laser vaporization could be desorbed and detected by mass spectrometry. There was the Sussex group's *Chemical Communications* paper on the separation of C_{60} and C_{70} from the fullerite and their carbon-13 NMR spectra. The IBM group's Raman spectra of purified solid films, and their NMR spectrum of the sublimed fullerite appeared shortly afterwards. Measured in terms of publication submission dates, Kroto,

The publication of the *Nature* paper was a tremendous personal triumph for Huffman (pictured here on the left) and Krätschmer.

Hare, Abdul-Sada, and Taylor had beaten Johnson, Bethune, and Meijer to the NMR spectrum of buckminsterfullerene by a mere ten days. But also very close behind were Whetten and Diederich and their colleagues at UCLA and the large group of scientists pulled together by Smalley at Rice.

With the sample of soot they had obtained from Krätschmer, the UCLA chemists had repeated the separation step. They had played around with the conditions in their own carbon evaporator and had increased the yield of the fullerite to an incredible *14 per cent* of the soot. After trying various procedures, they had finally hit on column chromatography using alumina as the best way of separating C_{60} and C_{70} from the fullerite, and had measured the FAB mass spectra, carbon-13 NMR, ultra-violet/visible, and infra-red spectra of the purified materials. All the results confirmed those reported by the Heidelberg/Tucson group.

The group of chemists and materials scientists at Rice University had also moved very quickly. In a paper received by the *Journal of Physical Chemistry* only one day after receipt of the contribution from Whetten and Diederich and their colleagues, O'Brien, Curl, Smalley, and 14 other Rice scientists described a new C_{60} 'generator' they had constructed. This was a scaled-up version of a carbon evaporator, producing soot by the vaporization of six-millimetre

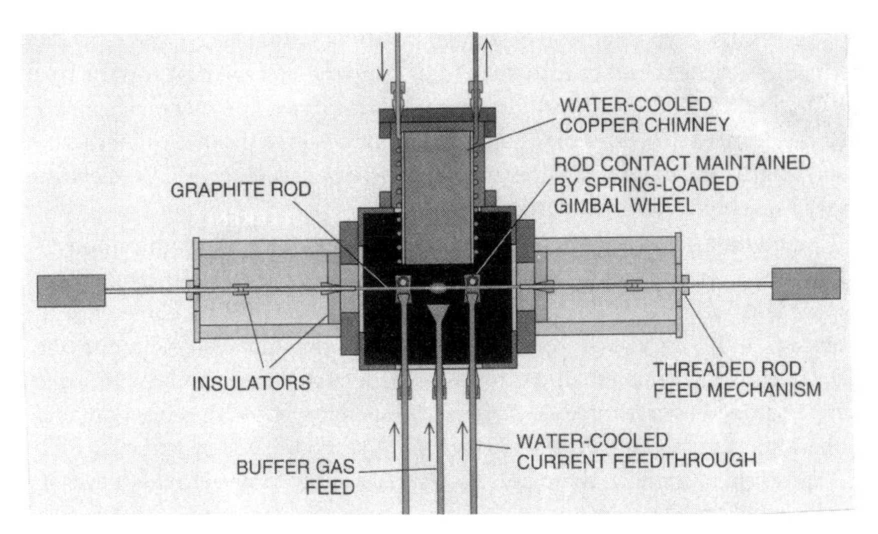

The Rice group constructed this 'fullerene factory', designed to produce C_{60} on a large scale by the evaporation of six millimetre diameter graphite rods in a contact arc. The resulting carbon soot was collected on the inside of the water-cooled copper chimney and extracted using boiling toluene. Up to ten grams of soot could be produced in a few hours' operation, yielding one gram of fullerite. Adapted from 'Fullerenes' by Robert F. Curl and Richard E. Smalley. Copyright © 1991 by Scientific American Inc. All rights reserved.

diameter graphite rods in a 'contact arc'. As in the Heidelberg evaporator, one rod was sharpened to a point (using a pencil sharpener) and spring-loaded against the other, which was machined flat. The difference was that under the evaporation conditions used by the Rice group, up to ten grams of soot could be produced in a few hours of operation. Extraction with boiling toluene gave fullerite with a yield of about ten per cent.

Smalley referred to this paper as the Rice group's first contribution to the 'feeding frenzy'. It also contained reports of the carbon-13 NMR and infra-red spectra of the fullerite in solution, together with the first measurements of the electrochemical properties of C_{60}. These latter measurements revealed that C_{60} is readily reduced in solution to produce at least two negatively charged ionic forms, promising a new class of fulleride and fulleronium salts. However, perhaps most notable at this early stage was the first reported chemical reaction of C_{60} to produce $C_{60}H_{36}$ and $C_{60}H_{18}$ using standard laboratory reagents. Thus began a new era of 'round' or three-dimensional chemistry.

The STM images of fullerite obtained by the IBM group were published in *Nature* on December 13. Equivalent images obtained by Jeffrey Wragg, J. E. Chamberlain, and H. W. White at the Department of Physics and Astronomy at the University of Missouri, using a sample of the fullerite provided by Krätschmer and Huffman, were published in the same issue.

The results from all these experimental studies established the essential correctness of the soccer ball structure for C_{60} and the squeezed sphere structure for C_{70}. But the failure of the single-crystal X-ray diffraction experiments meant that the detailed structures of C_{60} and C_{70}, expressed in terms of carbon–carbon bond lengths and bond angles, hadn't been properly nailed down. The scientists moved quickly to dot this particular *i*.

There was a way of getting at the bond lengths of C_{60} through carbon-13 NMR spectroscopy of the solid, and the group at IBM attempted these experiments shortly after Bethune returned from the ISSPIC conference in Konstanz. With assistance from Jesse Salem, another Almaden colleague, the IBM scientists assembled an arc fullerene generator, enabling them to make samples large enough for solid-state NMR spectroscopy. Nino Yannoni and Robert Johnson found that, much to their surprise, their NMR spectrum of solid C_{60} consisted of a single, sharp line, very similar in appearance to the spectrum of C_{60} in solution.

For molecules oriented randomly in a powdered solid, there is inevitably a spread of distances between the nuclei of atoms in neighbouring molecules. There is therefore a variation in the extent of shielding of the nuclei from an external magnetic field, giving rise to characteristically broad, asymmetric NMR signals. The fact that the room temperature NMR spectrum of solid C_{60} showed a sharp line indicated that the soccer ball molecules were rotating very rapidly, averaging out the effects of the spread of distances between the nuclei that would have otherwise broadened the signal.

Cooling the solid down to the temperature at which nitrogen gas freezes to a liquid (77 kelvin), Yannoni found the kind of broadness and asymmetry in the signal that might have been expected from the start. A moment's reflection produced an all-too-obvious explanation. It had been the rapid rotation of the soccer ball C_{60} molecules that had so frustrated the earlier attempts to measure a single-crystal X-ray diffraction pattern. The C_{60} molecules simply spin around too fast in the lattice to provide definitive atomic positions, and the expected spread of distances between atoms in neighbouring molecules is averaged out. Similarly, the STM images that had been obtained independently by several groups should in principle have hinted at the detailed positions of the atoms within the carbon spheres, but also suffered from the same rotational averaging.

Further studies of the temperature dependence of the NMR spectrum gave evidence for two distinct phases in the solid. The IBM group coined the term 'rotator phase' to describe the room temperature behaviour of solid C_{60} in which the molecules rotate some 20 billion times per second. Cooling the solid reduces this rate of rotation until a transition occurs to a 'ratchet phase' in which, instead of rotating smoothly, the C_{60} molecules click from one orientation to the next. Three thousand miles away, on the east coast of the US, a group at AT&T Bell Laboratories at Murray Hill in New Jersey was reaching the same conclusions.

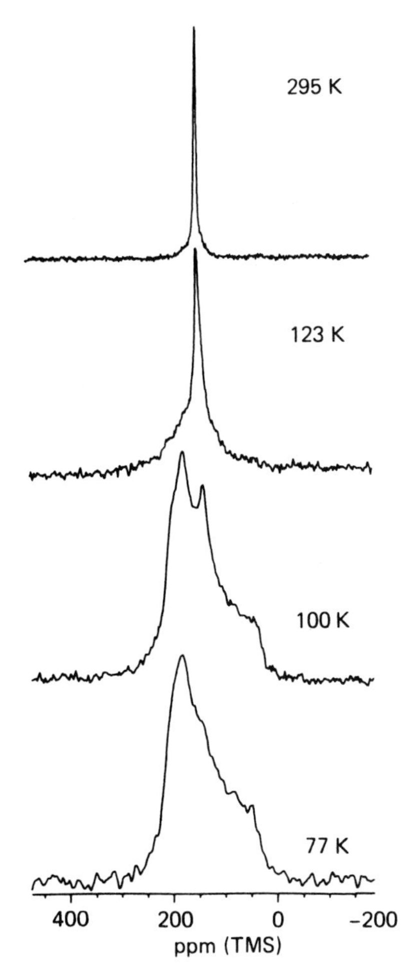

The one-line carbon-13 NMR spectrum of C_{60} in the solid state was expected to be broadened due to the inevitable spread of distances between atoms in neighbouring molecules, and hence the spread in shielding of the nuclei from an applied magnetic field. However, when IBM's Nino Yannoni measured the NMR spectrum of a purified sample of solid C_{60}, he found a single sharp line. Cooling the sample to 77 kelvin broadened the signal in the way anticipated. These, and further experiments, demonstrated that at room temperature (295 kelvin), the C_{60} molecules in the solid are *rotating* some 20 billion times per second. Small wonder that single crystal X-ray diffraction or STM imaging of fullerite failed to show defined atomic positions. Adapted, with permission, from Yannoni, C. S., Johnson, R. D., Meijer, G., Bethune, D. S., and Salem, J. R. (1991). *Journal of Physical Chemistry*, **95**, 9. Copyright (1991) American Chemical Society.

To measure the carbon–carbon bond lengths of C_{60}, the IBM group applied special NMR techniques to a sample of the solid they had enriched with carbon-13, cooled to freeze out the rotations. Signals resulting from the magnetic coupling of carbon-13 nuclei sitting adjacent to one another in a molecule depend sensitively on the distance between them. For C_{60}, it was first necessary to enrich the sample with additional carbon-13 in order to increase the probability that some molecules would form with at least two adjacent carbon-13 nuclei. The experiments yielded two distances, corresponding to carbon–carbon bond lengths of 0.140 and 0.145 nanometres. The shorter of the two was presumed to be characteristic of the bond lengths of the 30 carbon–carbon double bonds that interconnect the pentagons in the soccer ball structure. Such is the symmetry of C_{60} that these bond lengths are all that is required to specify completely its overall dimensions (diameter 0.71 nanometres) and its moment of inertia.

Bethune's colleague Bob Johnson was also very keen to apply another elaborate NMR technique to C_{70}. This technique, called the two-dimensional incredible natural abundance double quantum-transfer experiment (2D INADEQUATE, for short), enables the determination of the explicit connectivity of the carbon atoms in a molecule. Applied to C_{70}, 2D INADEQUATE NMR would allow the five signals first reported by the Sussex group to be unambiguously assigned to specific carbon atoms in the structure, and identify which of these carbon atoms were bonded together. The experiment was a 'textbook' example of the application of the technique, and the results fully supported both the squeezed sphere structure first proposed by the original bucky pioneers in 1985 and the assignments of the NMR lines given by the Sussex group.

Interestingly, Johnson's experiments had also involved samples enriched with carbon-13 (for the same reasons as before), but he found that the carbon-13 nuclei were distributed randomly in the structures of the C_{70} molecules. This suggested that the molecules were being assembled in the arc discharge from completely atomized carbon—building up the structure atom by atom—natural physical and chemical processes completing a task that had defied the best efforts of synthetic chemists.

The first definitive X-ray crystal structure of the soccer ball framework was reported early in 1991 by Joel Hawkins and his colleagues at the University of California at Berkeley. These scientists chose to overcome the problem of rotation-induced disorder in crystalline C_{60} by breaking the molecule's spherical symmetry. By making a chemical derivative of C_{60} which would effectively introduce an inflexible 'handle', they hoped that the substance would crystallize with the handles lined up, restricting any rotational motion and so producing defined crystal planes. After many false starts, they found a suitable derivative.

The Berkeley chemists reacted C_{60} with osmium tetroxide (OsO_4) in pyridine solution to produce (among other products) a 1:1 adduct $C_{60}(OsO_4)(pyridine)_2$. In this reaction, the metal oxide adds to one of C_{60}'s carbon–carbon double

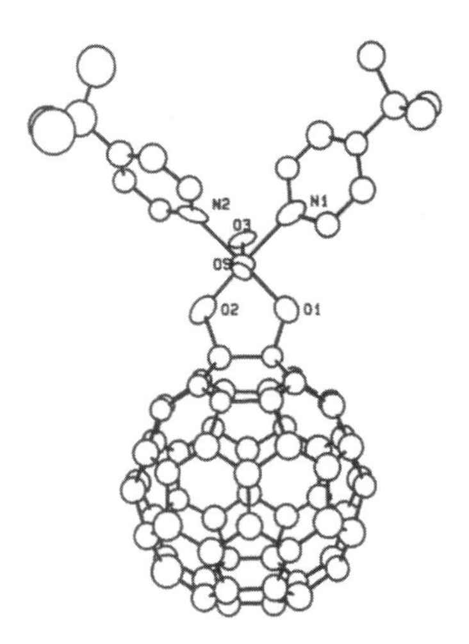

By making a chemical derivative of buckminsterfullerene, Hawkins and his colleagues at Berkeley were able to break its spherical symmetry and prevent the soccer ball framework from rotating. This allowed them to measure the X-ray diffraction pattern of the derivative, giving them detailed information on the carbon–carbon bond lengths and angles in the soccer ball. This is the resulting 'ball-and-stick' structure of $C_{60}(OsO_4)$(4-*tert*-butylpyridine)$_2$ deduced from the diffraction pattern. Adapted, with permission, from Hawkins, Joel M., Meyer Axel, Lewis, Timothy A., Loren, Stefan, and Hollander, Frederick J. (1991). *Science*, **252**, 312.

bonds, producing a new pentagonal bonding arrangement in which a single carbon-carbon bond is 'bridged' by an O-Os-O group. The two pyridine molecules attach themselves to the other side of the OsO_4 unit.

Subsequent exchange of the two pyridine molecules by two molecules of 4-*tert*-butylpyridine produced $C_{60}(OsO_4)$ (4-*tert*-butylpyridine)$_2$ which, as Hawkins and his colleagues discovered, does crystallize to give an ordered crystal structure suitable for X-ray analysis. Its X-ray diffraction pattern gave the detailed positions of all the carbon atoms in the soccer ball framework, with carbon–carbon bond lengths of 0.1386 and 0.1434 nanometres.

With the publication of this structure, the very last piece in the jigsaw puzzle of evidence was pressed firmly into place. The serious business of exploring the new field of fullerene science could now begin in earnest.

PART III

From substance to science

13

Chemistry of the spheres

To say that the fledgeling field of fullerene science exploded in 1991 is to indulge in the English art of understatement. By the end of 1990, the annual production of fullerene-related experimental and theoretical research papers had risen to 50. By the end of 1991, this figure had shot up to over 600. At the beginning of 1993, the US Patent Office received more correspondence on fullerenes than on *all other subjects combined*. It was not an explosion: it was a big bang.

The task of keeping abreast of the literature became more and more difficult. At Rice University, Smalley struggled to maintain a bibliography of research papers, review articles, and even popular articles that he called *The almost (but never quite) complete buckminsterfullerene bibliography*. He passed the stewardship of this database of fullerene publications to the Arizona Fullerene Consortium, organized by Huffman, at the end of 1992. Another bibliography was established in early 1991 by physics professor John Fischer at the University of Pennsylvania. Both are available to anyone with access to the Internet electronic mail network.

The fascination of the fullerenes has two primary sources. The obvious attraction is that buckminsterfullerene and its relatives represent a whole new class of chemical and material substance. In the months following publication of the Krätschmer, Lamb, Fostiropoulos, and Huffman paper, virtually *any* research on the fullerenes was by definition novel. But the big bang was also given a hefty dose of inflation by the relative ease with which fullerite could be produced in the laboratory. By mid-1991, it was no longer necessary even to possess a carbon evaporator. Fred Wudl and his colleagues at the University of California at Santa Barbara developed a simple 'benchtop' reactor (total cost: $700 parts and labour) which employed the same principles to produce up to 500 milligrams of fullerite a day.

Those with no wish to get involved in synthesis and who possessed a sizeable budget for consumables could simply buy as much buckminsterfullerene or fullerite as they needed. The first commercial supplies of buckminsterfullerene became available towards the end of 1990. The supplier was Materials and Electrochemical Research Corporation based in Tucson, manufacturing fullerite using the carbon evaporation process under license from the University of Arizona. The price was $1250 per gram of fullerite. In 1992, both buckminsterfullerene and fullerite were listed for the first time in the catalogue of the Aldrich Chemical Company, an international company specializing in the supply of laboratory reagents and fine chemicals. The catalogue price for 25 milligrams of C_{60} was £143.60: 25 milligrams of fullerite could be purchased for £47.20. Buckminsterfullerene was not just totally new and exciting, it was readily *accessible.*

During the early stages of the big bang, the main preoccupation of the new breed of fullerene specialists was to characterize the new substance in both molecular and material terms. The emphasis began to shift away from the use of spectroscopic analysis as a means of proving the soccer ball structure of C_{60} and effort was instead focused on probing its properties. It was time to start working out what buckminsterfullerene could *do.*

Many scientists drew analogies with the discovery of benzene. In 1825, Michael Faraday decided to take a closer look at the clear, sweet-smelling liquid that tended to collect at the bottom of the containers of gas which were delivered to the Royal Institution by his brother Robert, who worked for the London Gas Company. From this liquid, he isolated a new chemical substance and used his extraordinary talents as an experimental scientist to deduce its chemical formula: C_6H_6. The substance later came to be known as benzene. For 40 years the structure of benzene remained a complete mystery. In 1865, the German chemist August Kekulé provided the solution.* Benzene is a *cyclic* molecule; it consists of a ring of six carbon atoms with each carbon atom bonded to a single hydrogen atom. This discovery transformed chemistry.

No longer constrained to 'one-dimensional' molecules made up of chains of carbon atoms, chemists were quick to exploit the possibilities offered by two-dimensional structures containing rings. Aromatic chemistry—the chemistry of molecules containing benzene and related structures—is now a vast sub-discipline of organic chemistry. The discovery of buckminsterfullerene, 120 years later, has taken the chemistry of carbon compounds from two dimensions into the three dimensions of the sphere.

It had always been assumed that buckminsterfullerene would be an aromatic molecule; a kind of three-dimensional benzene with its pi-bond electrons delocalized over the whole 60-atom sphere. The earlier Hückel calculations had predicted that C_{60} would have aromatic character, with 60 long bonds roughly

*This discovery is not without its controversial aspects. See Sources and notes, p. 274.

corresponding to single carbon–carbon bonds and 30 short bonds roughly corresponding to stronger double bonds. More sophisticated calculations supported these predictions.

The experimental results from NMR spectroscopy and diffraction studies appeared at first to bear this out. Buckminsterfullerene was found to possess two different kinds of carbon–carbon bond in the right proportion. The NMR signals from the carbon nuclei in both C_{60} and C_{70} were found to lie in a range typical of carbon nuclei in aromatic molecules. What these results could not reveal was the extent of delocalization, or the *degree* of aromaticity. It was important to find an answer for the simple reason that the chemical reactivity of any molecule is governed by its electronic structure, and chemists eager to take their first exploratory steps in the chemistry of the fullerenes needed to know what kinds of molecules they were dealing with.

A direct route to a molecule's aromaticity is available through the measurement of its magnetic properties, particularly its response to an applied magnetic field. Under the influence of such a field, a tiny electronic current may be established by the movement of electrons through the molecule. For molecules with no unpaired electrons, this induced electronic current sets up a small magnetic field which opposes the applied field, and these molecules are classed as 'diamagnetic'. The size of the induced magnetization is referred to as the magnetic susceptibility of the molecule.

For aromatic molecules, the delocalized electrons can be imagined to be completely free to move around the ring. In an applied magnetic field, a strong 'ring current' is established which can give rise to a large magnetic susceptibility. Separating out the different contributions to the overall susceptibility is no easy task and relies heavily on theory, but it can be done. The measurement of the contribution from the ring current can therefore be used to assess the degree of aromaticity of a molecule.

For C_{60}, calculations based on the simple assumption that the molecules possess 60 pi-bond electrons delocalized over a sphere with a diameter of 0.7 nanometres predicted (not surprisingly) a very large contribution to the magnetic susceptibility from the ring current, some 40 times larger than the equivalent contribution found in benzene. However, more sophisticated theory predicted almost the exact opposite, suggesting that C_{60} is only very weakly diamagnetic and therefore not aromatic at all.

Experimental measurements of the magnetic susceptibility of small samples of solid C_{60} were reported in March 1991 by Robert Haddon and his colleagues at AT&T Bell Laboratories. They found that buckminsterfullerene gives a vanishingly small ring current contribution to the susceptibility, with an induced current *less* than that found in benzene. The conclusion at the time was that C_{60} is not aromatic. In contrast, C_{70} showed a large susceptibility consistent with a high degree of electron delocalization. This left the fullerenes somewhat out in the cold. With C_{70} aromatic but C_{60} not, it was obviously not going to be easy to give an unambiguous classification for the fullerenes as a family. It turned out, however, that the lack of any significant magnetic susceptibility in C_{60} is due to a cancellation of ring currents in the molecule which are in themselves substantial. In the year following the publication of the Bell group's assertion that C_{60} is not aromatic, the scientists devised a way to calculate the ring currents directly. Strong diamagnetic currents in the six-membered rings were found to be almost exactly cancelled by strong 'paramagnetic' currents (which add to the applied field) in the five-membered rings, leaving a small overall susceptibility. Thus, the prevailing view today is that C_{60} *is* aromatic, but with most unusual magnetic properties.

Buckminsterfullerene had also long been assumed to be a very stable, unreactive molecule, primarily because it could survive the torrid conditions inside a machine like AP2 at the expense of other, less symmetrical fullerenes and because it showed a marked reluctance to enter into chemical relationships with radical scavengers such as oxygen or nitric oxide.

But this turned out to be not quite the case. The molecule *does* show a tendency to react with oxygen over long-ish periods, and studies of its electrochemical properties revealed that it is a mild oxidizing agent (it will oxidize other molecules by removing electrons from them).

When faced with something totally new, organic chemists (like all other scientists) immediately try to find ways to compare it with things they already know. This is a matter of categorizing or pigeonholing the new so that they can tackle it by drawing on their wealth of knowledge and understanding of the category as a whole. In 1991, buckminsterfullerene's perceived lack of aromaticity, its high affinity for electrons and simple molecular models gave chemists enough clues to get them started on their first exploratory steps in fullerene chemistry.

It became very difficult for scientists to avoid superlatives when talking or writing about buckminsterfullerene, such was the grip it had on their imaginations. If C_{60} is not a 'super-aromatic' molecule, then it must be a super-something, they reasoned. But what? The localization of the double bonds in C_{60} suggested that it can be classified as a super-alkene: essentially a collection of 30 carbon–carbon double bonds characteristic of alkene molecules such as ethene, $H_2C{=}CH_2$, wrapped into a ball.

Irrespective of the subsequent revision of C_{60}'s status as an aromatic molecule, 'super-alkene' proved to be quite a useful classification. Alkenes are known to undergo addition reactions with radicals and with halogen molecules such as fluorine, chlorine and bromine. Electron deficient alkenes will react with a variety of electron-rich molecules. Alkenes can be reduced by the catalytic addition of hydrogen, and they can be oxidized. Alkenes can also be polymerized. When Smalley and his colleagues showed that C_{60} could be reacted with standard laboratory reagents to produce $C_{60}H_{18}$ and $C_{60}H_{36}$, they were demonstrating a reaction characteristic of alkenes.

Reports of the solution-phase reactions of C_{60} with a variety of radicals were published in the latter half of 1991. Paul Krusic and Edel Wasserman at Du Pont's Central Research and Development Department in Wilmington, Delaware, and Petra Keizer, John Morton, and Keith Preston at NRC's Steacie Institute for Molecular Sciences in Ottawa demonstrated that C_{60} reacts with benzyl and methyl radicals. A benzyl radical is basically a toluene molecule with a hydrogen atom pulled off and is denoted $PhCH_2^{\bullet}$, where Ph stands for phenyl (C_6H_5) and the little dot denotes an unpaired electron. Similarly, a methyl radical is methane with a hydrogen atom pulled off and is denoted $CH_3^{\bullet}$.

These are radical *addition* reactions, the electron-rich radicals adding to the double bonds of the electron-poor C_{60} molecule. The chemists found that the addition of one benzyl radical produces a new radical 'adduct', $(PhCH_2)C_{60}^{\bullet}$. Further addition produces $(PhCH_2)_2C_{60}$, $(PhCH_2)_3C_{60}^{\bullet}$, and so on all the way to at least $(PhCH_2)_{15}C_{60}^{\bullet}$. Adducts with an even number of benzyl groups have all their electrons paired and so should be reasonably stable, but the chemists discovered that the radicals $(PhCH_2)_3C_{60}^{\bullet}$ and $(PhCH_2)_5C_{60}^{\bullet}$ are also conspicuously stable. They concluded from this that the radicals tend to add at points around the circumference of what is essentially a corannulene unit within C_{60}, allowing the lone unpaired electron to be stabilized by delocalizing it over the carbon atoms that make up the central pentagon. Even so, the delocalized electron would still represent quite a target for further chemical attack, were it not for the rather bulky benzyl groups which protect it by simply getting in the way.

Despite the relative ease with which radical adducts of C_{60} could be produced, these reactions do not represent a straightforward means of producing fullerene derivatives. The problem is stopping the reaction once it's got started. When it reacts with radicals, buckminsterfullerene behaves rather like a sponge. Successive additions produce a mixture of molecules and radicals that can individually be detected, but which cannot be easily separated. This is nowhere more apparent than in the reaction of C_{60} with methyl radicals. Mass spectrometric studies showed that this reaction in benzene solution produces a broad range of products which can be assigned the general formula $(CH_3)_nC_{60}$, with n between 1–34.

Further experiments by the Du Pont group made use of other radicals to demonstrate the generality of C_{60} addition reactions. Their investigations also revealed that the radical adducts $RC_{60}^{\bullet}$ (where R is some organic group) tend to form dimers in solution. These are molecules $RC_{60}—C_{60}R$ with a weak chemical bond linking two C_{60} groups, the first example of direct bonding between two carbon spheres.

One of the other superlatives associated with buckminsterfullerene and its derivatives was related to the perceived potential of such substances as 'superlubricants'. Graphite is well known for its lubricity, and is still used today as an additive in some commercial greases. Buckminsterfullerene seemed at first sight

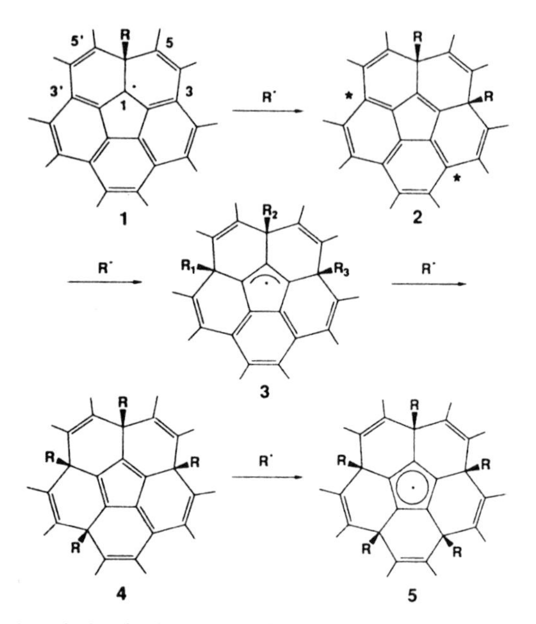

Benzyl and methyl radicals tend to add to C_{60} around the circumference of one of its corannulene units. Further addition of two more radicals completes the circuit and allows the lone unpaired electron to be delocalized around the central pentagon, producing an intermediate radical favoured through its lower energy and increased stability compared to alternative structures. The methyl-substituted radical is nevertheless quite reactive, and reacts further to give larger and larger products, up to $(CH_3)_{34}C_{60}$. Although the unpaired electron in $(PhCH_2)_5C_{60}$ represents an attractive target for further chemical reaction, the bulky benzyl groups prevent this by simply getting in the way.

to be ideally suited for use as a lubricant additive—a form of 'round' graphite consisting of 'molecular ball bearings'. Who could resist the appeal of the 'oil with balls'?

But this was before the scientists became fully aware of the nature of the chemical reactivity of C_{60}, at which point its suitability as a commercial lubricant additive was considerably diminished. Faint hopes were kept alive, however, by the prospects afforded by a fully fluorinated derivative, $C_{60}F_{60}$. Fluorinated polymers are widely used as non-stick coatings (Teflon) and sealants and are known for their inertness. Perhaps 'Teflon balls' would retain the basic lubrication properties anticipated for C_{60} without the inconvenience of chemical instability.

Initial signs did not look promising. Studies by the group at Sussex in collaboration with John Holloway and Eric Hope at the University of Leicester and John Langley at the University of Southampton showed that in the presence of solvents such as methyl alcohol or benzene, $C_{60}F_{60}$ is very susceptible to attack by water. The fluorine atoms are substituted by hydroxyl

(OH) groups, producing hydroxy fullerenes (or fullerols) and hydrogen fluoride. In itself, this reactivity is bad news, but any lubricant additive which reacts with water to release highly corrosive hydrogen fluoride is unlikely to find its way into many car engines. However, some uncertainty remains over the degree of fluorination of the C_{60} examined in these studies, and further experiments by other scientists are suggesting that all hope for Teflon balls should not yet be abandoned.

Aside from any commercial implications, the reactions of C_{60} with fluorine, chlorine, and bromine are important because the resulting halide products can be useful precursors to a variety of substituted derivatives. These reactions also involve addition to the carbon–carbon double bonds and, initially, similar problems were encountered as with the radical addition reactions. The halogens would tend to add to C_{60} in an uncontrolled manner, producing a range of products that were hard to separate and characterize.

However, Kroto, Walton, Taylor, and their Sussex colleagues Paul Birkett and Peter Hitchcock were able to tame the reaction of C_{60} with bromine, producing $C_{60}Br_6$ and $C_{60}Br_8$ which they could isolate and properly characterize. X-ray crystallography revealed the locations of the bromine atoms on the soccer ball framework. At about the same time, Wasserman and his Du Pont colleagues produced $C_{60}Br_{24}$ and determined its structure by X-ray crystallography. The Du Pont group found that the 24 bromine atoms wrap symmetrically around the outside of the C_{60} framework, two to each of 12 double bonds, shielding the 18 remaining double bonds from further attack.

The utility of halogenated fullerenes as intermediates in the production of other fullerene derivatives has been demonstrated by several groups. Examples include substitution reactions in which halogen atoms are replaced by other groups such as methoxy (OCH_3) and phenyl groups. The molecule $Ph_{12}C_{60}$ was produced from the reaction of C_{60} with bromine and benzene in the presence of an iron trichloride catalyst. George Olah and his colleagues at the University of Southern California in Los Angeles produced a range of phenyl-substituted fullerenes up to $Ph_{22}C_{60}$ by reacting polychlorinated C_{60} with benzene in the presence of aluminium trichloride. The direct reaction of C_{60} or C_{60}/C_{70} mixtures with benzene in the presence of an aluminium trichloride catalyst results in the addition of C_6H_6 across the fullerene double bonds, a phenyl group adding to one carbon atom and a hydrogen atom adding to the other. The products include $Ph_{12}H_{12}C_{60}$.

For Fred Wudl and his colleagues at the University of California at Santa Barbara, it was the distribution of pentagons that offered a vital clue to the reactivity patterns of C_{60}. There is a molecule, called pyracyclene, which consists of two pentagons and two hexagons fused together in a symmetrical manner, with a carbon–carbon double bond in the centre. The properties and reactivity of pyracyclene are well known. To Wudl, buckminsterfullerene looked like a collection of pyracyclene units.

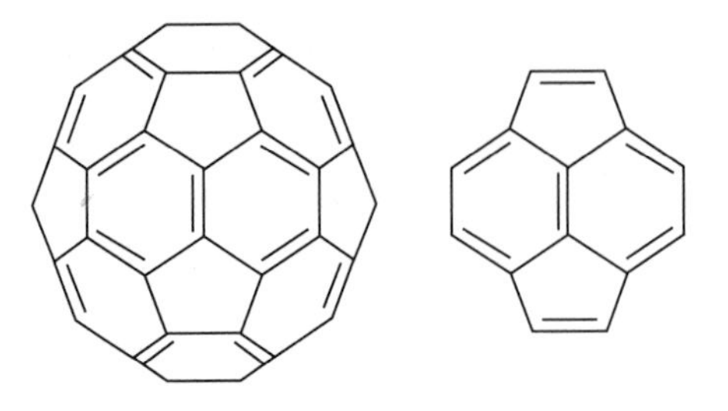

To Fred Wudl and his colleagues in Santa Barbara, buckminsterfullerene (shown left), with its 30 double bonds and 60 single bonds, looked just like a collection of pyracyclene molecules. Pyracyclene (shown right) consists of two pentagons and two hexagons fused together, with a carbon–carbon double bond in the centre. By drawing on what they knew of the chemistry of pyracyclene, the scientists could begin the task of exploring the chemistry of C_{60}.

The chemists' first attempts to react C_{60} with a variety of chemicals led mostly to 'intractable mixtures', the black, carbonized gunk familiar to all synthetic chemists trying to break new ground (and, indeed, to all students of organic chemistry lab class who, no matter how hard they tried, couldn't reconcile the black gunk with the 'pure white powder' they were supposed to be getting). Despite the intractable nature of the products from these early experiments, it was clear that C_{60} had all the properties they would expect of electrophiles (molecules that like to react with electron rich chemicals), dienophiles (molecules that like to react with conjugated carbon–carbon double bonds), and dipolarophiles (molecules that like to react with dipolar molecules in which there is a distinct separation of charge—a positive bit and a negative bit). It was whilst Wudl's colleague Toshiyasu Suzuki was examining the reactions of C_{60} with dipolar molecules called diazoalkanes that they received some interesting news from François Diederich.

The Santa Barbara chemists had sent Diederich a sample of what they believed contained 'higher fullerenes'—fullerenes larger in carbon number than C_{70}—for Diederich to analyse using laser desorption mass spectrometry. Diederich discovered that the sample did indeed contain small quantities of the higher fullerenes, but he also found some $C_{70}O$. The reason this molecule had been overlooked earlier was because $C_{70}O$ has an ultra-violet absorption spectrum almost identical to that of C_{70}. Addition of a single oxygen atom clearly didn't have much effect on the electronic states responsible for the absorption, so Diederich concluded that $C_{70}O$ must also have a very similar geometrical structure. Specifically, he assumed that $C_{70}O$ had the structure of an 'oxofulleroid': the double bond between two pentagons had gone, with the gap between them bridged by two single bonds to the oxygen atom.

There were similarities between this oxofulleroid structure and the kinds of products to be expected from the reactions between C_{60} and the diazoalkanes.

Wudl speculated that it might be possible to make fulleroid derivatives of C_{60}. These would be 'inflated' or 'expanded' versions of buckminsterfullerene which would share many of the same physical properties. Suzuki put Wudl's idea to the test by reacting C_{60} with diphenyldiazomethane. This time nature was kind to the Santa Barbara group. The result was a *single* product (not an intractable mixture) isolated in greater than 40 per cent yield. Diphenylfulleroid-60 (written Ph_2C_{61}) became the first in a series of fulleroid derivatives. Wudl, Suzuki and their colleagues Q. 'Chan' Li, Kishan Khemani, and Örn Almarsson announced their discovery in a paper published in the journal *Science* in November 1991, 'back-to-back' with the Du Pont chemists' announcement of their success with radical addition reactions.

What made this work so important at the time is that it really did open up the organic chemistry of the fullerenes. The chemistry of aromatic molecules is so broad because they can be reacted with other molecules in so many different ways. The aromatic molecules contain carbon–hydrogen bonds that are fairly easily replaced by bonds to all manner of reactive groups (called functional groups) which are the heart and soul of organic chemistry. The fullerenes have no carbon–hydrogen bonds, so their systematic inflation to fulleroid derivatives represents a key method for introducing functional groups in a controlled manner, whilst retaining many of the properties of the parent fullerene.

Wudl is not only a professor of chemistry and physics at Santa Barbara, he is also Associate Director of the University's Institute of Polymers and Organic Solids. Making new polymers out of buckminsterfullerene was therefore high on Wudl's list of objectives.

In principle, there are two ways in which fullerenes can be incorporated into polymeric molecules. The fullerenes may be part of the 'backbone' of the polymer, looking not unlike beads strung on a thread. In fact, Wudl has coined the term 'pearl necklace' polymers to describe these. In the alternative structure, the fullerenes dangle from the backbone as pendant groups, in what Wudl has called 'charm bracelet' polymers.

The Santa Barbara group has explored synthetic approaches to both types of polymer through the fulleroids. An early success was the preparation of bifulleroids: molecules with two fulleroid groups joined together by various bridging groups, representing a small section of pearl necklace polymer. Unfortunately, these molecules are not very soluble, making it rather difficult to react them further and add more pearls. However, the chemists may be able once again to draw on their long experience and chemically modify the molecules to make them more soluble.

The propensity for C_{60} to react with radicals suggested yet another approach to the preparation of fullerene-containing polymers. Douglas Loy and Roger Assink, two chemists at Sandia National Laboratories in Albuquerque, New Mexico, speculated that reactions of C_{60} with diradicals (radicals with two unpaired electrons) might yield polymeric structures. Thus, a general diradical

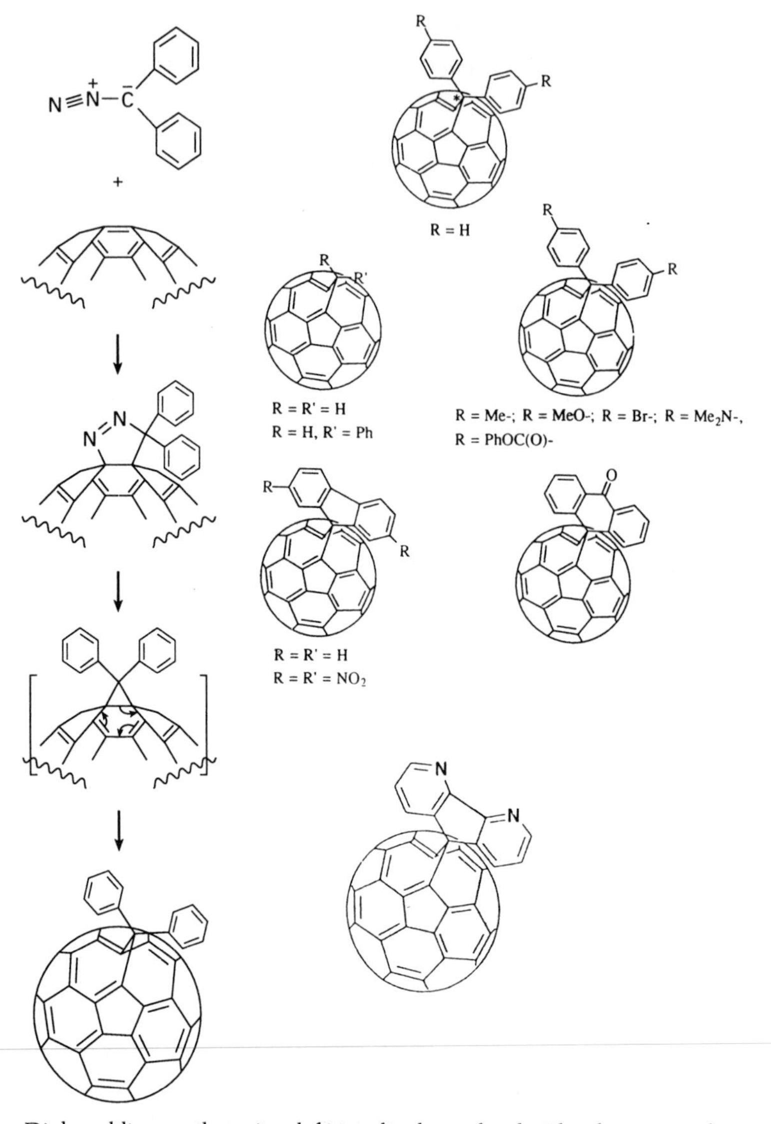

Diphenyldiazomethane (top left) is a dipolar molecule. The electrons in the CNN group are distributed so that the carbon atom carries a formal negative charge and the central nitrogen atom a formal positive charge. To an electron-deficient double bond in C_{60}, the concentration of negative charge looks very attractive. The two molecules therefore combine in a reaction which produces an unstable intermediate with a new five-membered ring. Loss of a nitrogen molecule (N_2) from this ring gives another unstable intermediate which quickly rearranges by opening up the soccer ball framework. The result is a substituted fulleroid. This reaction represents an important way of introducing substituents onto the carbon cage, illustrated on the right by the range of fulleroids that have been synthesized so far.

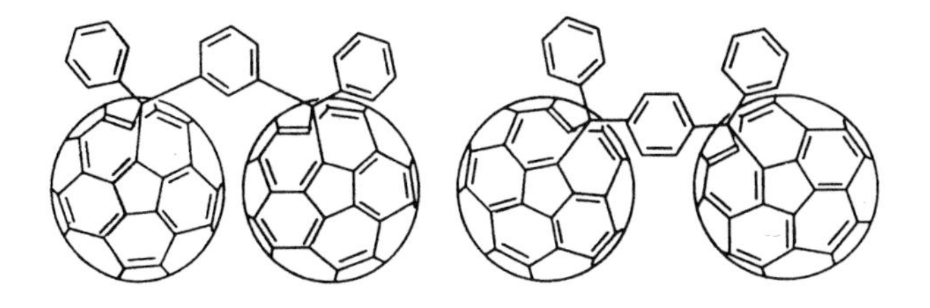

Wudl's approach to making C_{60}-containing polymers is to string fulleroid molecules together like pearls on a necklace. These two difulleroids represent small sections of pearl necklace polymer with two pearls each. Unfortunately, these molecules are not very soluble, making them poor choices for further reactions to produce large polymers. Wudl and his colleagues in Santa Barbara are hoping to modify them, adding chemical groups specifically designed to increase the molecules' solubility.

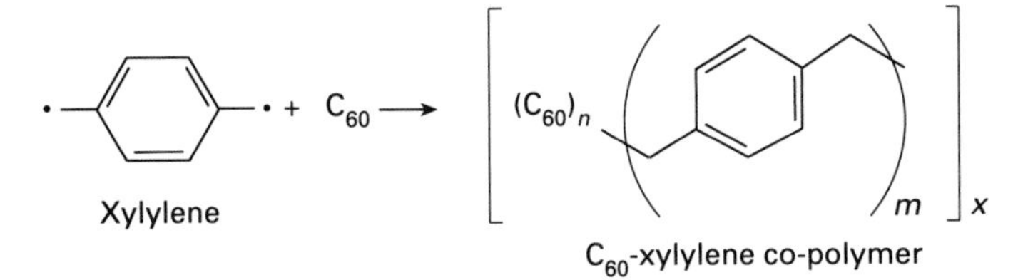

The molecule xylylene can be drawn with two different electron 'configurations'. In one of these, the molecule has the appearance of a benzene derivative, with two CH_2 groups at either end each with a single, unpaired electron (note that the hydrogen atoms are not shown in this diagram). In this configuration, xylylene acts as a diradical, and can react with C_{60} to produce a co-polymer.

denoted ·R· might react with C_{60} to produce another diradical ·R—C_{60}·, which would react with another ·R· to give ·R—C_{60}—R· and so on to produce a long chain polymer.

They chose to try this out with the molecule xylylene, whose electrons can adopt a configuration in which the molecule has an unpaired electron—a dangling bond—at either end. Slipping xylylene units in between the C_{60} molecules did indeed produce the first C_{60}-containing *copolymer*, with a ratio of just over three C_{60}s for each xylylene unit. Loy and Assink took this to indicate that the solid polymeric material that precipitated from their solution was cross-linked, meaning that the long chains were also joined together cross-wise by chemical bonds much like two sides of a ladder are joined by its rungs.

Such is the variety and applicability of polymers that they are virtually synonymous with modern living. If fullerene-containing polymers are found to have different or unusual properties, there are already potentially vast markets where they might find applications. In addition, the chemicals industry is continually on the look out for more cost-effective catalysts and, although it is certainly far too soon to get overly excited about the prospects for fullerenes, there are some encouraging signs. A polymer formed from C_{60} and palladium atoms has been shown to catalyse hydrogenation reactions.

Dividing chemistry into neat sub-disciplines—physical, organic, inorganic—may be convenient for administrative or educational purposes, but it hides the enormous overlap that exists between them. Organometallic chemistry, the study of metal compounds containing organic substituents, is one such inter-sub-disciplinary subject that requires its practitioners to range freely over aspects of inorganic and organic chemistry. When Hawkins and his colleagues synthesized an osmylated C_{60} derivative in order to measure the first definitive X-ray diffraction pattern of the soccer ball framework, they were synthesizing an organometallic compound.

The organometallic chemists drew the same lessons from their early forays into fullerene chemistry that the organic chemists were beginning to assimilate: the reactivity of C_{60} is comparable to that of a particularly electron-deficient alkene. Examples from the comprehensive literature on organometallic chemistry suggested that it should be possible to attach a variety of metal-containing substituents directly to the outside of the C_{60} framework. These new compounds would differ from the osmylated C_{60} in that the metal atom would be bonded directly to the carbon atoms of the cage.

The first to report the synthesis of such a metal-fullerene compound was another group at Du Pont's Central Research and Development Department. Paul Fagan, Joseph Calabrese, and Brian Malone produced a platinum–C_{60} derivative and confirmed its structure using X-ray crystallography. They quickly followed this with further reports of the synthesis of ruthenium, palladium, and nickel compounds. Alan Balch and his colleagues at the University of California at Davis produced iridium–C_{60} and iridium–C_{70} compounds.

In all these products, the metal atoms are attached to two carbon atoms of the cage. These are the carbon atoms which bridge the two pentagons in the basic pyracyclene unit; the atoms involved in bonding to the substituent in the fulleroid derivatives. This simply demonstrates that the same chemical principles are involved in both types of reaction. The carbon–carbon double bond linking any two pentagons together is the shorter of the bonds in C_{60}, indicating that it has the greater double bond character and reflecting the molecule's desire to avoid incorporating double bonds within the pentagons themselves. It is also an electron deficient bond, and presents the most attractive target for electron-rich metal-containing reagents. The organometallic derivatives differ from the fulleroids in that the two carbon atoms remain connected by a single bond, forming a triangular bonding arrangement with the metal atom at one apex.

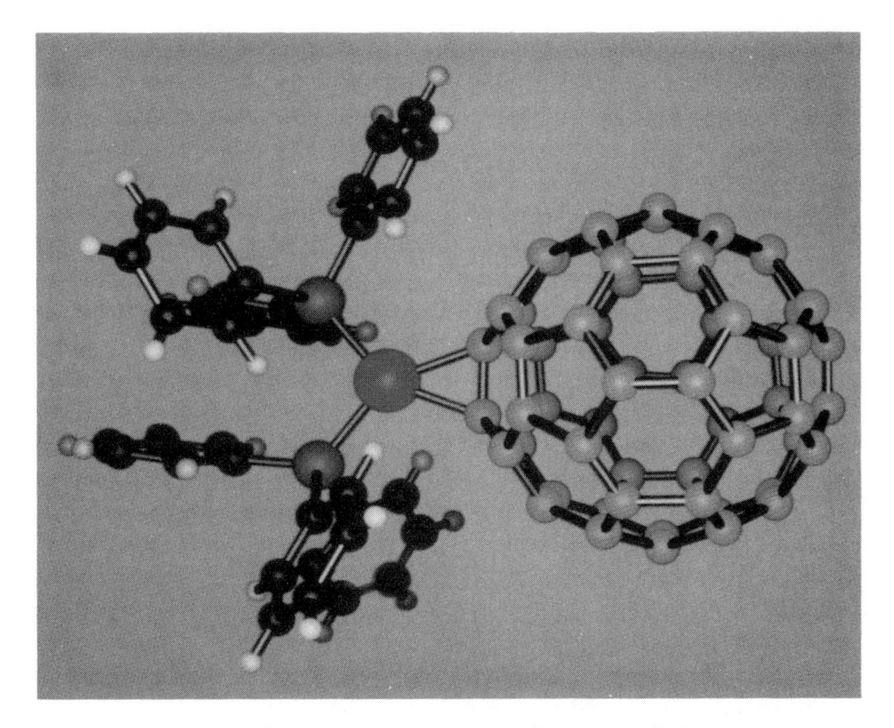

The structure of a platinum-containing derivative of C_{60}, written $(\eta^2\text{-}C_{60})Pt(Ph_3P)_2$, deduced from X-ray crystallography. In this molecule, the large metal atom is bonded directly to the soccer ball framework. The atom is attached to two carbon atoms that connect two pentagons, forming a triangular arrangement of chemical bonds. Reproduced with permission from Fagan, Paul J., Calabrese, Joseph C., and Malone, Brian. (1992). *Accounts of Chemical Research*, **25**, 134. Copyright (1992) American Chemical Society

These reactions are not limited to the addition of a single substituent. Derivatives of C_{60} with up to six metal-containing groups have been made and characterized by Fagan and his colleagues. The chemists discovered that with each successive addition, the electron affinity of the resulting product is markedly reduced compared to the starting material. The most probable cause of this is that the addition of a substituent removes a double bond from conjugation, changing the pattern of pi-bond energy levels and, in particular, pushing up the energy of the low-lying empty levels. The higher the energy of these levels, the less attractive they appear to a potential electron donor.

The end result is a clear benefit to the chemist. Adding one organometallic substituent reduces the reactivity of the product towards further addition, allowing singly-substituted derivatives to be isolated.

As with the fulleroids, the importance of these studies is that they demonstrate practical methods for producing fullerene derivatives based on a relatively straightforward understanding of the chemical properties of this new

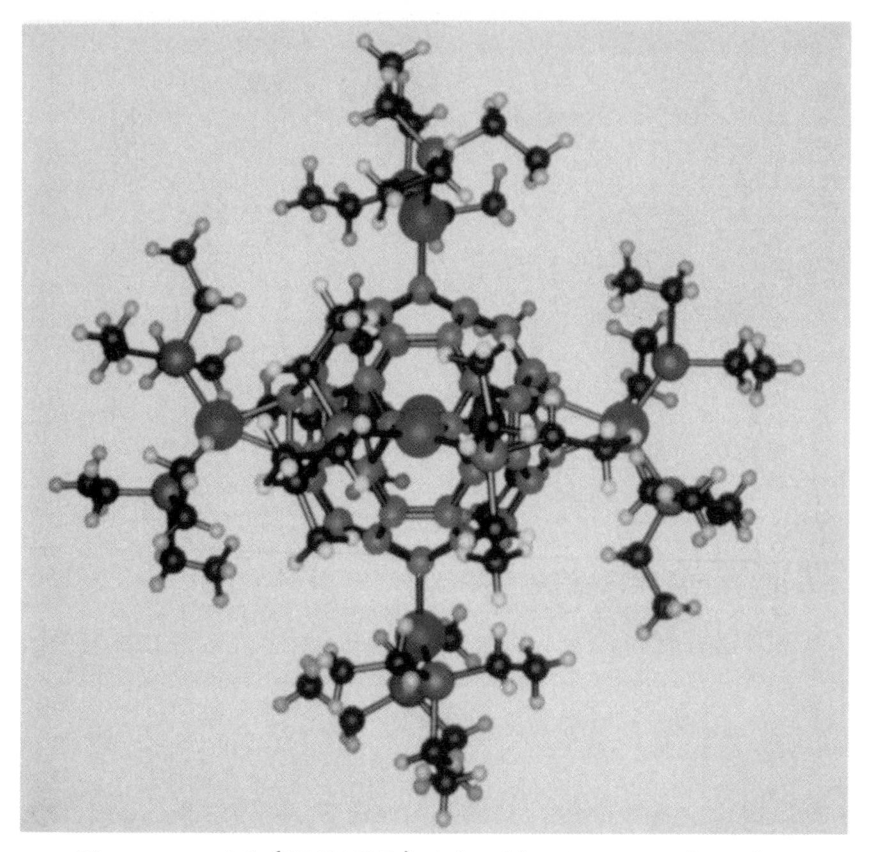

The structure of $C_{60}\{[(C_2H_5)_3P]_2Pt\}_6$ deduced from X-ray crystallography. The metal-containing substituents are arranged symmetrically around the central soccer ball framework, pointing towards the corners of an octahedron. Reproduced with permission from Fagan, Paul J., Calabrese, Joseph C., and Malone, Brian. (1992). *Accounts of Chemical Research*, **25**, 134. Copyright (1992) American Chemical Society.

class of carbon molecules. By showing how C_{60} behaves, analogy with molecules having similar reactivity patterns allows chemists to select synthetic approaches with a much enhanced probability of success.

All the above examples from the growing literature on fullerene chemistry are based primarily on conventional synthetic approaches to exploring the chemistry of what are, in essence, novel three-dimensional super-alkenes. They are examples of chemistry on the *outside*. And yet what makes the fullerenes so new and so special is the prospect of forming them with atoms trapped on the *inside*.

Within days of realizing what it was they were dealing with in September 1985, the team of scientists at Rice were looking for (and finding) evidence for

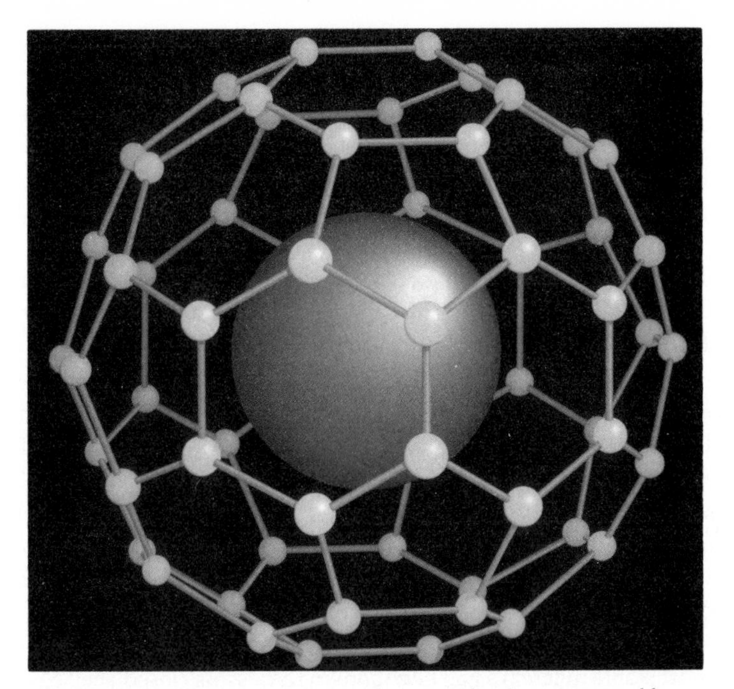

What makes the fullerenes so new and so special is the prospect of forming them with atoms on the *inside*. This computer-generated image shows an atom inside a C_{60} cage. Surprisingly, metal-containing fullerenes do not follow the same patterns of stability as the equivalent 'empty' fullerenes. A lanthanum atom encapsulated inside C_{82} turned out to be more stable than LaC_{60}.

fullerenes with lanthanum atoms incorporated on the inside. With the announcement of the discovery of a way to produce bulk quantities of fullerenes in September 1990, the obvious next step was to adapt it somewhow to the production and isolation of metal-encapsulating fullerenes.

During the following months, it was Smalley and his colleagues at Rice who reported the most significant progress. They developed a simple apparatus which they used to vaporize graphite at high temperatures using a laser, thereby generating large quantities of C_{60} and C_{70}. During the summer of 1991, they repeated these experiments with composite rods formed from powdered graphite and lanthanum oxide. Mass spectral analysis in Smalley's FT-ICR apparatus showed that the scientists had indeed formed lanthanum-containing fullerenes in large (milligram) quantities. This was confirmed using the methods they had developed to study these molecules a few years previously. Laser fragmentation of LaC_{60} was both difficult and proceeded in successive steps involving the loss of C_2, before breaking up completely at LaC_{36}. The 36-atom cage is the theoretical minimum that can be shrink-wrapped onto a lanthanum atom. By this time, fullerenes with metals stuck to the outside were known to behave very differently.

The LaC_{60} was present in the solid deposit in quite low proportion (a few per cent), together with a little LaC_{74} and LaC_{82}. Apart from the usual sooty material, the remainder of the solid was composed of 'empty' C_{60} and C_{70}. The lanthanum-containing fullerenes had survived temperatures of 1500 kelvin inside a furnace and exposure to air as the solid samples had been transferred to the FT-ICR spectrometer. There could be no doubt that the Rice group had successfully produced metal-encapsulating fullerenes in significant quantities. Now all they had to do was find a way to separate them from the rest of the soot.

They tried extracting the fullerenes in boiling toluene. Subsequent analysis showed that, apart from C_{60}, C_{70}, and a little C_{84} and C_{96}, there was only one lanthanum-containing fullerene left. Rather to their surprise, this was LaC_{82}. That the others (especially LaC_{60}) had disappeared was perhaps disappointing but also rather interesting. Clearly the reactivity of the fullerenes *had* been affected by incorporating metal atoms on the inside, with LaC_{82} now showing unexpected stability relative to the others.

The earlier Hückel calculations again provided a clue to this surprising result. All the most stable fullerenes had been shown to possess closed-shell electronic structures. It seemed very likely that the interaction between an encapsulated metal atom and a fullerene cage resulted in the transfer of one or more electrons into the pi-bond orbitals of the cage. With a stable fullerene, these electrons would have no choice but to go into empty levels. In the case of LaC_{60}, electrons donated from the lanthanum atom would have to go into the three low-lying empty levels. Electrons donated to these levels would remain unpaired, rendering LaC_{60} chemically reactive and therefore somewhat unstable.

If, on the other hand, a lanthanum atom were to become encapsulated inside a fullerene cage which possessed an open electronic shell, then the electrons could pair up with those of the fullerene, closing the shell and producing a much less reactive molecule. What was needed were fullerenes that were a few electrons short of a closed-shell configuration. Smalley and his co-workers suggested that this could explain the special stability of LaC_{82}: donation of two electrons from the La atom to C_{82} would produce $La^{2+}C_{82}^{2-}$, a molecule with a closed electronic shell.

This proposal had a pleasing symmetry to it. Stable fullerenes make unstable metal-encapsulating products. Stable metal-encapsulating products are formed from unstable empty fullerenes. But whilst there is a ring of truth to it, the explanation of the stability of LaC_{82} is not quite that simple. Further experiments by the fullerene research group at IBM showed that the lanthanum atom in LaC_{82} is present in the La^{3+} form, implying that it transfers three (not two) electrons to the carbon cage. This would produce a molecule best described as $La^{3+} + C_{82}^{3-}$, spoiling the simple picture by leaving one cage

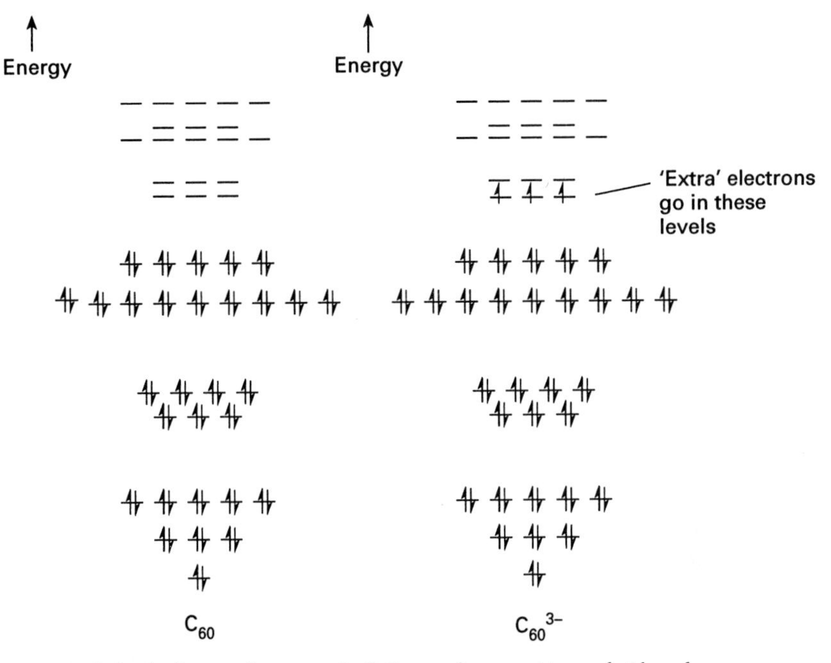

Hückel calculations for soccer ball C_{60} produce a pattern of pi-bond energy levels similar to that shown on the left. When all 60 pi-bond electrons have been assigned to these levels, filling them from the bottom up with two to each level, all the electrons are paired and the molecule has a closed electronic shell. Any electrons donated from a metal atom encapsulated inside the cage must therefore go into the lowest-energy empty levels. Because of an important principle of quantum mechanics, when faced with levels of equal energy the electrons prefer to remain unpaired, giving the pattern on the right, rendering the molecule chemically reactive and therefore relatively unstable.

electron unpaired. Smalley has speculated that this lone electron may not be transferred to the cage but is tucked away on the inside where it does little to affect the overall stability of the molecule. In contrast, Mark Ross, a scientist at the Naval Research Laboratory in Washington DC, has suggested that the special nature of LaC_{82} might be more to do with its greater solubility in the solvents used to extract it from soot than its greater stability.

Irrespective of why, the LaC_{82} had certainly survived the effects of prolonged boiling in toluene solution and exposure to Houston's mid-August air, so it seemed reasonable to suppose that its production could be scaled up. The Rice scientists managed to achieve just this using their carbon arc 'C_{60}-generator' fitted with special graphite rods whose centres had been drilled out and filled with lanthanum oxide. The IBM group used a closely related technique to produce milligram quantities of solvent-extracted LaC_{82} in about two per cent yield, mixed with empty C_{60} and C_{70}. Many other groups have subsequently used a similar approach, although everybody tends to employ slightly different conditions and procedures, often making it difficult to draw firm conclusions when the experiments don't work out.

Unfortunately, the metal-encapsulating fullerenes are not very soluble in those solvents used to extract and purify the empty fullerenes and for a long time attempts to separate metal-encapsulating fullerenes from their reaction mixtures were unsuccessful. But researchers in two Japanese laboratories have now reported the successful isolation and full characterization of LaC_{82}.

Having demonstrated that metal-encapsulating fullerenes could be produced in milligram quantities, scientists were quick to add to the list of examples. In addition to lanthanum, potassium, and caesium, fullerenes containing scandium, titanium, yttrium, zirconium, cerium, samarium, europium, gadolinium, terbium, holmium, hafnium, and uranium have also been produced, as have fullerenes with one, two, and three enclosed metal atoms. A variety of metal-encapsulating fullerenes with cages ranging from C_{28} to C_{82} have been detected in quantity. Among the fullerenes with two metal atoms inside, C_{80} appears to be the preferred cage size, and La_2C_{80}, Ce_2C_{80}, and Tb_2C_{80} have all been detected. Attempts to produce fullerenes containing aluminium, iron, nickel, copper, silver, and gold atoms have failed: nobody is quite sure why. Fullerenes containing helium and neon atoms have also been detected. Smalley and his colleagues have produced fullerenes with one or more carbon atoms in the cage replaced by boron atoms. The fullerene research group at IBM has now produced purified samples of fullerenes containing one and two scandium atoms.

The compound La_2C_{80} was first produced and reported by Whetten and his colleagues at UCLA. When Smalley read about the results of this work, he was initially sceptical. He believed that the UCLA chemists might have had problems with some uranium contamination of their lanthanum oxide powder, and that what they had really seen in their mass spectrum was some kind of uranium oxide-C_{80} complex. To check this out, the group at Rice tried to produce such complexes directly by laser vaporizing graphite rods impregnated with uranium oxide.

These experiments appeared at first to be a failure. The mass spectrum showed a series of peaks resembling that of the empty fullerenes, with no peaks that might correspond to uranium-containing products. But when the scientists checked the mass calibration of their spectrometer, they discovered (much to their surprise) that the peaks actually corresponded to a *series* of uranium-encapsulating fullerenes. Even more surprising was the observation that among these, UC_{28} appeared to be remarkably stable.

This result was remarkable because uranium is a very large atom and C_{28} is one of the smallest possible fullerenes, suggesting that the one shouldn't fit too comfortably inside the other. The explanation, like the explanations given in virtually all the cases where special stability is seen, relates to structure. The C_{28} cage has 12 pentagons and four hexagons, with the pentagons arranged in four sets of fused triplets. Each group of pentagons is positioned at the vertex of an

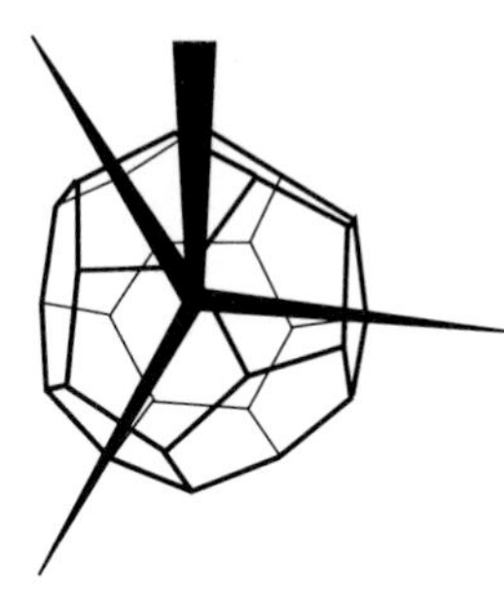

This structure of C_{28}, with its pentagons in a tetrahedral arrangement of fused triplets, looks like a 'super' carbon atom. If C_{28} molecules can be joined together in a tetrahedral array like the carbon atoms in diamond, then this will represent yet another new form of carbon.

imaginary tetrahedron. The cage has an open electronic shell, and needs four electrons to close it. Uranium likes to form compounds in which it shares four of its outermost electrons. Consequently, as Rice theoretician Gustavo Scuseria pointed out, UC_{28} might be stable because the central uranium atom is *chemically bonded* to each of the four groups of pentagons in the C_{28} cage, producing a molecule with a closed electronic shell.

The ability of the tetrahedral C_{28} to tie up its dangling bonds and close its electronic shell by bonding with uranium in this way suggests that the hydrocarbon molecule $C_{28}H_4$ should also be particularly stable, a possibility raised by Kroto in a paper published in 1987. The tetrahedral arrangement of pentagon triplets hints at a further possibility. Carbon atoms themselves adopt a tetrahedral bonding arrangement in diamond. Could C_{28} behave like a giant carbon atom, forming a similar tetrahedral lattice? It if can, then this would represent yet another new form of carbon.

At the time that the Rice scientists were demonstrating that lanthanum-encapsulating fullerenes can be produced in large quantities, Smalley became convinced of the need for a new kind of succinct notation to describe this new class of molecules and avoid confusion between fullerenes with metal atoms on the inside and on the outside. The symbolism LaC_{82} was not sufficiently explicit. But inventing a new symbolism is no easy task. Modern science had found an extraordinarily large number of ways of employing Greek and Cyrillic characters, subscripts and superscripts, bold fonts and italic fonts. There is a limited number of ways of writing a chemical formula: all had been used before and variants of these were simply unsatisfactory.

Then Ori Cheshnovsky, a chemist from Tel Aviv University who was visiting Rice that summer, made a suggestion. Why not use the '@' symbol to indicate encapsulation? After all, it was a little-used but common enough symbol, and seemed really quite evocative of something small inside a cage. Smalley was pleased. Combined with a set of brackets, the @ symbol would be all they needed to cover the range of fullerene complexes they were likely to encounter in the future. The lanthanum-encapsulating fulleride LaC_{82} would be written ($La@C_{82}$). A potassium fulleride formed from C_{60} with two potassium atoms on the outside and one on the inside would be written explicitly as $K_2(K@C_{60})$. Even buckminsterfullerene itself could be written ($@C_{60}$), to avoid any potential confusion with other non-fullerene C_{60} molecules that might one day be synthesized.

Like all new forms of notation, this one has its supporters and detractors. Most scientists active in fullerene research now use the @ symbol to denote encapsulation, although they rarely follow Smalley's suggested use of brackets. In context, K_3C_{60} and $K@C_{60}$ are reasonably clear and unambiguous, and brackets can always be used where this becomes absolutely necessary. Very few reading about C_{60} will need further reminding that this formula refers to a soccer ball arrangement of 60 carbon atoms.

Realization of the full promise of the metal-encapsulating fullerenes may be quick to follow the first successful separation of $La@C_{82}$ and the scandium-containing fullerenes. Solid films of pure metal-encapsulating fullerenes will have properties that can only be guessed at. Whereas the chemistry of the empty fullerenes is now being written, the exploration of the chemistry of metal-encapsulating fullerenes has yet to begin. It is already clear that these chemistries will be quite different, the transfer of electrons from the central metal atom to the carbon cage changing the chemical rules which determine the molecules' stability and reactivity.

The above barely does justice to the extraordinary amount of research that has been published on the chemistry of the fullerenes since September 1990. The synthesis of C_{60}–glucose derivatives in late 1992 by François Diederich and Andrea Vasella at the Swiss Federal Technical Institute in Zürich opens up a new fullerene biochemistry. For every reaction involving unexcited buckminsterfullerene, there are, potentially, parallel reactions involving C_{60} which has absorbed one or more photons of light, and significant strides in fullerene photochemistry have been reported. Given that C_{60} and the other fullerenes were first detected with the aid of mass spectrometry of their positive (and, later, negative) ions, it should not be too surprising to discover that an extensive fullerene ion chemistry is also in the making.

The new era of 'round' chemistry is well underway.

14

Superconducting fullerides

A tiny, cylindrical pellet of superconducting material floats magically in the space a few millimetres above the surface of a permanent magnet, seemingly indifferent to the force of gravity that is supposed to drag down all matter within its grasp. This is superconducting levitation. The pellet is made from a material that, when cooled below a certain critical temperature, loses all electrical resistance and becomes a superconductor. Once it is conducting a 'supercurrent', the material sternly resists the penetration of an external magnetic field. The unseen magnetic field from the magnet repels the superconductor with a force that is greater than the force of gravity. The cylinder is caught in this balance of forces and hovers above the surface, tugged down by gravity and pushed up by magnetic repulsion.

Under normal conditions, metals exhibit electrical resistance because their crystal lattices contain defects and because the ions that make up their lattices vibrate. Interactions between the charge-carrying electrons that constitute an electric current and both crystal defects and lattice vibrations can cause the electrons to be knocked off course, weakening (or resisting) the flow of current. This is the electrical equivalent of friction.

Theory says that a perfect, defect-free lattice with the minimum of vibration allowed by the laws of quantum mechanics should show effectively no electrical resistance, but such a lattice can be formed only at zero kelvin, the absolute zero of temperature. Theory also says that absolute zero can be approached (and is approached very closely these days) but never reached, since this would require an infinite amount of energy. Thus, the idea of zero electrical resistance was regarded much like the idea of perpetual motion: nice but wholly impracticable.

At least, this was the accepted view when, in 1911, Heike Kammerling Onnes measured the electrical resistance of mercury cooled to 4.2 kelvin, the temperature at which helium condenses to a liquid. Onnes was a pioneer of low-temperature physics at the University of Leiden in the Netherlands. In 1908 he had finally found a way of condensing helium to liquid form, thereby achieving

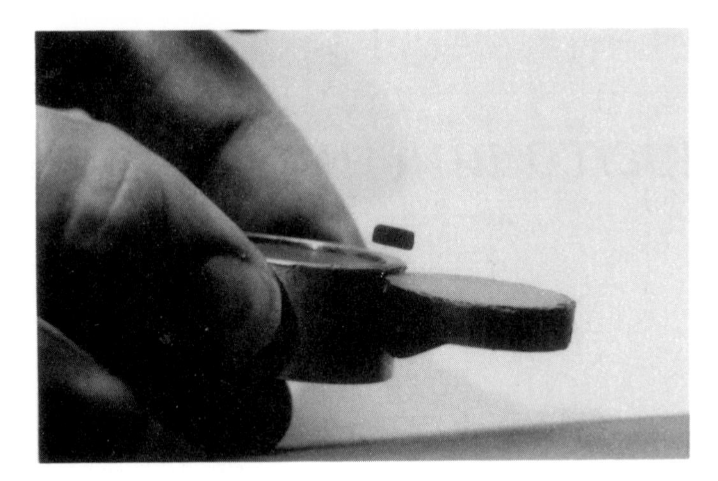

Superconducting levitation. The hovering pellet is made from a material that, when cooled below a certain critical temperature, loses all electrical resistance. Once in its superconducting state, the material sternly resists the penetration of an applied magnetic field. The field from below the permanent magnet repels the pellet with a force greater than the force of gravity, with the result that the pellet hovers magically in the space above the surface of the magnet. From A. Guinier and R. Jullien (1989). *The Solid State.* Courtesy of the University of Birmingham Superconductivity Research Group.

the lowest temperatures ever reached on earth, and it was whilst he was exploring the properties of different materials at these low temperatures that he made an unexpected discovery.

At 4.2 kelvin, the electrical resistance of mercury didn't just fall to low levels, as expected, it disappeared completely. The result was an electrical conductor with no resistance: a superconductor. Once established, an electrical current would travel around a ring of superconducting material for ever, with no loss of strength. A current of several hundred amps established in a ring of lead cooled to 7.2 kelvin was subsequently shown to persist for two and a half years.

In the absence of a good explanation for this effect, experimentalists continued to probe the properties of the new superconductors in the hope that they would turn up a few clues. In 1933, F. Walther Meissner and R. Ochsenfeld discovered a key property that was to become one of the most recognizable signatures of superconductivity. When placed in a weak-to-moderate magnetic field and cooled below its superconducting 'transition' temperature, a superconducting material expels the field from its interior, at the same time setting up its own supercurrent. The magnetic lines of force are forced to flow around the material rather than through it.

There was no simple explanation for this effect (the Meissner effect) in terms of classical theories of interactions between electric and magnetic forces. The

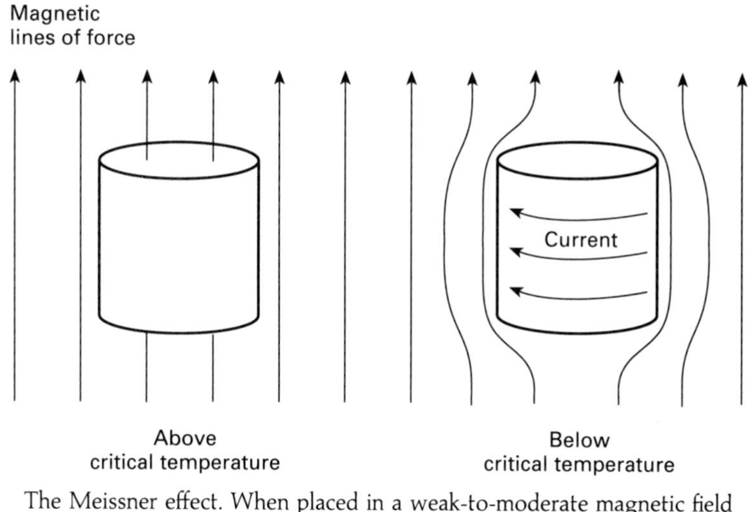

The Meissner effect. When placed in a weak-to-moderate magnetic field and cooled below its superconducting transition temperature, a superconductor expels the field from its interior, forcing the magnetic lines of force to flow around it.

explanation had to lie instead in quantum mechanics. Here was a decidedly quantum phenomenon clearly measurable on a classical, macroscopic scale.

The explanation was finally worked out in the 1950s. American physicist Leon Cooper showed that to understand superconductivity, we have to modify slightly the way we think about electrons moving through a lattice of metal ions. We learn early on that like charges repel each other, and yet Cooper said that electrons in a superconductor experience a mutual *attraction*, albeit a very weak attraction. In essence, what happens is that one electron passing near a positively charged ion exerts an attractive force which pulls the ion out of position, distorting the lattice slightly. The electron moves on, but the distorted lattice continues to vibrate. This vibration produces a region of excess positive charge in the lattice, weakly attracting a second electron. The end result is that both electrons move through the lattice in a co-operative way, as a 'Cooper pair'. On the scale of distance set by the ions that make up the lattice, the electrons in such a pair are quite far apart, and their mutual attraction (really a reduced repulsion) is easily destroyed by thermal agitation. Hence the need for very low temperatures.

Whereas the resistance towards the flow of single electrons stubbornly remains at low temperatures, for some metals the resistance towards the flow of Cooper pairs does not, and these metals can become superconductors. The reason for this difference lies in the quantum mechanics of the electrons and the lattice, and a successful quantum mechanical theory of superconductivity was eventually developed in 1956 by John Bardeen, Leon Cooper, and J. Robert Schrieffer.

The story goes that Schrieffer, who was Bardeen's Ph.D. student at the University of Illinois, was having such difficulty finding the right theoretical description that he was seriously thinking of giving up. Bardeen had to go to Stockholm to collect a Nobel Prize (which he shared with Walter Brattain and William Shockley for their invention of the transistor) and urged Schrieffer to keep trying for a little while longer. This was good advice. Schrieffer guessed the right solution to the problem and the resulting Bardeen–Cooper–Schrieffer (BCS) theory of superconductivity was found to explain all the available experimental data. In 1972, Bardeen found himself travelling to Stockholm again, to collect another Nobel Prize which he shared with Cooper and Schrieffer for their work on the BCS theory.

The key to understanding how superconductivity works lies in the dramatically different properties of electrons when they become paired through interactions with lattice vibrations. The pairing changes the basic quantum rules governing the electrons' behaviour, allowing them to 'condense' into a single energy state. Once in this state, large numbers of electrons can move coherently through the lattice in a manner more accurately imagined as a macroscopic wave, rather than particle motion. These waves of Cooper pairs are not thrown off course by crystal defects or lattice vibrations, but sweep around or over them. There is therefore nothing to resist or impede their flow.

For a long time, finding commercial applications for this rather marvellous phenomenon was hampered by the need to cool the superconducting material to the temperature of liquid helium. This was not impractical but it was (and still is) rather costly. Of course, for applications where the total investment costs greatly exceed the cost of liquid helium refrigeration, the requirement for very low temperatures becomes less of a problem. This is the case for the most widely discussed application of superconductors: high-speed, magnetically levitating trains (maglevs). The experimental system developed by the Japanese National Railways uses low-temperature superconducting magnets which levitate the train as it passes over aluminium coils built into the track. The greatest costs are associated with the track, with the refrigeration system accounting for only one per cent of the total investment.

The costs of liquid helium refrigeration drew the attentions of experimentalists to the challenge of finding metals or metal alloys that would superconduct at higher temperatures. It was believed that more economically viable applications could be found if the transition temperature could be raised at least to 77 kelvin, the temperature at which gaseous nitrogen condenses to a liquid. Refrigeration systems based around liquid nitrogen can be operated at a fraction of the cost of equivalent systems based on liquid helium.

As the scientists discovered more examples of superconductivity in metals and metal alloys, the record transition temperature was pushed higher and higher. But after 60 years of experimentation, the best they could achieve was a transition temperature of 23 kelvin, obtained in 1972 for a niobium–germanium alloy, Nb_3Ge. This was far short of the 77 kelvin target.

However, broadening the scope of application is not just about having high transition temperatures, but depends also on the response of the material to higher magnetic field strengths. For a material to exclude an external magnetic field completely would actually require an infinite density of Cooper pairs at the surface, which is obviously impossible. Instead, the field does penetrate the surface slightly, and is cancelled out by the magnetic field generated by the flowing current. But if the applied field is increased beyond a certain critical level, the material is overwhelmed. The field penetrates the body of the material and superconductivity ceases. For the conventional metal superconductors the field strengths at which this occurs are very low, restricting applications to relatively low powers. These are called Type-I superconductors.

The metal alloy superconductors were found to behave differently, retaining their superconducting properties at much higher magnetic field strengths (and thus relatively higher powers). These Type-II superconductors do not yield to the influence of the magnetic field so readily. Above certain critical field strengths, the lines of force will penetrate the material but only through narrow, cylindrical regions called flux tubes. Those parts of the material not penetrated by a flux tube will continue to superconduct. At higher and higher field strengths, more and more flux tubes are formed and eventually these combine to turn off the superconductivity altogether.

An alloy of niobium and tin, Nb_3Sn, retains its superconducting properties at quite high magnetic field strengths and is consequently used in many commercial superconducting magnets. Its transition temperature is only 18 kelvin, however, and so liquid helium refrigeration is still required. Despite this restriction, the development of metal alloy superconductors has led to a number of small-scale commercial applications. Amongst the most significant of these is magnetic resonance imaging—used for medical diagnosis—and some 4000-odd magnetic resonance imaging machines are currently in operation. Other applications in industry and computers have either been extensively discussed or progressed to various pilot projects.

The nature of superconductivity research changed dramatically in late 1986, when IBM scientists Georg Bednorz and Alex Müller discovered that an inorganic ceramic material containing lanthanum, barium, and copper oxide becomes superconducting at 30 kelvin. Within a matter of months, the record for a superconducting transition temperature was pushed up to an incredible 93 kelvin by Maw-Kuen Wu and James Ashburn at the University of Alabama and Paul Chu at the University of Houston. They produced another copper oxide ceramic with the lanthanum atoms replaced by smaller yttrium atoms.

This was very much a watershed. The yttrium–barium–copper oxide ceramic showed superconducting properties well above the temperature of liquid nitrogen, suddenly making a wide range of small-scale applications potentially realizable. In 1988, the transition temperature was pushed to 125 kelvin, set by a thallium–barium–calcium–copper oxide ceramic produced by Stuart Parkin

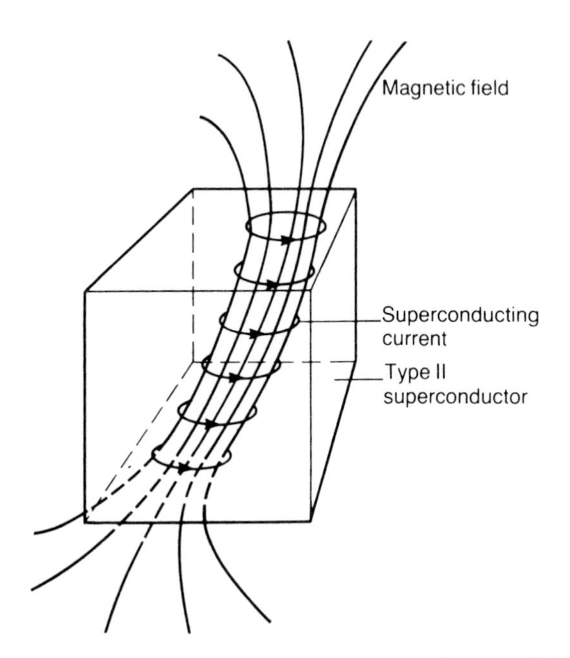

In strong magnetic fields, the superconductor may be overwhelmed and penetrated by the lines of force. In a Type-I superconductor, this signals the end of its superconducting properties. However, the metal alloy (Type-II) superconductors behave differently. The magnetic field penetrates the material but only in narrow, cylindrical regions called flux tubes. Those parts of the material not penetrated by a flux tube will continue to superconduct.

and his colleagues at IBM's Almaden Research Center. This record stood firm for six years and has only recently been overtaken by a mercury–barium–calcium–copper oxide ceramic which becomes superconducting at 133 kelvin.

The 1986 discovery of a new class of high-temperature superconductors heralded nothing short of a revolution in materials science. The astonishing properties of these ceramics (which are normally *insulators* at room temperature) seem to be at odds with the scientists' understanding of how superconductivity is supposed to work. Theoreticians have become divided into three camps. There are those who believe that the properties of the new superconductors can be explained using the conventional BCS theory supplemented by a novel mechanism for pairing the electrons which, nevertheless, still involves lattice vibrations. There are those who believe BCS theory with a new (non-vibrational) pairing mechanism is needed. And there are those who believe that only a new theoretical description will provide an adequate explanation.

The theoretical explanation of high-temperature superconductivity is only one aspect of the field that still remains unresolved and controversial. The crystal and electronic structures of the copper oxide ceramics are very

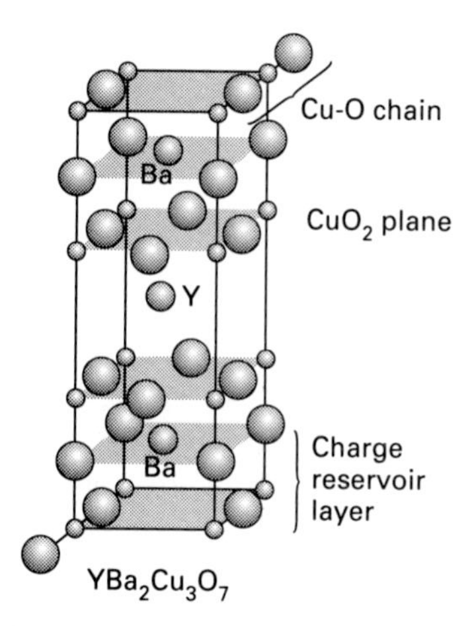

This structure of the yttrium–barium–copper oxide ceramic $YBa_2Cu_3O_7$ shows how the copper oxide units form continuous planes. These planes can act as 'highways' along which the charge-carrying electrons move.

complicated, making progress towards understanding the mechanisms of superconductivity difficult to secure. It is clear that the nature of the plane of copper oxide molecules in the lattice is important—with the planes providing 'highways' along which the charge carriers can travel. The function of the other metals or metal oxides is simply to donate electrons into orbitals located on the copper atoms or to pull electrons out of orbitals located on the oxygen atoms.

Donating electrons produces an excess, giving an *n*-type superconductor in which the carriers of the current are the negatively charged electrons. Pulling electrons off the oxygen atoms produces positively charged 'holes', giving a *p*-type superconductor in which the holes are envisaged as the carriers: electrons hopping in one direction to fill the holes producing a net flow of holes in the opposite direction. Beyond these simple facts, the superconducting properties of the copper oxide ceramics are extremely sensitive to factors that affect their detailed electronic structures. Today, not all these factors are fully understood.

The importance of high-temperature superconductors lies in the giant step they represent towards practical applications. These are Type-II superconductors, and their high transition temperatures coupled with a certain robustness to moderate external magnetic fields opens up a number of possibilities. With liquid nitrogen refrigeration, certain low-power applications in low-loss microwave and electronic devices and frictionless bearings are now very close to realization. To be sure, there are still many problems to be overcome. The copper oxide ceramics can be chemically unstable and are not very flexible. They are brittle and not easily worked into the variety of shapes (such as wires) needed to fit them into electrical devices. There is also a limit to their ability to superconduct large currents and operate in the intense magetic fields typical of large commercial electromagnets. Part of the reason for this last problem is the granular nature of the ceramic materials. Poor crystallinity gives rise to 'weak links' which reduce the efficiency of conduction. And although the ceramics bear up against the invasion of intense magnetic fields, the passage of current through the material tends to shift the flux tubes around, giving an effect known as flux creep which is manifested as a kind of electrical resistance.

Slow progress is being made, however. The Japanese Sumitomo Corporation is one of a number of commercial companies that has solved some of the problems and is now manufacturing superconducting wires and tapes from copper oxide ceramics. The company has reported making 114-metre length superconducting wires that can carry currents of up to 11 300 amps per square centimetre at 77 kelvin. Although the wires cannot maintain this current capacity in the presence of even moderate magnetic fields, they are at least approaching the performance standards of the power industry.

As always, some compromise will probably be necessary. Flux creep can be stopped by freezing the flux tubes at temperatures below about 30 kelvin. This will cost more than liquid nitrogen refrigeration, but should be nowhere near as expensive as the liquid helium refrigeration required for the low-temperature metal alloy superconductors. Tsuneo Nakahara, vice chairman and deputy chief executive officer of Sumitomo, has estimated that the value of the market for these 'compromise' high-temperature superconductors will reach nearly $4 billion per year by 2000.

The development of new high-temperature superconductors based on inorganic ceramic materials rather than metals or metal alloys demonstrates a possibility and poses a question. What about superconductors based on organic molecules? In 1964, William Little, a physicist at Stanford University in California, thought that it should be possible to make superconductors out of organic molecules. At least, he had calculated that certain organic polymers should be superconducting with transition temperatures *above room temperature*. Such polymers have never been made, and room-temperature superconductors still represent a 'Holy Grail' of materials science.

However, organic superconductors became a reality in 1981, when Klaus Bechgaard at the Oersted Institute in Copenhagen prepared $(TMTSF)_2 PF_6$, the hexafluorophosphate salt of tetramethyltetraselenafulvalene which, as Denis Jérôme at the University of Paris at Orsay discovered, becomes superconducting at 0.9 kelvin under high pressures. By 1990, the record transition temperature for an organic superconductor stood at 13 kelvin.

Then there was buckminsterfullerene. Shortly after the September 1990 announcement of a method of bulk synthesis of C_{60}, Krishnan Raghavachari at AT&T Bell Laboratories in New Jersey hurried along to the office of his colleague, Robert Haddon. In 1986, these two Bell scientists had made a small wager, which Haddon had recorded on the blackboard in his office. Haddon had bet that: 'C_{60} will be in a bottle by the end of 1990', his confidence based on the presumption that organic chemists would find clever ways to assemble C_{60}, atom by atom. He had gone on to speculate that C_{240} would be in a bottle by the end of 1992. Raghavachari was disappointed to find that the wager had long since been erased.

This was how Haddon heard the news of the breakthrough achieved by the Heidelberg/Tucson group. A little over two weeks later, he gave an informal presentation to his Bell colleagues in the Materials Discussion Group, predicting the arrival of the first three-dimensional organic 'metal', and the next organic superconductor.

Haddon reasoned that the spherical C_{60} molecules form a close-packed arrangement that is roughly uniform in three dimensions. The molecules pack together as tiny spheres, with any one sphere in close contact with other spheres along all three dimensions in the lattice (except at the surface). This network of contacted spheres suggested an array of intersecting highways along which electrons could conduct electricity through the crystal lattice. Thin films of solid C_{60} and C_{70} presented possibilities as three-dimensional organic conductors.

But a thin film of C_{60} by itself wouldn't do. Unlike a solid metal, in which the electrons associated with the metal atoms are readily released to become mobile charge carriers, all the electrons in C_{60} are tied up in chemical bonds. What was needed to make a conductor out of solid C_{60} was a way to introduce the charge carriers. Some form of *doping* was implied.

The striking spherical symmetry of soccer ball C_{60} affects its ability to donate and accept electrons. In terms of energy, hydrocarbon compounds made up of one or more of the basic benzene-type hexagons are in general equally happy to donate electrons as well as accept them. Graphite, with its flat planes of hexagons, has similar properties as an electron donor or acceptor. But as the planes are distorted towards a spherical symmetry by introducing pentagonal defects, the energy states are likewise affected with the result that buckminsterfullerene much prefers to accept electrons rather than donate them.

The answer, then, was to introduce charge carriers by doping the thin film with atoms that would gladly feed C_{60}'s appetite for electrons. Suitable donors would give up electrons to the C_{60} molecules, creating an excess of electrons that would then become free to act as charge carriers, moving through the lattice by hopping from one soccer ball C_{60} to the next. It seemed that it should be relatively straightforward to make *n*-type organic conductors based on C_{60}. There was more. Haddon also believed that such organic conductors could become superconducting under the right conditions.

Elementary chemistry suggested suitable candidates for the dopant atoms. Alkali metal atoms readily give up an electron to form stable ionic compounds with electron acceptors. Sodium chloride—$Na^{+}Cl^{-}$—or common table salt, is an example. Simple estimates based on the size of a cage of 60 carbon atoms suggested that the smaller alkali metal atoms should be able to slip into the empty gaps (or interstices) between the soccer ball C_{60} molecules without disrupting the crystal structure. At his informal presentation on October 18, 1990, Haddon suggested lithium.

Haddon set to work that same afternoon. He tried doping C_{60} with sodium using an electrochemical method similar to the kind that had been used to produce the TMTSF salts and other organic superconductors. However, these early experiments never yielded useful materials. Haddon and his colleagues Donald Murphy and Matthew Rosseinsky then tried bulk doping techniques. In December, Haddon learned of the work of another Bell colleague, Arthur Hebard, and immediately saw in Hebard's approach a way to make thin films of doped C_{60} and measure their conductivity.

Using Hebard's approach, Haddon, Murphy, and Rosseinsky constructed a small vacuum apparatus in which they placed a glass substrate coated first with thin strips of silver and then with a thin film of solid C_{60} or C_{70} deposited over the top. The silver strips were connected to wires for the conductivity measurements. In the bottom of the apparatus they placed a small sample of solid alkali metal, which they then heated for several hours. The metal slowly vaporized and its atoms penetrated the fullerene films, taking up positions in the interstices of the crystal lattice and donating their electrons to the fullerene molecules. Rosseinsky ran the first experiments using this apparatus towards the end of January, 1991. These were a complete success. Thin films of C_{60} and C_{70} doped with the alkali metal atoms lithium, sodium, potassium, rubidium, and caesium all became conducting, with the potassium 'fullerides' showing the optimum conductivity.

The conductivity was observed to increase to a maximum and then fall off with increasing exposure of the thin films to the alkali metal vapours. Over this time, the C_{60} films changed colour from yellow to magenta. This suggested that the chemical composition of the doped films was changing with time, with more and more metal atoms penetrating the lattice and increasing the ratio of metal atoms to fullerene molecules.

The scientists could explain the changes in conductivity with the aid of the many Hückel calculations that had been reported for soccer ball C_{60}. These calculations indicated that lying not far above the highest energy filled levels would be three empty levels of equal energy. It is into these three levels that any electrons donated to C_{60} are expected to go, just as for the metal-encapsulating C_{60}, except that here the alkali metal atoms remain on the outside of the soccer ball framework. The scientists argued that there would be at least two kinds of alkali metal fulleride salts, M_3C_{60} and M_6C_{60} (where M represents Li, Na, K, Rb, or Cs). In the M_3C_{60} salt, each metal atom donates one electron into each of the three low-lying empty levels in C_{60}, half-filling them and producing a C_{60}^{3-} ion. In the solid, the low-lying levels overlap and combine to form a conduction 'band'. The electrons entering this band would be free to act as charge carriers under the influence of an applied electric field. Thin films of M_3C_{60} would be conducting.

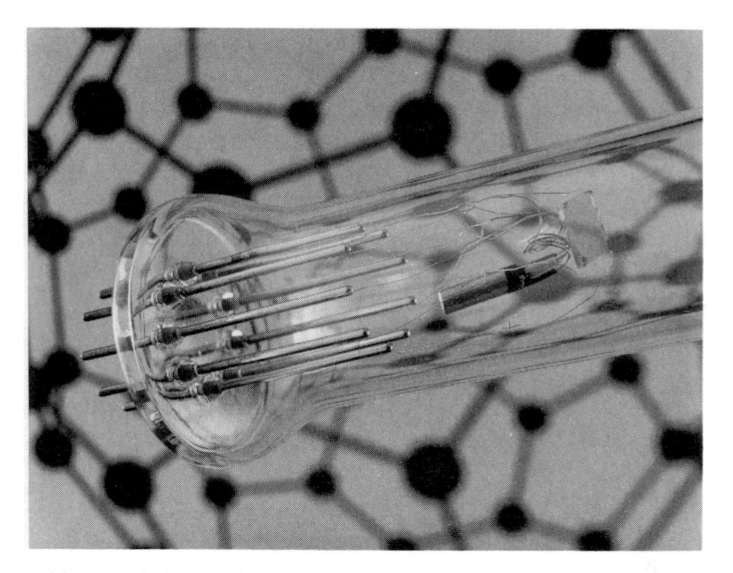

Haddon and his colleagues used this apparatus to measure the conductivities of the alkali metal doped fullerenes. A thin film of C_{60} or C_{70} was first deposited on the small rectangular glass substrate seen inside the tube. This film was then doped by exposing it to alkali metal vapour. Wires connect the four corners of the substrate to the outside world through metal pins, and enable the conductivity of the doped film to be measured. The cylindrical object positioned just above the substrate is a sensitive thermometer. Reproduced with permission from Haddon, R. C. (1992). *Accounts of Chemical Research*, **25**, 127. Copyright (1992) American Chemical Society.

Increasing the proportion of metal atoms absorbed into the crystal lattice eventually leads to M_6C_{60}, in which the six donated electrons must go into the three empty levels. With six electrons in three levels, all the electrons are paired and therefore unavailable to act as charge carriers. Thin films of M_6C_{60} would be insulating. The maximum in conductivity observed by the Bell team therefore corresponded to the production of doped films with the compositions of the M_3C_{60} salts. Continued doping produced more and more M_6C_{60}, reducing the conductivity. The scientists also discovered an intermediate M_4C_{60} phase, which is insulating.

Of all the alkali metal fullerides investigated by the team at Bell Laboratories, the potassium fulleride salt K_3C_{60} was found to exhibit the highest conductivity. The scientists announced their results in a paper published in *Nature* towards the end of March, 1991. However, by this time they had already made a rather more startling discovery.

Haddon and his colleague Si Glarum, a pioneer of a special microwave technique for measuring the onset of superconductivity, had tried to find evidence for superconductivity in samples of fullerides prepared using

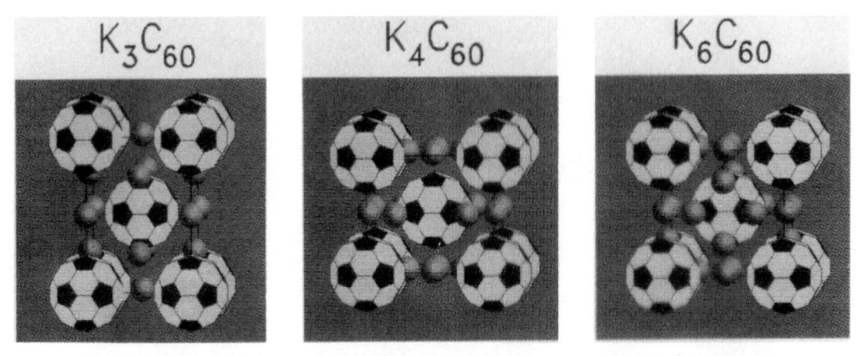

These structures show the different spatial arrangements of potassium atoms and soccer ball C_{60} molecules in K_3C_{60}, K_4C_{60}, and K_6C_{60}. With three unpaired electrons to act as mobile charge carriers, the K_3C_{60} phase is conducting. In K_6C_{60}, the six electrons donated to C_{60} are paired and therefore unavailable to act as charge carriers. This phase is insulating. The K_4C_{60} phase is also insulating. Reproduced with permission from Haddon, R. C. (1992). *Accounts of Chemical Research,* **25**, 127. Copyright (1992) American Chemical Society.

electrochemical methods, and had failed. In February 1991, Haddon tried the doping method using alkali metal vapour. Initial measurements seemed unpromising, and they abandoned these experiments around the end of the month. But in March, they decided to go back to their earlier samples and check them again using Glarum's microwave technique. On March 13, 1991, Glarum found strong evidence that their sample of potassium-doped C_{60} became superconducting at 18 kelvin. At the time this was the highest transition temperature for an organic superconductor.

The Bell scientists quickly proved that this was the case. Rosseinsky prepared a sample of potassium-doped C_{60} of nominal composition K_3C_{60} and within a week Tom Palstra and Arthur Ramirez had confirmed its superconducting properties through magnetization measurements and observation of a Meissner effect. Further modifications to their apparatus allowed them to carry out measurements at liquid helium temperatures, at which they observed a transition to zero electrical resistance. The Bell scientists drafted a paper describing their results which they submitted to *Nature* on March 26. Initial discovery to submission of the paper had taken less than two weeks.

Haddon contacted Philip Ball at the London office of *Nature* to tell him the news. Ball was bewildered. He called Laura Garwin, head of physical sciences on the editorial staff at *Nature* and asked her what was the most outrageous property she could think of for soccer ball C_{60}. Without hesitation, she replied: 'It's a superconductor?'

The Bell scientists published their results in the April 18 issue of *Nature,* and publicly announced their findings the same week during a special evening session of the American Chemical Society national meeting in Atlanta. This discovery sent further shockwaves through the already burgeoning field of fullerene science. Within a short time, Katsumi Tanigaki and his colleagues at the NEC Corporation's Fundamental Research Laboratories in Japan had pushed the transition temperature for a fulleride superconductor to a record 33 kelvin using caesium and rubidium dopants to produce Cs_2RbC_{60}. (A subsequent announcement of superconductivity at 45 kelvin in fullerides doped with potassium–thallium and rubidium–thallium alloys was later retracted.)

As with the ceramic superconductors, the theoretical explanation for superconductivity in the fullerides is still a subject for debate. Some theorists favour conventional BCS theory combined with electron-pairing mechanisms involving lattice vibrations or even internal vibrations of the fulleride ions themselves. Others reject BCS theory as inadequate and offer their own alternatives.

Perhaps most importantly, the discovery of fulleride superconductors opens up an enormous range of possibilities for chemists seeking to push transition temperatures ever higher. The fullerides are readily produced by doping thin films of fullerenes with a variety of alkali metals. Different fullerenes and different dopants would appear to offer an almost limitless series of combinations.

In principle, the fulleride superconductors have all the same potential applications as the copper oxide ceramics (they are also Type-II superconductors). One important difference is that the fullerides are composed predominantly of carbon, a relatively light element. With sufficiently high transition temperatures, fulleride superconductors could find applications in lightweight electric motors and electromagnets. However, their reactivity with air and moisture presents a problem. Powdered samples of K_3C_{60} are pyrophoric (they burn spontaneously in contact with air) and the thin films rapidly lose their conductivity. These are problems that do not seem to be insurmountable: appropriate chemical modification of the fullerenes may produce more robust materials.

It took 60 years of research to find a conventional metal alloy superconductor with a transition temperature of 23 kelvin. Only two years after the discovery of copper oxide superconductors, the record for a transition temperature was set at 125 kelvin. Commercial applications for the ceramic superconductors have been slow in coming, but they are coming nevertheless.

It took just six months to exploit the spherical geometry and electronic structure of the fullerenes to produce molecular conductors and high-temperature superconductors. Now admittedly, nature has a way of frustrating the scientists' eternal optimism, and no doubt there will be many obstacles to the commercial exploitation of fulleride superconductors. But it is still very early days. What might come out of future research on superconducting fullerides is anybody's guess.

15

Shifting the carbon paradigm

The term 'revolutionary' is used rather loosely in modern science. In the western society of the late twentieth century, the pace of change experienced in one life-span has not only become incredibly fast; it has also become entirely familiar. Today's expectations are such that yesterday's wonderful new discovery is already yesterday's news. As the scientists clamour for our attention, we hear much talk of revolutionary discoveries that will profoundly change our way of thinking and maybe even our lives. Apparently, we live in a constant state of revolutionary turmoil.

It is easy (and frequently all too necessary) for scientists to slip into such hyperbole. But what kind of scale should we use to distinguish an 'ordinary' discovery from a 'revolutionary' one? And were the discoveries of buckminsterfullerene and of fullerite revolutionary?

Although it is always dangerous to generalize, and especially so in science, we can recognize essentially two types of scientific activity. The first, which can be called 'normal' science, fits reasonably comfortably with the old saying about scientific progress deriving from 99 per cent perspiration and one per cent inspiration. Normal science is the everyday problem-solving that scientists attempt within the boundaries of currently accepted knowledge and understanding—the current 'world view'—reached largely by consensus within the scientific community.

The science philosopher and historian Thomas Kuhn coined the term 'paradigm' to describe the key example (a key experiment perhaps) which characterizes a particular world view. Today, it is generally understood that to speak of the accepted paradigm is to speak of the entire framework of accumulated knowledge and understanding of a particular area of scientific endeavour, or some self-contained sub-discipline of it. The paradigm defines the world view, and contains the rules by which we should attempt to devise new ways of probing nature both theoretically and experimentally. It also defines its own boundaries; the limits of application beyond which the science of the paradigm has no meaning or relevance.

The second type of scientific activity is 'revolutionary' science. In revolutionary science, one paradigm is replaced by another. The new paradigm may dramatically change the way we attempt to understand nature and gain new knowledge. Not surprisingly, the science of a 'paradigm-shift' is truly exciting.

If it is true that normal science *defines* the very nature of the experiments carried out within the boundaries of the accepted paradigm, then it is legitimate to ask how these boundaries can ever be transcended. To a certain extent, technological innovation provides one route. The development of a new instrument capable of revealing new facts about nature will necessarily take us beyond the boundaries of the accepted paradigm and into uncharted waters. But it is often the case that scientists will be eager to use the new technology not to prove that the current paradigm is inadequate, but to try to prove the very opposite. Scientists rarely set out specifically to falsify a theory or an interpretation. This would be an entirely negative activity. Rather, they set out to demonstrate that the theory or interpretation (or its rival if it has one) is essentially correct. They set out to *confirm* predictions and if possible add to the weight of evidence in favour of an already accepted or partly accepted world view. They set out to prove what they think they already know. This is a much more positive activity, and people on the whole like to be positive about their work.

But there is another way that revolution can come from normal science. This is where blind chance or pure accident emerges as one of the most important means of securing progress. Few sane scientists are going consciously to step beyond the boundaries of what the community regards as 'sound' science, or science within the confines of the accepted paradigm. Few will seek to challenge a world view which has been accepted by them and their peers as the 'truth'. But, occasionally, in seeking to show one thing they may accidentally stumble over something quite unexpected. This totally unexpected observation may pull them up sharp. Scientists blessed with open minds may see in this accidental discovery facts which do not fit the accepted paradigm. Thus is a discovery made which, according to the current world view, should quite simply be impossible.

In September 1985, there were two, and only two, known crystalline forms of carbon—diamond and graphite. This, if you like, was the carbon paradigm. When Kroto set off on the road to Texas, he was looking for interstellar molecules made of long chains of carbon atoms. What he, Smalley, Curl, Heath, and O'Brien found was buckminsterfullerene. The scientists were sufficiently open-minded to recognize the significance of their serendipitous discovery, and each was prepared to put his scientific credibility on the line by proclaiming that the old carbon paradigm was no longer valid. They had transcended the boundaries of the accepted world view which said that there could be only two allotropes of carbon.

Serendipity and scientific progress. This scientist is so engrossed with a problem defined in the context of the existing paradigm that he doesn't notice a great paradigm-shifting discovery until he literally falls over it. Chance gives the open-minded scientist an opportunity to transcend the boundaries of the existing paradigm and push back the frontiers of science in new and unexpected directions. Cartoon by Peter Schrank. Reproduced with permission from Baggott, Jim *New Scientist*, March 3, 1990.

At this stage, however, the shift in the carbon paradigm was still relatively small. Yes, there were these marvellous closed-cage structures representative of a new form of carbon but, even if they were real, they were being formed in small quantities under rather extreme conditions. The old carbon paradigm could be shoehorned to accept a few novel molecules, provided they remained curiosities—quite literally 'odd balls' formed with a low probability which the chemists had captured in their time of flight mass spectrometer through a lucky break.

Then Krätschmer, Lamb, Fostiropoulos, and Huffman shifted the paradigm considerably further. Connecting the camel humps with buckminsterfullerene was no less than wild speculation. It was a very bold (and, as Krätschmer initially believed, rather foolhardy) man that stood up and speculated in this fashion. But Huffman's bold leap was in the right direction. Part of the physicists' great difficulty had involved transcending the existing carbon paradigm. They could not easily bring themselves to believe that the fullerenes could form spontaneously in their simple experiment in anything other than tiny quantities. The carbon paradigm wouldn't allow it.

With the arrival of a new crystalline form of carbon in September 1990, it would have been tempting to conclude that the paradigm shift was complete. With production yields in a carbon evaporator of up to 14 per cent it was obvious that fullerite represented an important new stable allotrope of carbon. No matter how they were forming in the arc discharge, the drive to produce closed, spheroidal structures—thereby eliminating the dangling bonds—was clearly a powerful one. For the scientists intimately involved in these discoveries, the world of carbon chemistry and materials science had changed for good in September 1990.

But even now the paradigm shift was not complete. Nature still had some more surprises that were to challenge the old paradigm in ways that would have been unthinkable just a few years before.

In all the excitement, it was easy to forget that C_{60} and C_{70} were only the most stable members of a potentially limitless series of new molecules. When Diederich and Whetten and their colleagues at UCLA first began to produce fullerite, they noted that some five per cent of the solvent-extracted material consisted of fullerenes larger than C_{60} and C_{70}. Detailed examination of this five per cent revealed it to be a mixture of mainly C_{76} and small amounts of C_{78} (in two different forms), C_{84}, C_{90}, C_{94}, and $C_{70}O$. Mass spectrometric analysis of the intractable material that wouldn't dissolve in toluene showed that it contained fullerenes in the range C_{100} to C_{250}.

The UCLA chemists went to work on the five per cent extractable material. Repeated chromatography allowed them to isolate and purify small (milligram) samples of C_{76}, two forms of C_{78}, and C_{84}. To the magenta and deep red solutions of C_{60} and C_{70} they added bright yellow–green solutions of C_{76}, chestnut brown and golden yellow solutions of C_{78} (the colour of the solution depending on the structure of the fullerene) and olive green solutions of C_{84}. The fullerenes were nothing if not colourful.

But it was the structures of these 'higher' fullerenes that turned out to be most intriguing. The first to be studied in detail was C_{76}. This fullerene fitted none of the 'leapfrog' or 'cylinder' relationships that Fowler had developed as an aid to identifying the stable structures among the thousands of possible candidates. However, Fowler's colleague David Manolopoulos at the University of Nottingham had written a computer program that could automatically examine each of the possible structures in turn. The program first 'unwrapped' the structure into a spiral strip of pentagons and hexagons and then applied Hückel theory to calculate the relative energies of the pi-bond orbitals, allowing the most favourable to be selected.

The number of different possible geometrical structures that the fullerenes may possess increases enormously as the size of the cage increases beyond C_{70}, so it was necessary for Manolopoulos to restrict his search by looking only at those structures containing isolated pentagons. Applied to C_{76}, his program identified two suitable candidates, of which one was uniquely stable.

This most stable structure was also quite startling. It contains a band of fused hexagons which wind through the structure in a helical fashion, like a screw thread. This structure is therefore said to be *chiral*: it has different left- and right-handed forms that are mirror images of each other. For chemists, chiral structures consisting of helical arrangements have held a fascination (bordering on awe) ever since the discovery of the double helix structure of DNA by James Watson and Francis Crick in 1953. Chiral structures are ubiquitous in the biochemistry of living organisms.

Manolopoulos wrote up his C_{76} results in a paper which he sent in manuscript form to Whetten at UCLA. His timing was excellent. The predicted chiral structure had 19 different types of carbon atoms, expected to give 19 lines of equal intensity in a carbon-13 NMR spectrum. Diederich and Whetten had just measured a carbon-13 NMR spectrum of C_{76}, which had 19 lines. The discovery of the first chiral fullerene was announced in September 1991. Joel Hawkins and Axel Meyer at the University of California in Berkeley have since separated the two mirror-image forms of C_{76}.

With hindsight, the chiral structure of C_{76} is a logical extension of the general principles governing the relative stabilities of the fullerenes that scientists had been devising for some years. However, this did not prevent a fair degree of astonishment on the part of the UCLA chemists when they realized that this structure is actually found in nature. Further work on C_{78} revealed that one of the two forms they had isolated also has a chiral structure, suggesting that chirality might be a common structural feature in the higher fullerenes.

Fowler had earlier applied his leapfrog construction and cylinder formulas to deduce three candidate structures for C_{84}, of which one was chiral. Although the UCLA chemists were able to isolate C_{84}, they were unable to separate it further into its different forms (of which they believed there were at least two). A group of Japanese scientists at the Tokyo Metropolitan University subsequently reported the isolation and purification of C_{78}, C_{82}, and C_{84}, but they too were unable to separate the different forms of C_{84} present. In fact, their results were taken to indicate that there may be some sensitivity in the formation of some of the higher fullerenes to the experimental conditions. The UCLA group had found no evidence for C_{82}.

The next big surprise came from Japan. The unusual concentric spheres that Sumio Iijima had seen in his electron micrographs in 1980 had been used seven years later to argue in favour of Smalley and Kroto's icospiral nucleation mechanism. Iijima needed no convincing that he had made these strange fullerenes-within-fullerenes in the hot vapour from his carbon arc discharge, and

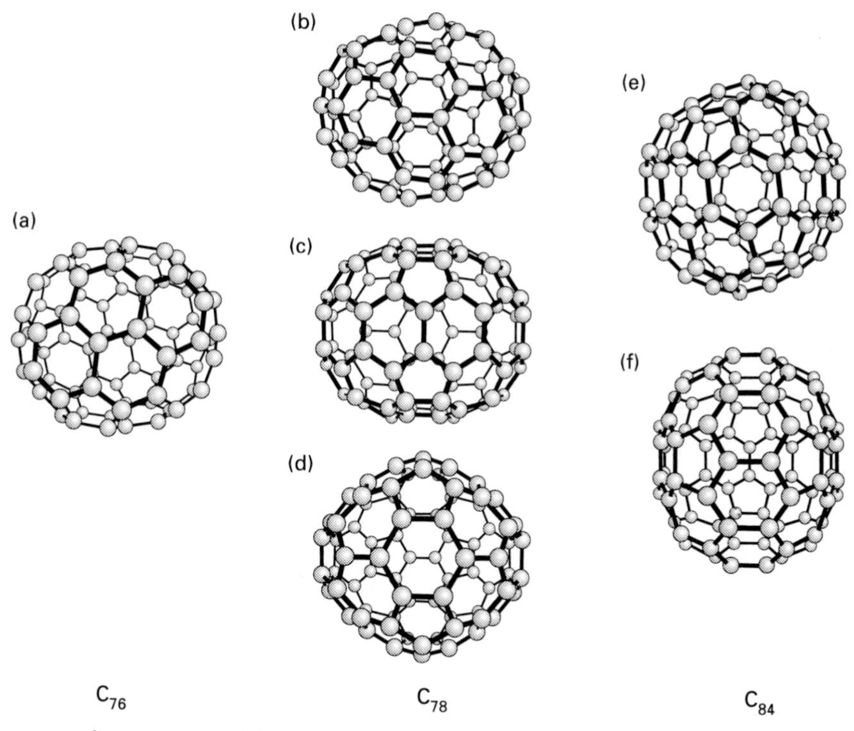

The structures of the higher fullerenes. The structure of C_{76} predicted by Manolopoulos was confirmed by Diederich, Whetten and their UCLA colleagues. It contains a band of fused hexagons which winds around the structure in a helical fashion, making C_{76} the first *chiral* fullerene. The topmost of the three forms of C_{78} also has a chiral structure. C_{84} exists in at least two different forms which have yet to be isolated. The structures given here were predicted by Fowler. Adapted, with permission, from P. W. Fowler and D. E. Manolopoulos, *An Atlas of Fullerenes*, to be published by Oxford University Press.

they had been deposited on his substrate surface along with amorphous carbon. The appearance of the Krätschmer, Lamb, Fostiropoulos, and Huffman paper in September 1990 simply added impetus to a research programme of his own that he was developing. He proposed to examine the different structures that graphite can adopt under the kinds of conditions obtained in a carbon arc evaporator, using electron microscopy.

As he varied the conditions in his own evaporator, he tried some experiments in which he kept the graphite electrodes a short distance apart, rather than bring them into contact as in what had by now become the standard method of fullerene synthesis. The gap between the electrodes was sufficiently small to allow the current to arc across them. As the positive electrode evaporated under a pressure of about 100 torr of argon, Iijima noticed that apart from the usual (presumably fullerene-containing) soot, fine needles of carbon were forming on the negative electrode.

The needles did not grow uniformly on the surface of the electrode, preferring to grow in bunches only on certain parts of it. They ranged in size from about four to 30 nanometres in diameter and up to one micrometre in length. When Iijima collected the needles and studied them using his electron microscope, he found that they were actually concentric *tubes*, with anywhere from two to 50 tubes-within-tubes. The average distance between neighbouring tubes in any given needle was 0.34 nanometres, the same as the distance between adjacent layers in the flat, chicken-wire structure of graphite.

Further electron microscopy studies showed that the tubes were composed of hexagons arranged in a helical spiral pattern. Imagine a long chain of hexagons wound into a cylinder in the manner of a screw thread. By winding at a certain pitch, it is possible for the hexagons in each new turn to fuse with those of the previous turn, forming a curved graphitic network. Tracing the chain around the tube reveals that it is wound in a helical fashion. The chiral structures of the (much smaller) cylindrical fullerenes provide obvious analogies.

In fact, with both ends of a single tube capped with hemispheres containing the requisite 12 pentagons necessary for closure, the family resemblence between the tubes and the fullerenes is apparent. Here was yet another form for a supposedly familiar element which only six years before was thought to be no longer capable of yielding any surprises.

When they appeared in Iijima's paper in *Nature* in November 1991, these tubular structures were fascinating, and clearly related in some way to the fullerenes, but they were perhaps not wholly novel. A report on the growth, structure, and properties of tubular graphite 'whiskers' had appeared some 30 years before. Roger Bacon, a physicist at the Research Laboratory of the National Carbon Company, a division of Union Carbide Corporation, had described the results of similar carbon arc experiments in a paper published in the *Journal of Applied Physics* in 1960. The most significant difference between these experiments and those of Iijima related to gas pressure: Bacon had employed a pressure of 92 atmospheres of argon.

As the positive electrode had evaporated in Bacon's experiment, a large cylindrical deposit had grown like a stalagmite on the negative electrode. The whiskers, which Bacon compared to a bundle of Christmas tinsel, were found in the core of this growth, and ranged in size from a fraction of a micrometre to over five micrometres in diameter and up to three centimetres in length. From a combination of limited-resolution electron microscopy and electron and X-ray diffraction, Bacon concluded that the whiskers had a scroll structure, with long sheets of graphite chicken wire rolled up rather like a newspaper.

Graphite cylinders formed this way would necessarily have overlapping edges. These would give steps which should have been clearly visible with the aid of Iijima's modern high-resolution electron microscope. Iijima saw no evidence for such steps and proposed an alternative spiralling growth

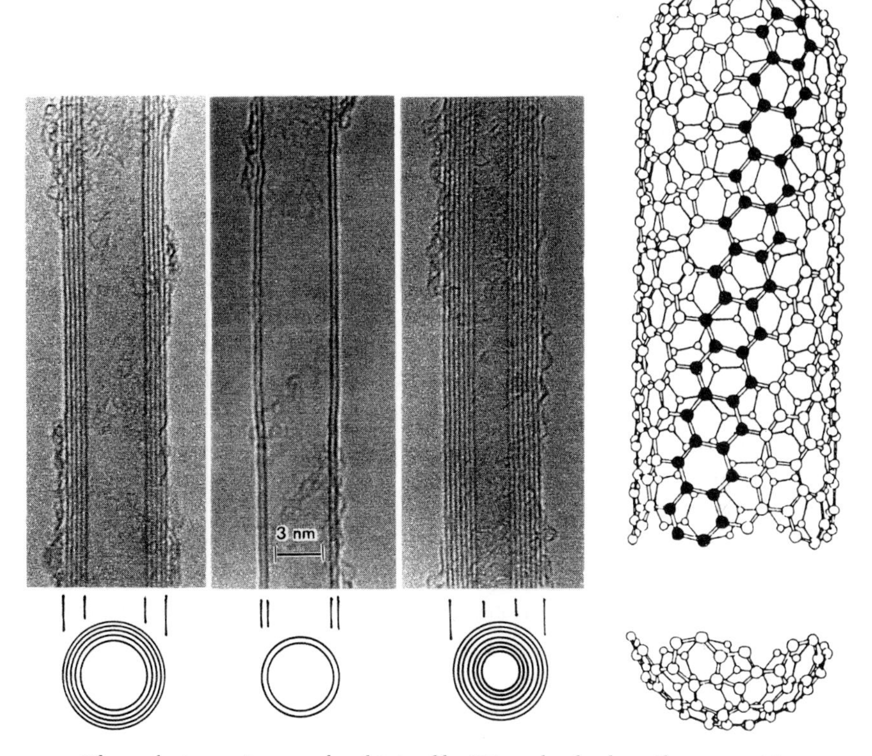

These electron micrographs obtained by Iijima clearly show the concentric nature of the carbon nanotubes. The family resemblence to the higher fullerenes is clearly seen in the schematic structure, which shows the hemispherical end caps (each containing six pentagons) and the helical arrangement of hexagons along the tube's length. Micrographs reproduced with permission from Iijima, Sumio (1991). *Nature*, **354**, 56. Copyright (1991) Macmillan Magazines Limited.

carbon atoms at the open ends. Some time later, Kroto and M. Endo suggested that the starting point could actually be a pentagonal face in the tube's hemispherical end cap.

Potential applications for Iijima's carbon tubes (which came to be known variously as *nanotubes* or *buckytubes*) were not hard to find. Bacon had noted the high degree of flexibility and tensile strength of his graphite whiskers, and these have now become standards against which the performance of carbon fibres is measured. The nanotubes, thought likely to possess properties similar to their larger cousins, therefore presented themselves as potentially useful structural elements in any nanometre-scale architecture. Indeed, subsequent theoretical calculations suggested that the nanotubes should be significantly stronger than carbon fibres (and, for that matter, any currently known material).

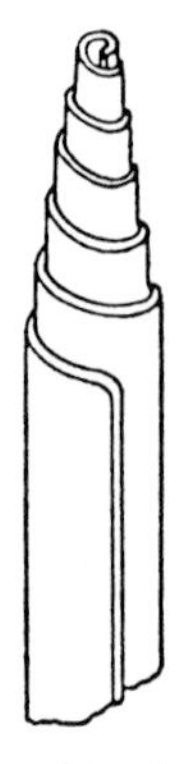

Bacon had surmised that his graphite 'whiskers' were being formed by rolling up flat graphite sheets rather like a newspaper. The resulting structure is expected to have distinct steps where the edge overlaps with the layer beneath it. Iijima saw no evidence for such steps in his electron micrographs, and concluded that the nanotubes are closed cylinders. Adapted with permission from Bacon, Roger (1960). *Journal of Applied Physics*, **31**, 283.

The tubes also had obvious implications for microelectronics (actually, nanoelectronics). If conducting or semiconducting nantotubes could be engineered to sufficient length, they could function as nanometre-scale electrical wires. Theoretical calculations of the electronic properties of the tubes predicted that their conductivity depends on their diameter and helical geometry. For tubes with diameters similar to those of the fullerenes, about a third of the different possible tube structures are predicted to have metallic properties, the remaining two-thirds having semiconducting or insulating properties. The electrical properties of ordinary materials are sensitive to the level of doping (as seen, for example, in the fullerides). A sensitive connection between conductivity in the nanotubes and their *geometry* was something totally new in solid state physics.

The other key difference between the experiments of Bacon and of Iijima, 30 years later, related to the yield of the tubes. Iijima produced the tubes on a small scale. Bacon's carbon stalagmites measured tens of centimetres in length, packed with whiskers which, on closer inspection with a higher resolution instrument, may have revealed themselves to be structurally identical to Iijima's carbon tubes. A large-scale synthesis of the nanotubes therefore seemed entirely feasible.

Such a synthesis was reported just eight months after Iijima's *Nature* paper appeared. Iijima's NEC colleagues Thomas Ebbesen and Pulickel Ajayan had also been studying different conditions in an arc discharge apparatus in an attempt to prepare fullerene derivatives. They used two graphite rods, one six millimetres in diameter and one nine millimetres in diameter. As they brought the rods close together under a pressure of 500 torr of helium, a current of typically 100 amps arced across the one millimetre gap. The smaller diameter rod evaporated. To maintain a steady current, the scientists maintained a constant distance between the two rods.

The result was a large cylindrical deposit on the larger rod, analogous to Bacon's stalagmite. About 75 per cent of the carbon originally in the smaller rod had been transferred to the deposit on the larger rod, the remainder contributing to the production of soot (and the fullerenes therein). The cylindrical deposit had a hard, grey metallic outer shell. Packed inside this shell were millions of nanotubes, representing about 25 per cent of the original carbon.

The core of the deposit appeared black and fibrous, with the fibres aligned along the direction the current had flowed from one electrode to the other. After separating this from the outer shell, and carefully grinding and smearing it into a thin film, the material took on a grey metallic lustre. Using high-resolution electron microscopy, Ebbesen and Ajayan confirmed that this material consisted of nanotubes, with the same structures that Iijima had reported. The tubes had diameters between two and 20 nanometres, with lengths up to several micrometres. The tips of the tubes were capped with pentagon-containing hemispheres.

Ebbesen and Ajayan carried out electrical measurements on the tubes in their bulk form, and found the material to have an average conductivity comparable to that of some semiconductors. Given that the broad range of tube diameters and geometries in the bulk material will give rise to a broad spread of conductivities, it is reasonable to suppose that many individual tubes are strong conductors.

Most of the theoretical predictions that had been made for the nanotubes related to individual tubes sizes. The method of bulk synthesis discovered (or re-discovered) by Ebbesen and Ajayan produced a very broad range of tube diameters with a widely varying number of tubes-within-tubes. The prospects for separating individual tubes according to size, and therefore testing some of the theoretical predictions, appeared pretty slim. But in June 1993, Iijima and a team of scientists from IBM and the California Institute of Technology announced two distinctly different ways of making large quantities of *single-walled* nanotubes about one nanometre in diameter.

Ebbesen and Ajayan did for the nanotubes what Krätschmer, Lamb, Fostiropoulos, and Huffman had done for the fullerenes. With a simple method for the bulk synthesis now available, and further methods for the preparation of single-walled nanotubes, scientists in laboratories around the world are investigating the novel properties of these new carbon structures. In the race to find the first commercial application for fullerene-related materials, the nanotubes provide strong competition for the superconducting fullerides.

Despite the novelty of the carbon nanotubes and their promise as potential elements in future nanoscale electronic devices, it is probably true to say that in

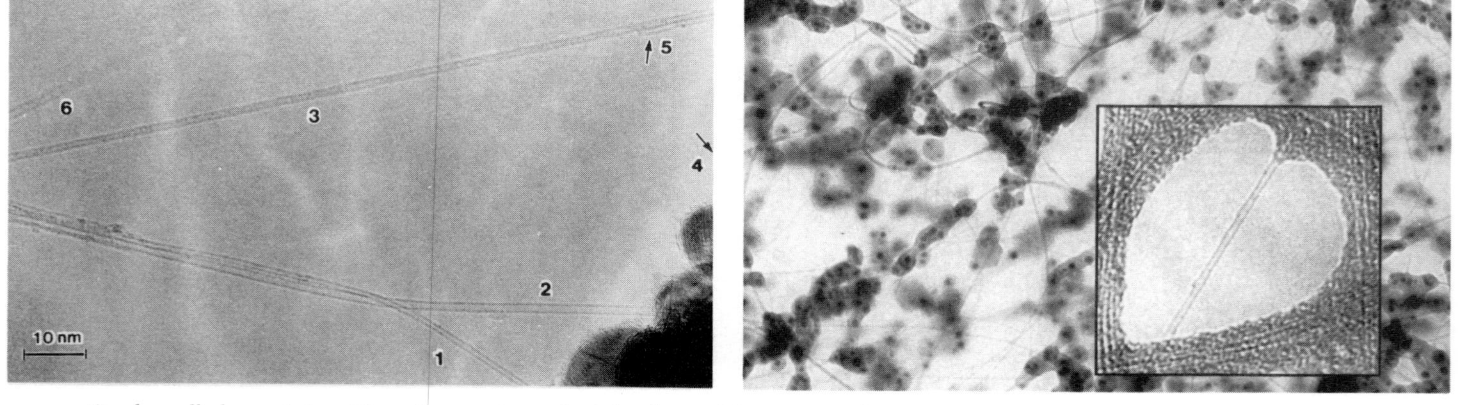

Single-walled nanotubes. The micrograph on the left, obtained by Iijima and Ichihashi, shows a number of long single-walled nanotubes. The tube labelled (1) has a diameter of 0.75 nanometres, (2) has a diameter of 1.37 nanometres. (3) is a long, straight tube and (5) is terminated by a hemispherical end cap. Reproduced with permission from Iijima, Sumio and Ichihashi, Toshinari. (1993). *Nature*, **363**, 603.

The micrograph on the right was obtained by Bethune and de Vries and their IBM colleagues, and C.H. Kiang at the California Institute of Technology. Using a different method, they were able to produce quantities of single-walled nanotubes which weave their way like threads through a jumble of carbon soot and cobalt clusters. The inset is a higher magnification image showing a single-walled nanotube with a diameter of 1.2 nanometres. Reproduced with permission from Bethune, D. S., Kiang, C. H., de Vries, M. J., Gorman, G., Savoy, R., Vazquez, J., and Beyers, R. (1993). *Nature*, **363**, 605. Copyright (1993) Macmillan Magazines Limited.

the minds of many scientists, the shift in the carbon paradigm was even now not quite complete. Basically, they still didn't get it. The growth mechanism for the tubes proposed by Iijima—perhaps starting from a pentagonal defect as proposed by Kroto and Endo—seemed entirely plausible but left some unanswered questions. Everybody knew that the flat, chicken-wire structure of graphite was the favoured form under 'normal' conditions, so how was it possible to generate such beautifully symmetrical structures—the fullerenes and the nanotubes—from the chaos of a carbon arc? Why did spheroidal or tubular forms of graphite appear to be favoured under 'unusual' conditions?

Of course, these were the same questions that the original five-a-side soccer team wrestled with way back in September 1985. The complete answers have yet to be found, but it is now understood that to find them we must first cast aside our prejudices and fully embrace the shift in the carbon paradigm. It is time to explode the myth of flat graphite.

What makes graphite flat? This seems such an odd question to be asking in the high-tech world of the late twentieth century. But it was obvious to the original researchers at Rice in 1985 that, at high temperatures in an inert gas atmosphere, fragments of graphitic chicken-wire would be inherently unstable with respect to any structure which could eliminate the dangling bonds. For such small fragments, consisting of tens or hundreds of carbon atoms, flatness was seen as a big disadvantage. Spheriodal structures would be energetically much more favourable.

This was all fine for small fragments of graphite generated in the extreme environment of a laser-produced plasma or electrical discharge, but obviously had nothing to do with the bulk form of graphite with which we are most familiar. Or did it?

Daniel Ugarte, a physicist at the Institute of Experimental Physics at the Federal Polytechnic in Lausanne, Switzerland, tried yet another approach to the study of the different forms that carbon can take when exposed to extreme conditions. He produced carbon soot in the usual way, examining the thin films using electron microscopy. However, rather than hold the dose (or number) of electrons in his microscope beam to modest values (as is usual in most microscopy studies), he purposefully increased it some 10–20 times higher than normal. His purpose was not only to obtain visual images of the films and their constituents, but also to see if their structures changed under intense electron bombardment.

In a field singled out by its collection of highly memorable images, the electron micrographs that Ugarte obtained were breathtaking. With prolonged electron beam irradiation, the thin tubes and polyhedral graphitic particles that constituted the bulk of the soot were steadily transformed into an assembly of spherical particles. Closer examination of these particles revealed them to be concentric carbon cages—one inside the other. The larger particles had diameters approaching 50 nanometres and consisted of roughly 70 fullerene shells.

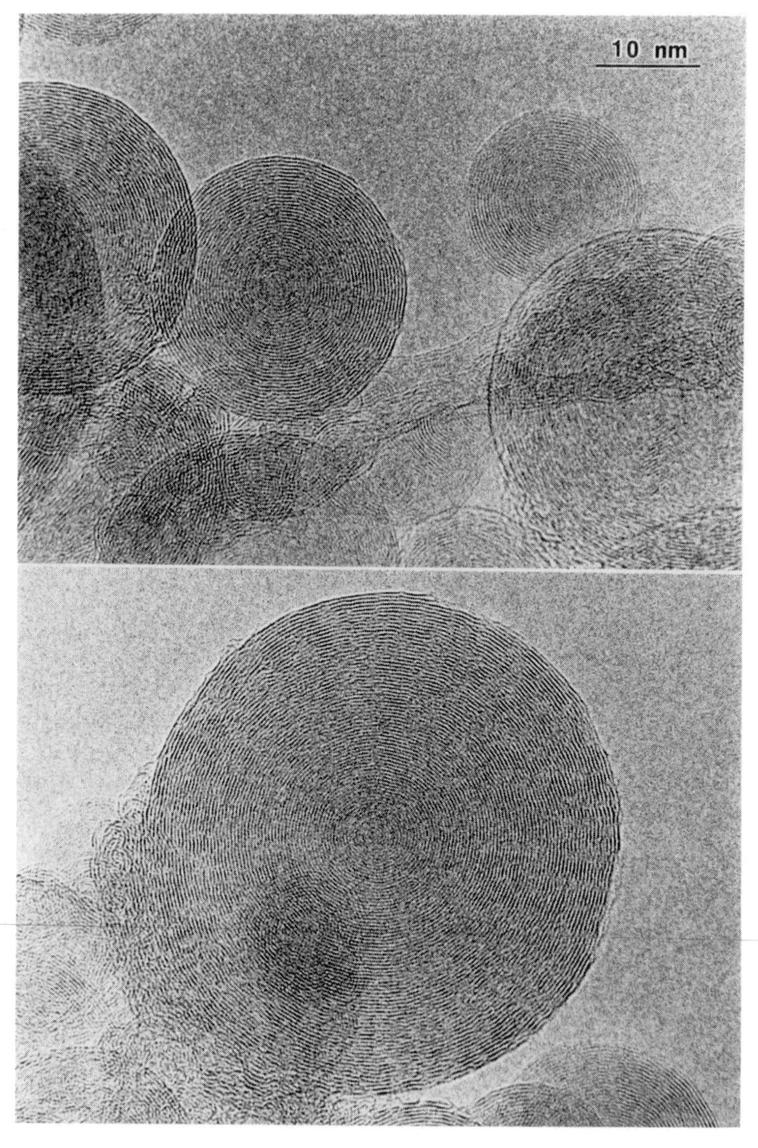

The ultimate fullerenes. These onion-like structures, formed from graphite by bombarding it with electrons, are nested fullerenes, one inside another. The largest of these 'hyperfullerenes' observed by Ugarte had dimensions approaching 50 nanometres and consisted of roughly 70 fullerene shells. Reproduced with permission from Ugarte, Daniel (1992). *Nature*, **359**, 707. Copyright (1992) Macmillan Magazines Limited.

These were Iijima's strange structures from 1980 all over again, except that Ugarte was producing them not from carbon vapour, but by electron bombardment of pre-formed carbon structures. The processes involved in these two kinds of experiments are not necessarily the same. Apart from heating the sample, bombardment with high-energy electrons can also excite electrons in the target structures, promoting rearrangements that might not be possible through thermal excitation.

Nevertheless, Ugarte's experiments had clear implications for the old carbon paradigm. Given the necessary encouragement and structural 'fluidity', flat graphite will spontaneously incorporate pentagons in its structure and curl up to form closed networks. The drive to eliminate the dangling bonds is not confined to small fragments of graphite chicken-wire; it is common also to larger-scale structures. With prolonged irradiation, Ugarte saw some evidence for particles several micrometres in diameter: spherical analogues of the longer nanotubes. Even larger, *macroscopic* particles seemed possible.

Ugarte had obtained his incredible micrographs by February 1992, and included them in a short paper in which he suggested that spheroidal shell structures might be the most stable form for carbon in a network of limited size (as opposed to the 'infinite' networks of flat graphite chicken-wire). But paradigms have a tremendous inertia, and scientists with fixed views strongly embedded in the current paradigm will simply dismiss challenging new ideas out of hand. So it was with Ugarte's challenge. Two weeks after submitting his paper to *Nature*, it was rejected by its referees. He tried again some months later, submitting much the same paper to *Physical Review Letters*. It was again rejected, with one referee comparing Ugarte's work to 'cold fusion'. Despite everything that had happened as the fullerene story had unfolded over the previous seven years, some scientists were still too deeply entrenched in the old way of looking at carbon.

Ugarte then met someone who introduced him to the members of Smalley's group at Rice. The Rice scientists were (not surprisingly after everything that had happened there) substantially more open-minded and receptive. They recognized the importance of Ugarte's work and encouraged him to send his paper as a preprint to everybody active in fullerene science, including Kroto, Krätschmer, and Whetten. Shortly afterwards, Ugarte met Kroto in Trieste. Kroto explained that he too was very enthusiastic about Ugarte's micrographs, and was surprised to hear of the difficulty that Ugrate had experienced in getting them published. He persuaded Ugarte not to give up, and agreed to endorse the paper on his behalf to *Nature*'s editors. The paper was eventually published in *Nature* in October 1992, Ugarte's stunning micrographs capturing the front cover.

Just as Iijima was able to pick out nanotubes consisting of only a few tubes-within-tubes, so Ugarte was able to identify smaller particles with just a few fullerenes-within-fullerenes. These particles are now often referred to as 'bucky onions', 'Russian dolls', or 'hyperfullerenes'. For example, in amongst the soot could be seen hyperfullerenes consisting of C_{60} inside C_{240} inside C_{540} inside C_{960}, confirmed by comparing the experimental images with computer simulations—calculated images based on idealized arrangements of the shell structures. Slight differences between the experimental and calculated images can be explained in terms of this idealization: in reality, the shells are not expected to persist with their faces aligned (as assumed in the calculations) and, as found in the X-ray crystallography of fullerite, rapid rotation of the spheres is likely.

Ugarte has since carried out further, systematic studies of all the different types of carbon structures that can be produced in an arc discharge. The relationships between the spherical structures and soot particles are still unclear. Ugarte discovered no evidence for the nautilus-like structures that might be expected to be produced by the icospiral nucleation mechanism, but then even the milder doses of more conventional electron microscopy may change the structures as they are being imaged. It is not impossible that, if they were present, the nautilus-like structures evolve under the gaze of the electron microscope into hyperfullerenes.

For every question answered by a piece of fundamental research, one or more new questions are inevitably raised. Research on the nanotubes and hyperfullerenes is no exception. But whilst Ugarte's work leaves many questions for future research one thing, at least, is clear. The dramatic images of the hyperfullerenes demonstrate that at high temperatures graphite prefers not to be flat. In a short article written to accompany Ugarte's *Nature* paper in October 1992, Kroto pondered on the myth of flat graphite, writing that:

> . . . a close look at graphite material shows that the effectively infinite flat sheets of graphite, which are presumed to form spontaneously, are actually rather elusive—to the point of non-existence.

He went on to compare the shift in the carbon paradigm with the shift in the 'shape-of-the-earth' paradigm that resulted from Christopher Columbus's voyage to the West Indies 500 years earlier. Could it be, he asked, that flat graphite had finally gone the same way as the flat earth?

The structural relationship between the fullerenes and the hyperfullerenes led Ugarte to speculate that fullerite might be only the first of a potentially vast new series of different solid forms of carbon. Isolating the multi-shelled spherical particles is not going to be easy. But if it can ever be done, the result will be various types of 'hyperfullerite', consisting of large onion-like graphitic particles organized in some close-packed arrangement. The physical properties of such materials can only be guessed, and they will certainly have some very interesting chemical properties.

But some things can be tried without waiting for a method of separation. The group at Rice University moved very quickly following their September 1985 discovery to see if they could incorporate metal atoms inside C_{60}, despite the fact that they could not generate and isolate bulk quantities of the material. In

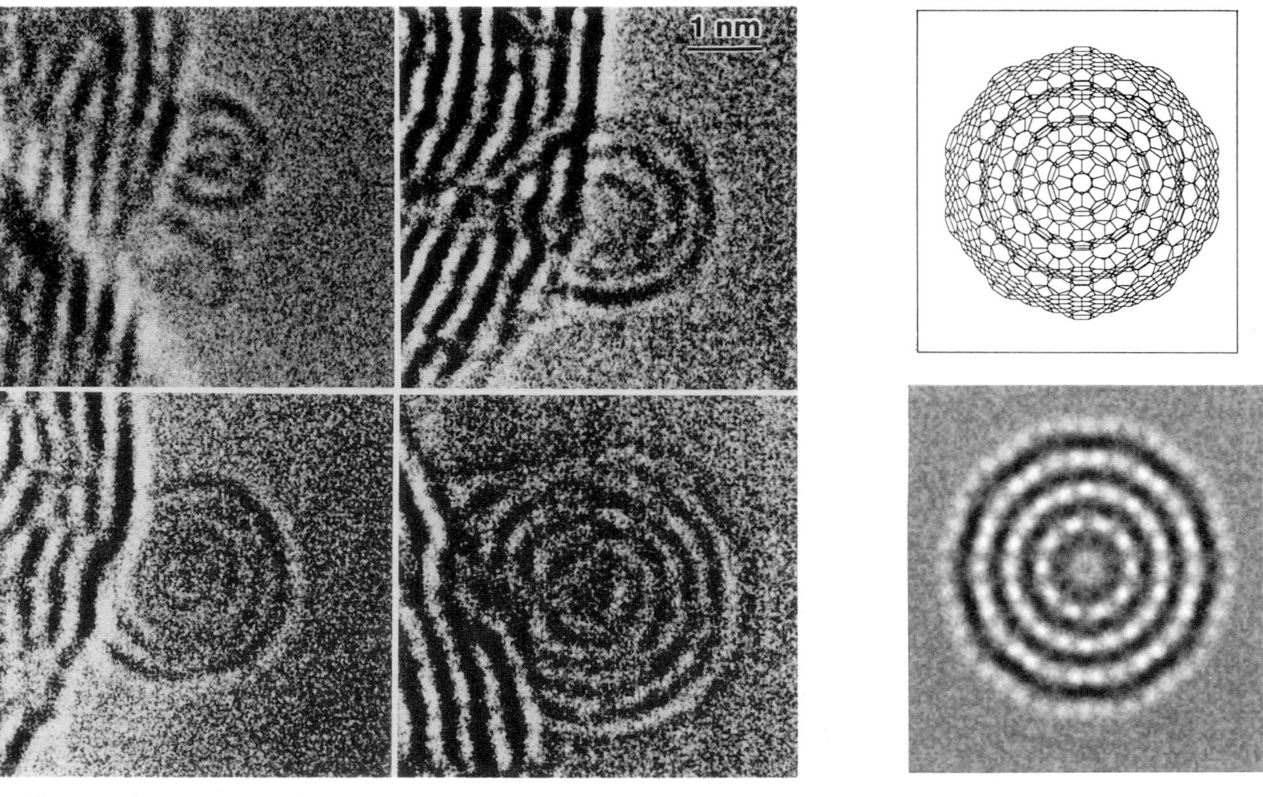

Smaller particles consisting of just a few fullerenes-within-fullerenes could also be identified in Ugarte's micrographs. Their constitution could be determined by comparing them with computer simulations of the micrographs based on idealized arrangements of the shell structures. The diagrams on the right show such a simulation for the hyperfullerene containing C_{60} inside C_{240} inside C_{540} inside C_{960}. Reproduced with permission from Ugarte, Daniel (1993). Preprint.

the same way, scientists have been eager to see what can be done to fill the voids in the centres of the hyperfullerenes.

In early 1993, Rodney Ruoff, Donald Lorents, Bryan Chan and Ripudaman Malhotra at SRI International in California, and Shekar Subramoney at Du Pont's Experimental Station in Delaware, reported the synthesis of hyperfullerenes containing single crystals of lanthanum carbide. They had simply packed the centre of one of their graphite rods with lanthanum oxide and repeated the standard procedure for making lanthanum-encapsulating fullerenes, except that they maintained a small gap between the rods as the current was passed.

Electron microscopy studies of the resulting deposit revealed nanotubes and hyperfullerenes, plus a large number of particles which appeared to contain metal crystals. Careful analysis of the images suggested that these were crystals of a specific form of lanthanum carbide. Left to its own devices, this form of lanthanum carbide reacts on exposure to moist air to give lanthanum oxide and acetylene. That the crystals formed inside the hyperfullerenes had remained unaffected by exposure to air in the laboratory for several days indicated that the nested fullerenes effectively isolate the crystals from their surroundings, forming a protective sheath. Ugarte has since shown how to fill and empty hyperfullerenes with tiny clusters of gold atoms.

Shortly after publication of the paper by Ebbesen and Ajayan announcing a method for the bulk synthesis of the nanotubes, scientists speculated that it might be possible to fill the tubes with material through capillary action. Tubes with their end caps removed would literally suck material inside like nanoscale pipettes. Could this really be possible? The answer, provided by Ajayan and Iijima in a paper published at the end of January 1993, was a resounding yes!

The NEC scientists produced bulk quantities of nanotubes using the Ebbesen–Ajayan method. After separating the tubes from the rest of the graphitic material, they sprinkled the sample with particles of molten lead produced by electron bombardment of a solid target placed close by in a vacuum. The particles were deposited randomly, with diameters between one and 15 nanometres. Some were deposited on or close to the end caps of some of the nanotubes. The scientists then annealed the sample by heating it in an oven, in air, for 30 minutes at about 700 kelvin (above the melting point of lead). Subsequent electron microscopy showed that some of the tubes had indeed drawn the molten metal inside their hollow centres.

The mechanism by which the lead-filled tubes are formed is not fully understood. However, the end caps of the tubes must first be broken before the liquid metal can be sucked inside by capillary action. The sides of the tubes appear to remain intact throughout, and the scientists found that the end caps also survived the annealing process intact if this was carried out in the absence of air. This suggests that in air, the end caps are vulnerable to chemical attack—subsequent research suggests a lead-catalysed reaction involving oxygen—which destroys successive carbon networks, exposing the inside of the tube to the outside world. This makes sense given that the end caps contain the

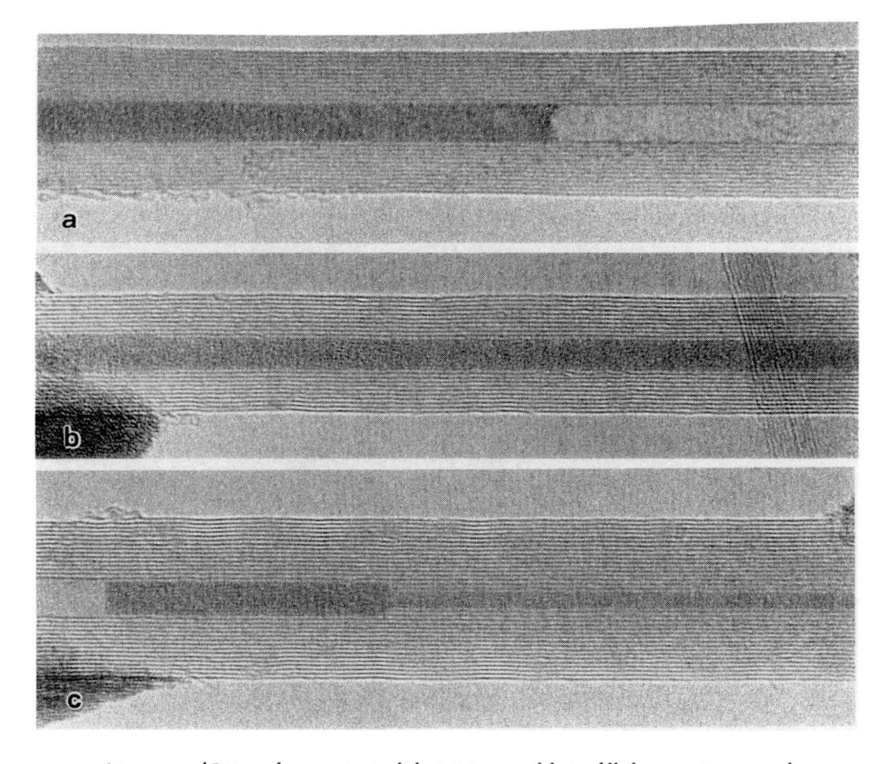

Ajayan and Iijima demonstrated that it is possible to fill the empty core of a nanotube with lead. Breaking one of the end caps through a chemical reaction with oxygen (possibly catalysed by lead) allows exposure of the inside of the tube to the outside world. Molten lead in the vicinity of an open tube may then be drawn along the tube's centre through capillary action. The lead then crystallizes under confinement inside the tube to a completely new form. The result is a potential molecular-scale wire. Reproduced with permission from Ajayan, P. M. and Iijima, Sumio (1993). *Nature*, **361**, 333. Copyright (1993) Macmillan Magazines Limited.

pentagonal defects required for closure, and are therefore likely to possess the kinds of super-alkene properties seen in buckminsterfullerene itself. Opening and thinning of nanotubes (stripping away outer tubes) has also been demonstrated using carbon dioxide.

Ajayan and Iijima can't be sure what form the lead takes when it crystallizes inside the tubes, although the evidence seems to rule out the possibility that it might be a form of lead oxide. There is a distinct possibility that with such geometrical confinement, the molten lead inside the tubes crystallizes to a completely new form. The narrow tubes have internal diameters of only 1.2 nanometers, room for only two or three lead atoms. There appears to be no evidence for incorporation of metal between the concentric layers, which is reasonable considering that the spacing between adjacent tubes is only 0.34 nanometres: hardly room for a single lead atom.

In fact, lead was not the scientists' first choice. They had tried several other low-melting-point substances without success and were somewhat surprised when they discovered that lead works so effectively. Theoretically, the large viscosity of molten lead should make it more difficult to get the material inside the tubes.

The lead-filled tubes present very visual examples of molecular-scale wires. The conductivity of these composite tubes no longer depends on the electrical properties of the carbon sheath. Indeed, if the empty tubes were insulating, then the composite tubes would be nanoscale versions of shielded cable. Furthermore, when cooled to about seven kelvin, lead becomes superconducting, opening up even more possibilities for applications in the emergent science of nanotechnology. Even the difficult process of separating the tubes may be made easier because of the considerably increased density of the filled tubes compared with the empty ones. At the present time, large-scale fabrication and separation of metal-filled nanowires seems very feasible.

There may, however, be even more immediate benefits for fundamental science. It is now possible to study in a systematic fashion the behaviour of materials confined to extremely small spaces. Investigation of the physical properties of materials squeezed into nanometre-size pores has so far been the preserve of the theoreticians. Now there is a way to measure these properties experimentally.

Where will it all lead? Nobody knows. The surprising nature of this once so unsurprising element means that further strange discoveries cannot be ruled out. Theoreticians have predicted that carbon structures with 'negative curvature' might also be stable. These are inside-out fullerenes incorporating seven-membered rings instead of five-membered rings. The theoreticians have urged experimentalists to look for these weird structures in the soot produced by a carbon arc discharge. So far, they haven't been found. But, with the simple carbon arc discharge now receiving such intense scrutiny, we should be ready for further surprises.

Whatever the future holds, our perception of the most humble of the common forms of carbon can never be the same again. The old carbon paradigm has finally given way to the new.

16

Still the last great problem in astronomy

The efforts of hundreds of scientists measuring, monitoring, classifying, categorizing, and recording the behaviour and properties of the fullerenes may have eroded the mystique of these marvellous new molecules, but some fairly central mysteries remain. How are such perfectly symmetrical molecules formed in the chaos of a high-temperature plasma or carbon arc discharge, seemingly at odds with a basic law of nature that says order does not come spontanteously from disorder? Given that our association with elemental carbon has a history spanning thousands of years, how come the fullerenes were not discovered until 1985? What roles, if any, do the fullerenes play in interstellar space?

Throughout all the feverish research activity that has brought forth a veritable panoply of fullerenes, fullerene derivatives, superconducting fullerides, nanotubes, and hyperfullerenes, speculation on these central questions has continued. However, unlike the small group of scientists closeted in Smalley's office in the months following the initial discovery in 1985, today's speculators at least have the benefit of the knowledge and understanding wrought from almost four years' intense study.

The mechanism by which the fullerenes are formed is still a matter of considerable speculation and debate. The icospiral nucleation mechanism proposed by the Rice–Sussex group in 1985 (referred to later by Smalley as the 'party line') was novel but it was also grounded in the old carbon paradigm. Fragments of graphite chicken wire would rearrange to incorporate pentagons in their structures, allowing them to buckle and at least partially close up to eliminate some of those unfavourable dangling bonds. But, the scientists had reasoned, graphite would much rather be flat. Dangling bonds notwithstanding, the incorporation of pentagons would be a relatively rare occurrence. Full closure to a fullerene cage would therefore be infrequent: for C_{60} they had estimated a one in ten thousand or one in a million chance. Instead, the structure would more often continue to grow, forming a nautilus-like shape on its way to becoming a tiny soot particle.

Buckminsterfullerene was seen as a survivor. Once formed by scant chance, its closed spherical structure and chemical inertness prevented it from reacting further. Conditions could be arranged inside AP2 which would cause the other fullerenes to be 'reacted away'—grown to particle sizes too large to register in the time-of-flight mass spectrometer, leaving only C_{60} (and a little C_{70}). At the time it was developed, the icospiral mechanism fit the known facts without rocking the existing paradigm more than absolutely necessary.

But this mechanism clearly could not explain the extraordinarily high yields of fullerenes that were eventually shown to be produced in carbon evaporators and arc discharges. Under the right conditions, full closure to a fullerene cage was not the exception, it was the rule. In arriving at the icospiral nucleation mechanism, the Rice–Sussex group had simply been too conservative. So, what was *really* going on?

Whatever was happening, one thing at least seemed to be clear. The fullerenes were *not* forming from small fragments of graphite chicken wire detached directly from the surface of the solid. A number of careful experiments with graphite samples enriched with carbon-13 (such as those reported by the IBM group) demonstrated that the fullerenes were being produced from *atoms*. If anything, this made the mechanism even more remarkable. These incredibly ordered, symmetrical structures were being assembled virtually atom by atom from random motions in the chaos of a high-temperature plasma or arc discharge.

Smalley found one significant clue to what was going on by looking at the conditions under which C_{60}, C_{70}, and small amounts of the higher fullerenes were formed with relative ease, and comparing them to conditions which appeared to disfavour the fullerenes. The IBM group had succeeded in producing milligram quantities of fullerenes from the laser vaporization of graphite. When Smalley learned this from Bethune on their journey from the ISSPIC conference in Konstanz, he was sorely irritated. Why had the IBM group succeeded where he and his colleagues at Rice had failed?

They quickly discovered the reason. All they needed to do was repeat the experiment with the graphite rod enclosed in a furnace. Laser vaporization of the graphite in the presence of several hundred torr of helium at a temperature of around 1300 kelvin produced soot containing up to 20 per cent fullerenes. The Rice group went on to make milligram quantities of the metal-encapsulating fullerenes using this method.

The effect of the high pressure of inert gas, identified to be so crucial to the success of the carbon evaporator experiments, is to confine the carbon atoms and fragments of graphitic material formed in the plasma and prevent them from drifting into cooler regions of gas. Smalley came to the conclusion that by keeping the fragments in the high-temperature region of gas around the rod, the necessary rearrangements could take place and the resulting pentagon-containing fragments could be thermalized (or annealed) through collisions.

Smalley's modified 'party line' mechanism for the spontaneous formation of buckminsterfullerene. As the carbon fragment grows through successive additions of atoms and small radicals, its structure is continually being annealed to give the lowest energy form. This form will be naturally curved through the incorporation of isolated pentagons. Further growth and annealing will nearly always produce soccer ball C_{60}. Adapted from 'Fullerenes' by Robert F. Curl and Richard E. Smalley.

Further reactions between the fragments and carbon radicals would add to the growing structures which, because of the high temperature, would rearrange wherever possible to ensure the number of isolated pentagons was maximized. Just as Smalley had discovered through his *eureka* experience with the last paper pentagon, the last carbon radicals would add to close the structure completely, producing a perfectly symmetrical soccer ball.

Smalley's mechanism is really an adaptation of the 'party line'. What makes it reasonable is that it combines the (unquestioned) drive to reduce the energy of the growing fragments by reducing the number of dangling bonds with the need for high temperatures to induce the necessary rearrangements. Without high temperatures, the rearrangements will be too slow and the fragments will continue to grow as flat or slightly buckled chicken wire, eventually producing soot but no fullerenes.

With high temperatures, however, Smalley's mechanism demands that the rearrangements nearly always produce structures with the maximum possible number of isolated pentagons. Soccer ball C_{60} is therefore formed in abundance because at every stage in the growth mechanism the annealing gives predominantly structures that are fragments of the soccer ball shape. Those few that are not go on to close into C_{70} and the higher fullerenes. There are parallels with the lesson from Ugarte's work on the hyperfullerenes. Given the right encouragement, flat graphite *will* rearrange to incorporate the pentagons required for closure into spherical structures.

While this all sounds very plausible, the mechanism is not without potential contradictions. A large number of theoretical calculations suggest that C_{70}—not C_{60}—is the more favourable structure in energy terms. But then the overall mechanism is a product not only of the energies of the individual species (atoms, radicals, molecules, fragments) involved in the reactions but also the energetics (and hence the relative speeds) of the reactions themselves. C_{70} could well be the more favourable structure, but if the reactions leading to C_{60} are significantly faster, then C_{60} molecules will be formed in greater numbers.

There have been alternative proposals. At the special 'fast-breaking events' session of the American Chemical Society symposium in Atlanta in April 1991, Jim Heath presented another view on fullerene formation. The basic difference between Heath's mechanism and Smalley's modified 'party line' lies in the way they treat the early stages of cluster growth. Rather than assemble the fullerenes from fragments that adopt the right curved shapes through annealing, Heath suggested that the route to the fullerenes involves transitions from linear carbon chains to monocyclic rings and from rings to closed cages.

Starting from atoms and small radicals, the first products of the condensation taking place in a plasma or arc discharge are the C_2–C_9 linear chains seen in abundance in the early cluster-beam experiments. These chains are reactive because they have dangling bonds at either end, but they do not grow to ever longer lengths (such as C_{33}). Beyond C_{10}, theoretical calculations predict that a ring is a more stable structure than a chain, eliminating the dangling bonds by joining the head to the tail.

The final stage in Heath's mechanism involves the transition from rings to closed cages. This may take place in one of two ways. Either the rings rearrange to form curled graphitic fragments, which are then annealed and grow to give fullerenes in much the same way as in Smalley's mechanism, or the rings rearrange to give geometrically closed structures directly. These will necessarily be the lower fullerenes, carbon cages with sizes smaller than C_{60} which then grow larger through subsequent additions of C_2 radicals. This last process is not exactly the reverse of the 'shrink wrap' photofragmentation mechanism, but it is qualitatively similar. On energy grounds, it is necessary that expansion of the lower fullerenes through successive C_2 additions takes place in such a way as to maximize the number of isolated pentagons.

It is this last aspect of Heath's mechanism that makes it somewhat unsatisfactory. For C_{60} (and C_{70}) to become dominant in the soot, there must be a sufficient concentration of small carbon radicals present in the plasma to convert the bulk of the smaller fullerenes into C_{60} and C_{70}. The formation of the smaller fullerenes will still require some annealing (otherwise they will simply fall apart again), suggesting that they may not be produced until a fairly late stage in the clustering process. There is therefore a need for quite large quantities of small radicals to be still available by the time the smaller fullerenes have been formed. This might be okay in an experiment where radicals are constantly being generated (in a carbon arc evaporator, for example), but is unlikely in a pulsed experiment using an apparatus like AP2.

It is also difficult to account for the formation of metal-encapsulating fullerenes using Heath's mechanism. Many of the smaller fullerenes that are supposed to grow to C_{60} by the addition of C_2 radicals will be too small to hold the metal atoms. If there is no way of getting the metal atoms incorporated

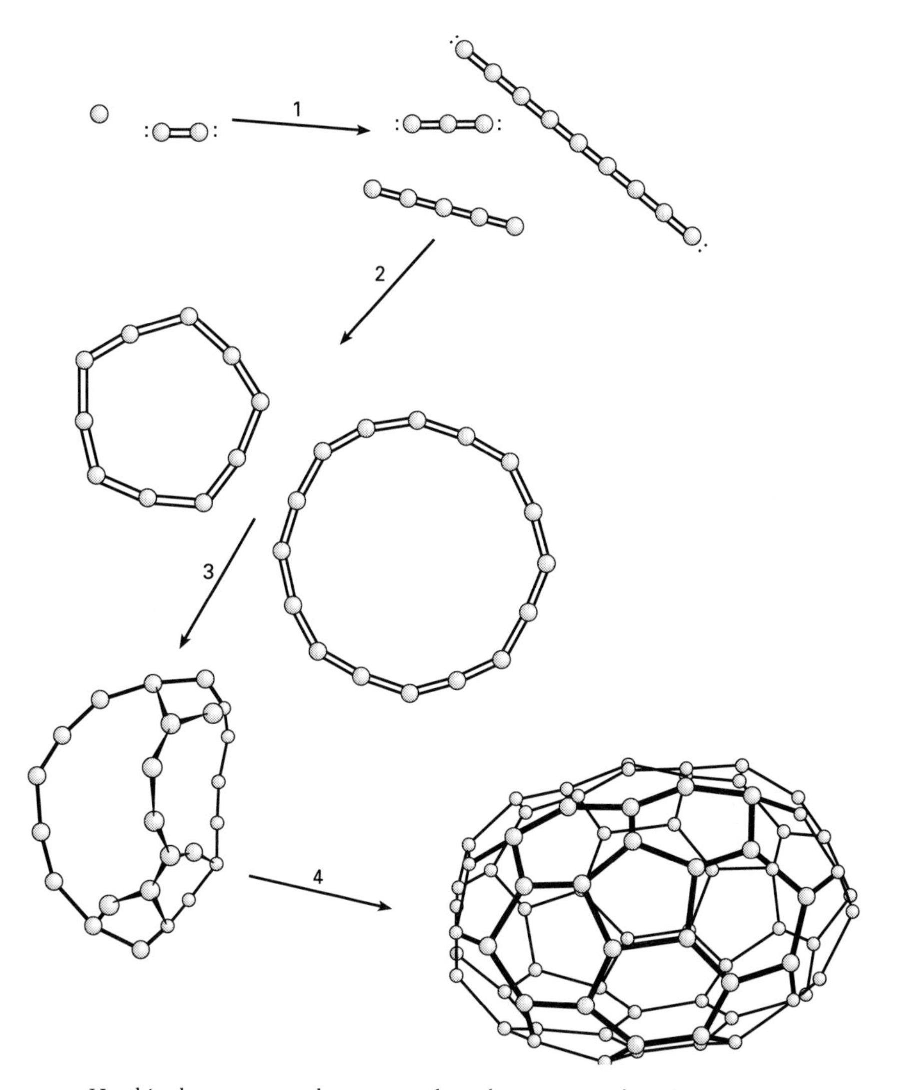

Heath's alternative mechanism involves the transition from linear carbon chains (C_n, where n is less than 10) to monocyclic rings (C_n, n between 9 and 21). Fullerenes are then formed by the further growth of these rings through unstable intermediates, which rearrange to give closed-cage structures. Adapted, with permission from Heath, James R. (1992). *Fullerenes*. (Hammond, George S. and Kuck, Valerie J. eds.). American Chemical Society, Washington D.C. Copyright (1992) American Chemical Society.

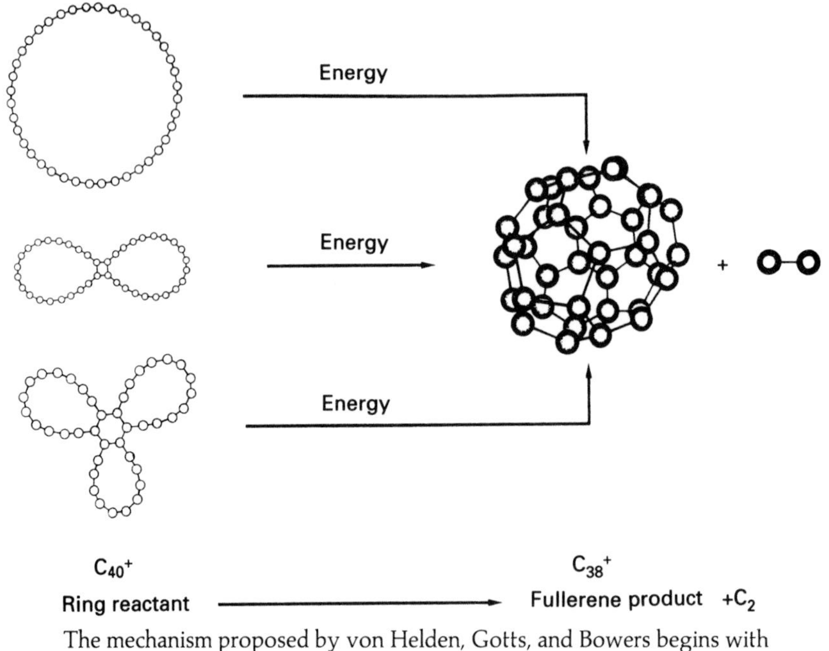

The mechanism proposed by von Helden, Gotts, and Bowers begins with linear chains which transform to monocyclic, bicyclic, and tricyclic planar rings. Through the annealing process, these rings then collapse into closed cages with the ejection of C_2 radicals or C atoms and C_3 radicals. This diagram shows the collapse of three different C_{40}^+ rings into a C_{38}^+ fullerene ion plus a C_2 radical. Adapted, with permission, from von Helden, Gert, Gotts, Nigel G., and Bowers, Michael T. (1993). *Nature*, **363**, 60. Copyright (1993) Macmillan Magazines Limited.

inside the smaller fullerenes, then there is in principle no easy way to get them inside by the time the fullerenes have grown.

In May 1993, Gert von Helden, Nigel Gotts, and Mike Bowers at the University of California at Santa Barbara presented experimental evidence in support of an alternative mechanism. This involves cluster growth to large sizes, followed by annealing and collapse of the large structures into fullerenes. The first stage of this mechanism is identical to Heath's: conversion of linear chains into planar rings, with the exception that these rings may be monocyclic, bicyclic, and even tricyclic. In the second stage of the mechanism, these large ring systems collapse into cages with the ejection of C_2 radicals or carbon atoms and C_3 radicals.

The involvement of carbon rings in these mechanisms is reminiscent of the earlier studies of Diederich, Rubin, and Knobler on C_{18}, C_{24}, and C_{30} produced from the decomposition of large carbon oxides. They had seen strong evidence for the spontaneous dimerization of two (presumably ring-shaped) C_{30} molecules to give C_{60}. In September 1992, Diederich and Whetten and their colleagues at UCLA reported evidence for *coalescence* reactions of C_{60} and C_{70}. In these reactions, several molecules of C_{60} fuse together to give a series of higher fullerenes grouped around C_{118}, C_{178}, C_{238}, and C_{298}. These are not quite simple multiples of 60, suggesting that elimination of C_2 radicals occurs to give what are presumably more stable structures. C_{70} similarly combines to give C_{138}, and C_{60} combines with C_{70} to give C_{128}.

Just to complicate matters, Huffman and Lamb and a group from the Naval Research Laboratory in Washington DC have described the results of reacting C_{60} and C_{70} with ozone, which produced *odd-numbered* carbon clusters C_{119}, C_{129}, and C_{139}. These are dimers of C_{60} and C_{70}, or the combination of C_{60} and C_{70}, less one carbon atom. The ionized forms appear to show behaviour similar to that of the large fullerenes. The scientists believe they have found a way to open the cage structures in a controlled manner, similar in many ways to the opening of the end caps of the nanotubes.

As far as the formation of C_{60} is concerned, the relative importance of reactions or rearrangements of pentagon-containing fragments, carbon rings, or the coalescence reactions of smaller fullerenes is just not known. It may be that they all have important roles to play. Given that there may be thousands of different structures involved in the mechanism—all contributing to fullerene production to a greater or lesser extent—the full details of the mechanism may never be known except in the most general terms.

Shortly after the initial discovery of buckminsterfullerene in 1985, the Rice–Sussex group had indulged in some pretty wild speculation. Connecting the formation of the fullerenes with the growth of soot particles had led the scientists to suggest that C_{60} is a ubiquitous molecule, present in small amounts in every candle flame. It had also led them into a good deal of trouble with the community of soot specialists. With hindsight, their enthusiasm for the stability and special nature of C_{60} seems at odds with their conservatism over the formation mechanism. The constraining influence of the old carbon paradigm had prevented them from giving full courage to their convictions.

The physical connection between C_{60} and soot was firmly established in 1991 by Jack Howard and his colleagues at the Massachusetts Institute of Technology (MIT). These scientists examined the contents of the sooting regions of benzene–oxygen flames and, depending on the conditions, found C_{60} and C_{70} in modest yield (about three grams per kilogram of fuel combusted). Homann and his colleagues at the Max Planck Institute for Physical Chemistry had earlier identified C_{60} in sooting benzene and acetylene flames using mass spectrometry. Four years later, the MIT scientists had the benefit of full infra-red and ultra-violet spectroscopic data for isolated C_{60} and C_{70}. By sucking the sooty material directly from the flames, extracting it with toluene and separating the extract using high-performance liquid chromatography, they were able to use the spectroscopic data to confirm that C_{60} and C_{70} were indeed present.

But this observation was not heralded as a vindication of the icospiral mechanism. The MIT scientists found that the yield of the fullerenes is not directly related to the yield of ordinary soot, suggesting that the fullerenes and soot are produced through two distinct mechanisms. Nevertheless, the Rice–Sussex group had been partly right in some of their wilder imaginings: under the right conditions, where there is soot, there are also fullerenes.

So, why hadn't the fullerenes been found a lot sooner? After all, fire is one of man's oldest discoveries. Why aren't the ceilings and walls of primitive cave dwellings covered with thick deposits of buckminsterfullerene? At the Materials Research Society symposium held in Boston in November 1990, Smalley pondered on this question.

Smalley argued that part of the reason why C_{60} does not occur naturally is related to its chemical reactivity. Despite initial evidence of inertness, C_{60} was found to burn quite happily in oxygen. Thus, although C_{60} might have been readily formed in the condensing vapour in the sooty regions of early cave fires, it was destroyed as it mixed with air and passed through the hottest parts of the flame. Not all the C_{60} could be expected to be consumed in this way, however, particularly if there was insufficient oxygen to burn all the fuel completely. Smalley was therefore led to speculate that the soccer ball molecules may have become trapped somehow inside soot particles. He believed that were it not for these processes, buckminsterfullerene would indeed have been discovered as a substance some tens of thousands of years ago.

A further reason was provided in May 1991 by the Sussex group. Samples of solid C_{60} exposed to light and air were found to degrade over time. On redissolving the samples in benzene, an insoluble reddish-brown deposit was produced. A study of the photochemistry revealed that the carbon cage molecules break up in the presence of trace amounts of ozone (O_3) to form a range of oxygenated products. Even if C_{60} is produced naturally and survives long enough to form a solid deposit, exposure to light and air will destroy it. The Sussex chemists had carried out a search for buckminsterfullerene in chimney soot, without success.

This seemed to settle the matter. The tendency for the closed-cage carbon molecules to burn, degrade, or become entrapped prevents the detection of naturally occurring fullerenes. Imagine the surprise, then, when it was announced in July 1992 that C_{60} and C_{70} had been found in samples of shungite, a coal-like rock from Shunga, a Russian town some 200 miles north-east of St Petersburg.

Peter Busek and Semeon Tsipursky at Arizona State University in Tempe and Robert Hettich at Oak Ridge National Laboratory in Tennessee had taken great care to ensure that they weren't being misled by what they were seeing. It was Tsipursky who noticed the honeycomb images reminiscent of the fullerenes in high-resolution electron micrographs of shungite. Mass spectrometry studies carried out at Arizona State University produced the characteristic C_{60}^- and C_{70}^- signals and further 'blind' tests by Hettich with samples of shungite, specially produced fullerenes, and blanks confirmed their findings.

How these fullerenes found their way into what are probably Precambrian rocks (making them about 600 or more million years old) remains a complete mystery. It is possible that the molecules survived whatever process was responsible for their production hundreds of millions of years ago and were deposited and buried along with the carbonaceous material that was eventually to form into rock. To survive this long, they would have had to have been completely screened from light and air. Alternatively, they may have been formed much more recently as secondary products. Busek and his Arizona State University colleagues Terry Daly, Peter Williams, and Charles Lewis have since found evidence for C_{60} and C_{70} in samples of rock from Sheep Mountain in Colorado which is named (rather appropriately) fulgurite. This is a glassy rock formed under the extreme conditions created when lightning strikes the ground. The ground melts and dendritic structures consisting of thin tubes of glass are produced. Temperatures can often exceed 2000 kelvin. The scientists reasoned that as part of this process, C_{60} and C_{70} were formed from carbonaceous material (pine needles and cones) present in nearby soil.

There was still the question of the carbon arc discharge. The carbon arcs used routinely to produce the fullerenes in high yields have been in use since the beginning of the nineteenth century. This simple fact explains Krätschmer's sense of disbelief when confronted with the evidence of his own experimental results. If producing C_{60} from graphite in an arc discharge or by resistive heating is so easy, how come nobody had found it before?

The answer again lies partly in the reactivity of C_{60}. The early carbon arcs were struck in air, and so any C_{60} produced would likely have been destroyed by reactions with oxygen. The chances of discovering the fullerenes in the soot from an arc discharge would have increased enormously with the introduction of vacuum technology, but the subtle interplay of temperature and pressure would have made them hard to find by accident.

It can be argued that the fullerenes were found in the 1980s because their detection required the sophisticated technology of machines like AP2. Without the singular results reported by the original five-a-side soccer team in 1985, there would have been no reason for Huffman to press a reluctant Krätschmer to revisit their earlier carbon evaporator experiments. Buckminsterfullerene had revealed tantalizing glimpses of itself in the form of the camel humps, and it is clear that Huffman, Krätschmer, and Sorg came very close to discovering this new form of carbon more than two years before the experiments with AP2. But identifying C_{60} in the soot from the Heidelberg evaporator required the scientists at least to have an idea of what it was they were looking for (no matter how crazy this idea seemed at the time).

Other elements of the wider discovery were also in place well before 1985. Iijima had examined the unusual patterns of concentric rings in his electron micrographs in 1980, but had not been able to interpret them. Had he been aware of the Hückel calculations carried out on soccer ball C_{60} by Osawa and others he might have made the connection much earlier, but he wasn't and he didn't. Bacon and many other scientists had studied the carbon whiskers formed in an arc discharge, but hadn't the instrumental resolution (or the inclination) to look for nanoscale buckytubes. Besides, without the background established by five years of experimental and theoretical research on buckminsterfullerene, and the big bang of September 1990 and its aftermath, it would have been difficult to know what to make of these things.

Whatever else can be said about Krätschmer, Huffman, and Sorg's experiments with the Heidelberg evaporator in 1982–83, it must be admitted that they were a failure. The physicists were looking for a form of particulate carbon that would explain the observed 217-nanometre interstellar band. What they ultimately found was buckminsterfullerene, but the distinctive camel hump spectrum of this molecule looks nothing like the 217-nanometre feature. Whatever it is that is causing this feature to appear in the light from distant stars, it is certainly not C_{60}.

But if C_{60} itself is not the answer, its discovery opens up a completely new range of possibilities. Ionized fullerenes, fullerene derivatives, metal-encapsulating fullerenes, and even the nanotubes and hyperfullerenes have been discussed as potential candidates for explaining the appearance of interstellar features in the ultra-violet, visible, and infra-red. However, too little is known at present about the spectroscopic properties of all these different possibilities to enable the scientists to make unambiguous assignments. There is a lot of speculation, and to a certain extent the astrochemists and physicists are spoilt for choice.

Huffman believes that we still don't understand carbon well enough to know how best to look for it in interstellar space. He has proposed to Krätschmer that they return to the original purpose of the carbon evaporator experiments they did with Sorg in 1982–83, and start again.

And what of the diffuse interstellar bands? It was the search for long chain polyynes and cyanopolyynes in the hot gases produced inside AP2, and their possible relevance to the mysterious diffuse bands, that had led the original soccer ball team into the arms of C_{60} in September 1985. Buckminsterfullerene is just too perfect; too attractive a possibility for it to be passed over as a serious candidate. Its spontaneous formation in large quantities under just the kinds of conditions thought to prevail in the expanding outer shells of carbon-rich red giant stars would seem to make its presence in the interstellar medium almost inevitable. So where is it?

In a review article published in March 1992, Kroto and Hare re-examined the role of of carbon in the galaxy in the light of the discovery of the fullerenes. If C_{60} (or an analogue) is to feature in interstellar space then it must survive with its soccer ball framework and its pi-electron orbital systems virtually intact. If it is able to survive, then it will be bathed in cosmic radiation with sufficient energy to ionize it. Ionized C_{60} has therefore been considered as a possibile explanation of the diffuse interstellar bands. Kroto and Jura have made a case for $C_{60}M^+$ ions, where M is one of the atoms commonly found in the interstellar medium, such as sodium, potassium, calcium, iron, sulfur, and oxygen. In these complex ions, transitions involving the transfer of electron charge are thought to correspond roughly in energy and appearance to the diffuse bands.

If the fullerenes are abundant in space in some form or another, then might it not be possible to detect them inside meteorites? This possibility was examined by de Vries, H. R. Wendt, and Hunziker at IBM in collaboration with E. Peterson and S. Chang at NASA's Ames Research Center early in 1991. They searched for the fullerenes in samples of carbonaceous material from the Murchison and Allende meteorites. They found plenty of polycyclic aromatic hydrocarbons, but no fullerenes. These results led the scientists to conclude that the presence of hydrogen in interstellar space and the gases surrounding carbon-rich red giants might preclude the synthesis of the fullerenes in these environments, the dangling bonds of the curled graphitic fragments becoming tied up with hydrogen atoms before they can grow further and close up to form spheroidal structures.

Such results have done little to shake Kroto's conviction that the fullerenes are abundant in space in one form or another. In 1992 it was reported that the application of very high pressures to samples of the fullerenes can compress the carbon spheres into the tetrahedral structure of diamond. Superman long ago impressed Lois Lane by squeezing diamonds out of coal, but doing this in the laboratory with the fullerenes has opened up yet another potential commercial application for this new form of carbon. Bits of diamond and structures looking remarkably like nanotubes and hyperfullerenes have also been found in meteorites, and it has been suggested that one mechanism by which the diamond might form inside these lumps of rock involves the compression of hyperfullerenes by shock waves.

For the time being, identifying the substance or substances responsible for the diffuse interstellar bands remains, for the chemist, still the last great problem in astronomy.

Epilogue

In retrospect, I was extremely fortunate that the editor [of *The inventions of Daedalus*], Michael Rodgers, was a tolerant and far-seeing man. He allowed all that heavy technical detail to remain. The book was intended to be fairly light-hearted, and a sterner editor would have put a blue pencil through the lot. My position among the originators of the hollow graphite molecule idea would then have been far weaker. As it was, I didn't have the gall to include my calculations on the critical temperature and pressure of giant fullerenes, as they are now called; though I tidied them up later for the Royal Society conference and its published proceedings. . . . I was concerned to discover how big a hollow graphite molecule could be made. It didn't occur to me to wonder how small one could be made, or I might just have stumbled over the C_{60} structure itself!

David Jones. Letter to the author. December 10, 1992

Appendix

Molecular spectroscopy: windows on the microworld

The various techiques of spectroscopy provide the scientist with windows through which the microscopic worlds of atoms and molecules can be studied. It is impossible to overstate the importance that spectroscopy has played in bringing us to our present level of knowledge and understanding of *all* the atomic and molecular sciences: physics, chemistry, and biology. Put simply, we learn how microscopic systems behave by watching how they interact with electromagnetic radiation. And scientists have worked out how to harness the full range of the spectrum from radio waves to gamma rays.

Towards the end of the last century, the physicist James Clerk Maxwell capitalized on the remarkable experimental and theoretical work of Michael Faraday to devise a new theory of electromagnetic radiation. In this theory, radiation consists of two sets of transverse waves—essentially, waves which move up and down at right angles to the direction in which they happen to be travelling. One set of waves represents an oscillating electric field and the other an oscillating magnetic field moving at right angles both to the electric field and the direction in which the wave is travelling.

The fundamental characteristics of a wave are its wavelength (the distance required for one complete up-and-down cycle), its frequency (the number of up and down cycles in a given time interval), and its speed. Electromagnetic radiation moves at only one speed—the speed of light. However, it can have widely different wavelengths and frequencies, varying from the radio region, through microwaves, infra-red to visible light, ultraviolet light, and beyond to X-rays and gamma-rays.

Although it is still used routinely today, Maxwell's theory enjoyed fame as complete explanation for the behaviour of radiation for only a relatively short time. In 1900, the German physicist Max Planck introduced a new concept into physics which was picked up and used to very great effect by Albert Einstein. This was the idea of the light-quantum. In essence, we now understand that radiation can be described either in terms of waves or in terms of light-quanta, small particles or 'packets' of energy which we know today as *photons*. The connection between these two apparently quite contradictory types of

ν Frequency (Hz)	λ Wavelength (m)	$h\nu$ Photon energy (eV)	$h\nu$ Photon energy (J)		Microscopic source	Detection	Artificial generation
10^{22}	10^{-13}	1 MeV — 10^{6}		γ-RAYS	Atomic nuclei	Geiger and scintillation counters	Accelerators
	1 Å — 10^{-10}; 1 nm — 10^{-9}	1 keV — 10^{3}	10^{-14}	X-RAYS	Inner electrons	Ionization chamber	X-ray tubes
10^{15}		10	10^{-18}	ULTRAVIOLET	Inner and outer electrons	Photoelectric Photomultiplier	Lasers
10^{14}	1 μ — 10^{-6}	1 eV — 10^{0}	10^{-19}	LIGHT	Outer electrons	Eye	Arcs
1 THz — 10^{12}		10^{-1}	10^{-20}	INFRARED	Molecular vibrations and rotations	Bolometer; Thermopile	Sparks; Lamps; Hot bodies
1 GHz — 10^{9}	1 cm — 10^{-2}; 21 cm H line; 1 m — 10^{0}	10^{-6}		MICROWAVES; Radar; UHF; VHF; TV; FM Radio	Electron spin; Nuclear spin	Crystal	Magnetron; Klystron; Travelling-wave tube
1 MHz — 10^{6}	10^{2}; 1 km — 10^{3}		10^{-27}	Broadcast; RADIOFREQUENCY		Electronic circuits	Electronic circuits
1 kHz — 10^{3}	10^{5}	10^{-11}		Power lines			AC generators

The electromagnetic spectrum. The scales on the left give energy of the radiation (in units of electron volts and joules), which increases in proportion to its frequency (in hertz or cycles per second). Also given are the wavelengths, which vary from kilometres (radio waves) to hundredths of nanometres (one nanometre is a billionth of a metre). Different possible sources of the radiation are given on the right. Note that as the energy of the radiation increases, finer and finer details of molecular and (ultimately) atomic behaviour are probed. Eugene Hecht, *Optics* (second edition), p. 69.

behaviour is expressed in a relationship first discovered by Planck. This relationship connects the frequency of the radiation with the energy of its photons. Increasing the frequency (decreasing the wavelength) implies photons of higher energy. Thus, radio-frequency radiation consists of photons of relatively low energy, whereas X-rays or gamma-rays consist of photons of very high energy.

A molecule can absorb electromagnetic radiation by interacting either with its electric or magnetic components. Most spectroscopy is concerned with the interaction of molecules with the electric part of electromagnetic radiation. When a molecule absorbs radiation, its internal energy increases. Radiation of different energy can excite different kinds of internal molecular motions. Visible and ultraviolet light, with wavelengths measured in hundreds of nanometres (billionths of a metre), can excite electrons in molecules to higher energy states in which the bonding between each molecule's constituent atoms is changed. The molecules may become more amenable to chemical reaction as a result. This is photochemistry, and occurs every time the sun shines on green grass.

Lower energy infra-red (heat) radiation, with wavelengths measured in micrometres (millionths of a metre) can excite different patterns of vibrational motions in molecules. Like a collection of tiny balls (the atoms) connected together by so many springs (the chemical bonds), a polyatomic molecule may undergo a complicated set of movements that are picturesquely called stretching, bending, scissoring, wagging, and twisting. Superimposed on top of this vibrational motion may be rotational motions; the molecules tumbling through space as they vibrate. Moving to still lower energies, the absorption of microwave radiation (wavelengths of centimetres or millimetres) by a molecule can lead to the excitation of pure rotational motion.

But these increases in energy cannot occur continuously as the energy (or frequency) of the radiation changes. Instead, energy is absorbed or emitted in sharply defined amounts called quanta, and these processes are descibed by quantum mechanics. In the first three decades of this century, physicists made a series of discoveries that led to a revolution in our understanding of the nature of the microscopic objects which constitute atoms and molecules: protons, neutrons, and electrons. Today we can perform experiments to demonstrate that these objects behave like tiny particles, with well-defined paths or trajectories through space. We can also perform experiments to show that these objects can behave like waves, with their peaks and troughs adding or cancelling to give interference effects.

These two seemingly contradictory types of behaviour, which are exactly analogous to the wave and particle (photon) descriptions of electromagnetic radiation, are to some extent reconciled in quantum mechanics, developed in the 1920s and 1930s to explain the microscopic world of sub-atomic particles. Although arguments about what quantum mechanics means continue to this day, it is widely accepted as one of the most successful scientific theories ever devised.

The dual wave–particle nature of sub-atomic objects gives rise to the quantization of energy in molecules. Only certain discrete molecular energy 'states' are possible, characterized by integer numbers (the quantum numbers). Transitions between these energy states take place in instantaneous quantum 'jumps'. Sometimes these transitions involve the absorption or emission of photons.

Each molecule possesses a ladder of quantized energy states. States of translational motion—the movement of the molecule in three-dimensional space—are so close together in energy as to be virtually continuous (and therefore 'classical'). Neglecting translation, the bottom rungs of the ladder (the lowest energy states) are pure rotational states. As we move up the ladder we begin to encounter vibrational states, with each vibrational state possessing its own set of rotational states. Moving higher still, we find electronic states with different patterns or distributions of electron 'waves' surrounding the atomic nuclei. Each electronic state has its own set of vibrational states which, in turn, have their own sets of rotational states. The spectroscopist is charged with the task of mapping out parts of this complex pattern of energy states to find out interesting things about the structure and properties of the molecule.

In order to excite a molecule's rotations, the molecule itself must first possess a so-called electric dipole moment—a separation of electric charge so that one part of it is slightly negatively charged and another slightly positively charged. The molecule will then act just like an antenna, 'picking up' radiation containing photons tuned to just the right frequency (and hence energy) to bridge the gap between two rungs of the ladder (see p. 264). By monitoring the frequencies of radiation that are absorbed, we discover the precise energies of the different rungs on the ladder. This is absorption spectroscopy.

If we use microwave radiation, we access only the lowest rungs—the rotational states—and perhaps a few very low-energy vibrational states. The map of rotational states can be used to derive important parameters such as the molecule's moment of interia (a measure of its resistance to rotational acceleration) and the lengths of its chemical bonds.

Molecules may acquire rotational energy not only by absorbing microwave radiation, but also by colliding with other molecules or with the surfaces of large objects (the walls of a container or dust grains in space). They may then give up some of their excess energy by *emitting* microwave radiation in quantum jumps from higher to lower rungs on the ladder. The pattern of emitted microwave frequencies also reflects the pattern of quantized rotational energy states characteristic of the molecule and complements (but is not necessarily identical to) the molecule's microwave absorption spectrum. Thus, if a rotationally excited molecule is present in interstellar space, it may emit microwave radiation with a characteristic pattern of frequencies which can be detected by a radio telescope. By measuring these frequencies and comparing them with data from the microwave absorption spectrum, astronomers can tell if the molecule is present in space.

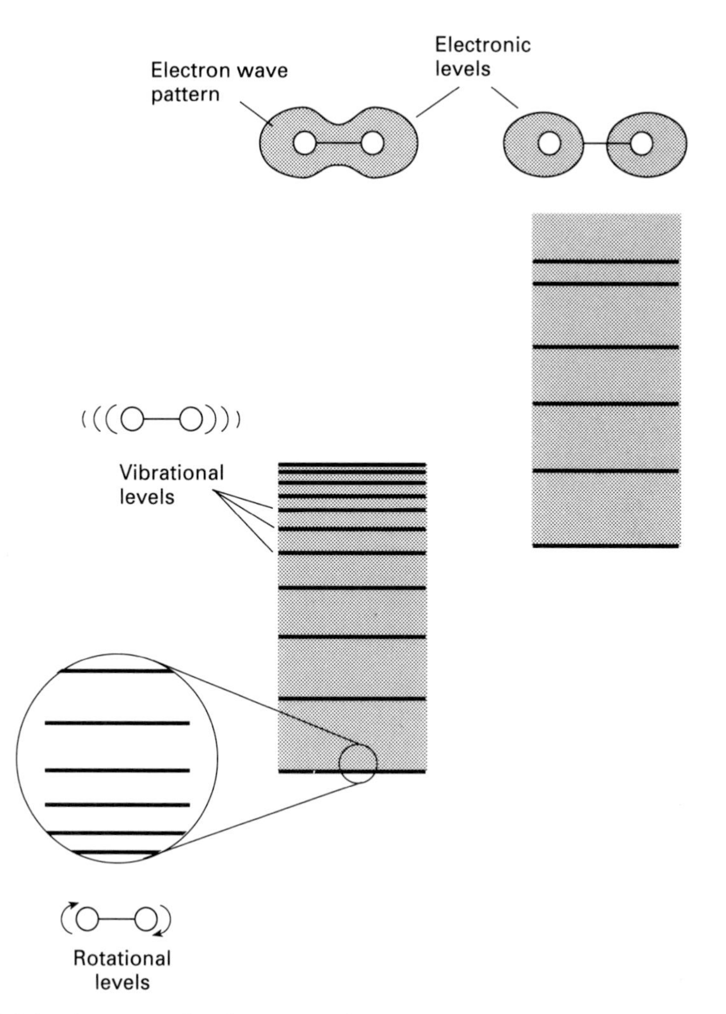

Molecules cannot absorb or emit radiation continuously. Their behaviour is instead governed by the principles of quantum mechanics, which breaks up the energy content of a molecule into discrete energy levels. These levels form a ladder. The bottom rungs of the ladder (the lowest energy levels) involve rotations—the molecules tumbling end over end as they move about. Higher up the ladder we find molecular vibrations: the chemical bonds in a molecule stretch, compress, and bend as though the molecule is made of a series of balls (the atoms) connected by springs (the bonds). Higher still, and we begin to excite electrons out of their stable 'wave patterns' in the molecule into new, higher energy ones. These new wave patterns may result in very different bonding in the molecule's electronically excited state, giving rise to photochemistry.

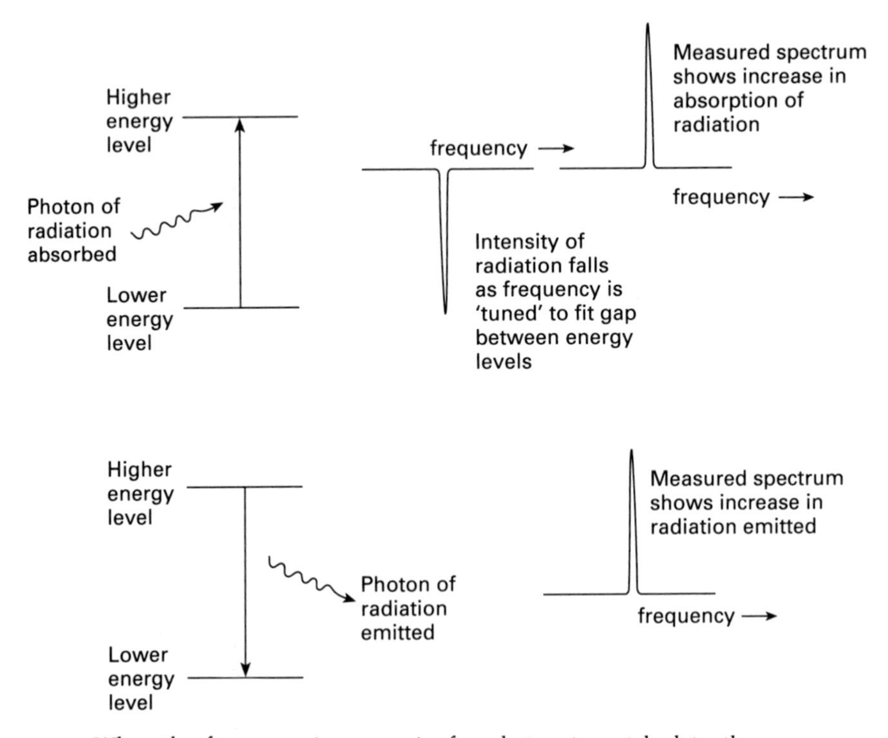

When the frequency (or energy) of a photon is matched to the gap between different rungs on a molecule's ladder of energy states, the molecule *might* absorb the photon. The actual process is governed by the rules of quantum mechanics, but if the photon is absorbed, the molecule gains energy and the radiation passing through the sample containing the molecules is diminished in intensity. If we measure the intensity of radiation passing through the sample, then we will find that it dips at the frequency at which the molecule absorbs. This constitutes a line in the molecule's *absorption spectrum*. Similarly if a molecule already in an excited state emits a photon, then this contributes a line in the molecule's *emission spectrum*. Measuring such spectra across a wide range of frequencies allows the spectroscopist to piece together bits of the molecule's ladder of energy states. This, in turn, reveals interesting information about the molecule's structure and properties.

Sources and notes

The intention of this book is to popularize a rather fascinating episode in late twentieth-century chemistry and materials science, and under no circumstances should it be regarded as an authoritative or scholarly contribution to the scientific literature on fullerenes. However, I recognize that, for whatever reasons of their own, some readers may be interested to know where my information has come from. I have purposefully avoided spoiling the flow of the narrative with endless references to my sources, and provide the relevant details here instead. From this section, the interested reader should be able to track back to the origin of virtually all my background material. The references are written according to the publisher's house-style, with (where relevant) the journal volume number in bold followed by the number of the first page of the article.

Chapter 1

The origin of the universe, the formation of stars, galaxies and planets, and the origin and evolution of life are favourite themes of science popularizers. John Gribbin's *Genesis*[1] is a very readable attempt to describe the theories of the origins of man and the universe that were prevalent in the early 1980s. It contains a useful account of early astrochemistry. Stephen Hawking's *A brief history of time*[2] is a more recent perspective on the big bang origin of the universe by a leading theoretical physicist.

Oparin's *Origin of life*,[3] first published in 1923, was subsequently expanded in a second edition and republished by Dover in 1953. Despite its age it is still well worth consulting. The detection of the first polyatomic molecule in interstellar space was reported by Townes and his colleagues in *Physical Review Letters*.[4] There followed a series of papers reporting the discovery of ever more complex molecules, including cyanoacetylene.[5] Fred Hoyle has long tended to associate himself with unfashionable theories, and *Lifecloud*[6] (with Chandra Wickramasinghe) is not regarded as a serious work by the majority of the scientific community. Although the ingredients for making glycine are found in interstellar space, of glycine itself there is no trace (yet). Hoyle and Wickramasinghe continue their crusade, most recently with the publication of their new book *Our place in the cosmos: the unfinished revolution*.[7]

The background to Kroto's early involvement in the search for interstellar cyanopolyynes is derived from a personal account, which was published in 1992 in the German chemistry journal *Angewante Chemie*,[8] various letters to[9,10] and an interview with[11] the author, and original scientific papers.[12,13] Walton's work on the synthesis of long chain polyynes is exemplified by his paper published in *Tetrahedron*.[14]

Early theories of the formation of interstellar molecules were presented by Herbst and Klemperer[15] and Dalgarno and Black.[16] The science of astro-chemistry around the end of the 1970s was reviewed by Kroto in his popular article in *New Scientist*[17] and his Tilden Lecture, published in *Chemical Society Reviews*.[18] The detection of $HC_{11}N$ was reported in *Nature* in 1982.[19]

An example of early measurements of carbon clusters in an arc discharge can be found in the paper by von Hintenberger *et al*.[20] The problems of identifying the origin of the diffuse interstellar bands in the visible are described in a review by Huffman[21] and in a recent popular book by Wynn-Williams.[22] Nigel Henbest's *Inside Science* on dust in space is a useful brief introduction to the subject.[23] Alec Douglas's proposal that linear carbon-chain molecules might be responsible for the diffuse bands was published in *Nature* in 1977.[24] Kroto's reference to the 'last great problem in astronomy' comes from a BBC *Horizon* programme which was screened in January 1992.[25]

Chapter 2

Most of the background to the early collaboration between Krätschmer and Huffman is derived from my interview with Krätschmer in Heidelberg,[26] a long telephone interview with Huffman,[27] and the physicists' own published accounts.[28–31] The book by Wynn-Williams[22] and the *Inside Science* by Henbest[23] are useful introductions to the properties and effects of interstellar dust. The review by Huffman[21] is not only comprehensive, it is also pertinent to the present tale. The 217-nanometre interstellar band was first reported in detail by Stecher.[32]

My description of Heidelberg and the Max Planck Institute for Nuclear Physics—'a nice place to do physics'—comes from observations made during my visit there in August 1992, combined with information extracted from a small city guide.[33] Krätschmer, Sorg, and Huffman's work on the matrix-isolated carbon clusters was published in 1985 in *Surface Science*.[34]

The carbyne story is actually a bit more complicated than appears from the text. In 1968, El Goresy and Donnay[35] reported the discovery of a new mineral, found in rock from the Ries Crater, which they claimed to be composed of pure carbon '. . . formed from hexagonal graphite by shock . . .'. They called this material 'chaoite', named for an American researcher of Chinese origin. This report was followed by similar claims from Soviet scientists A.M. Sladkov and Yu.P. Koudrayatsev, who first used the term 'carbyne'. Ten years later,

Whittaker presented evidence for a variety of carbynes, and suggested that chaoite was, in fact, a form of carbyne.[36] The supporting evidence was never very strong, and was called into question in the early 1980s. Some scientists went so far as to suggest that the stuff wasn't carbon at all, but layer silicates, implying that all the previous claimants had committed some serious errors in their analyses. The subject went quiet in the mid-1980s, but has now re-surfaced following the discovery of a form of carbon similar to chaoite alongside tiny grains of diamond in meteorites. This new form might have something to do with carbyne, but then again it might not.

El Goresy was actually on sabbatical leave when he made his discovery with Donnay, and his permanent work address is the Cosmophysics Department at the Max Planck Institute for Nuclear Physics in Heidelberg. I have benefited from comments on the carbyne story received from El Goresy through Fostiropoulos,[37] and some background information from Krätschmer.[38]

Chapter 3

The circumstances leading to Kroto's visit to Rice in 1984, and his various non-scientific pursuits, are derived from a personal interview,[11] his own account in *Angewante Chemie*,[8] and a letter to the author.[10] Curl has also given some background to the visit,[39,40] and his and Kroto's shared interest in Douglas's 1977 proposal concerning the diffuse interstellar bands.

My descriptions of Houston and the Rice campus come from observations made during a visit there in September 1992, a Smithsonian guide to historic America,[41] and a telephone interview with Smalley.[42] Although AP2 has changed somewhat in appearance since Kroto's first visit to Smalley's lab in 1984, my description of its current appearance is probably reasonably close. The work of Smalley and his colleagues on the development of cluster-beam techniques and their application to a variety of small molecules and clusters (including SiC_2) was published in a series of papers in the early 1980s.[43–47] The studies of semiconductor materials that contributed to Smalley's general lack of enthusiasm for the experiments on graphite were published in 1985 and 1986.[48,49]

The paper by Rohlfing, Cox, and Kaldor on carbon clusters was published in October 1984,[50] eight months after Kroto's visit to Rice. The published time-of-flight mass spectrum, with the C_{60} signal just slightly larger than the rest of the group of even-numbered clusters, is quite representative of the spectra that the Exxon group obtained during these first experiments with graphite.[51] The Exxon scientists therefore had no incentive to pursue further studies of C_{60} itself. The presentation of these results as a poster at the third ISSPIC conference in Berlin was recalled by Krätschmer.[26]

Smalley's reference to 'this silly game with graphite' comes from his personal account published in *The Sciences*.[52]

Chapters 4 and 5

Compiling a detailed chronology of the events which took place at Rice in September 1985 is made difficult by the general fallibility of long-term memory, and doubly difficult by the subsequent dispute that erupted between Smalley and Kroto (see the opening paragraph of chapter 8). In attempting to piece together the sequence of events for an article published in *New Scientist* in 1991,[53] I received conflicting accounts from Kroto[54–56] and Smalley.[57–60] Kroto[61] and Curl[62] have likened the situation to the Japanese story of *Rashomon*, in which witnesses to a crime all have different recollections of the events. Kroto has taken a special interest in what he calls 'the *Rashomon* factor', noting that Akira Kurosawa's film version focuses on the nature not of *objective* truth, but of *subjective* truth, or truth as it appears to others. In Kurosawa's *Rashomon*, nobody lies—all witnesses tell what they believe to be the truth: 'Five people interpret an action and each interpretation is different because, in the telling and in the retelling, the people reveal not the action but themselves.'[63] This applies not just to the witnesses, but to others—journalists and the authors of books—who attempt an objective presentation of the facts.

The differences between the scientists on the general matter of timing were eventually resolved by the discovery of the long-forgotten AP2 log book,[64,65] which also brought to light the (previously overlooked) roles played by Liu and Zhang. I first received copies of the relevant pages of the log book from O'Brien via Kroto. The entries in the log book largely confirmed the chronology initially supplied to me by Kroto.[9]

In May 1992, Smalley, Heath, Curl, and O'Brien used the log book and the original computer data files to reconstruct the full sequence of events, from the time graphite was first put into AP2 on August 23, 1985, to the receipt of their manuscript at the Washington office of *Nature* on September 13. The result is the 'AP2 chronology 1985'.[66] The data files themselves showed that, when replotted with the C_{60} signal on-scale, the special nature of C_{60} was already being revealed by AP2 on August 23.[67–69]

Kroto did not participate in the reconstruction exercise, preferring that each participant in the discovery should separately publish his own version of events, much as the witnesses in *Rashomon* tell their own stories. Consequently, Kroto does not endorse all the entries in the AP2 chronology.[10] Irreconcilable differences remain, with Smalley and Kroto continuing to tell their own, conflicting versions of the story. Taubes[70] has written of their dispute. Although in these chapters I have tried to accommodate the views of all involved, the version of events I have given is necessarily derivative, and should *not* be construed as 'official' or 'authorized'.

Insights into the working environment in the Smalley group are derived from my visit there in September 1992 and my interview with Mike Alford, a Smalley postdoctoral associate.[69] My paperback copy of Wolfe's *The right stuff* was published by Black Swan.[71] Although I have never carried out experiments on an apparatus like AP2, I have in the past been involved in similar laser experiments, and have drawn on my own experiences in embellishing the text. My description of the experiments, discussion, and general excitement is based largely on the AP2 chronology,[66] personal accounts, interviews and letters,[8,9,11,52,55–60,64,65] modified by specific comments on early drafts of these chapters received from Kroto,[10] Curl,[40] Smalley,[42] Heath,[67] and O'Brien.[68]

The results of experiments with hydrogen, nitrogen, H_2O, and D_2O were published in 1987.[72,73] By introducing methyl cyanide (CH_3CN) and ammonia (NH_3) into the helium buffer gas in subsequent experiments, the scientists were able directly to demonstrate the formation of cyanopolyynes inside AP2.

Marks's book on the work of Buckminster Fuller[74] is fascinating, and powerfully conveys the mixture of far-sighted genius and verbose eccentricity that so characterized the American inventor. The book is quite lavishly illustrated. Apart from the fateful picture of the Union Tank Car Company dome, it includes a delighful picture of Fuller himself popping up like the genie from Aladdin's lamp through a *pentagonal* hole in one of his domes. My description of Smalley's 'eureka' experience with sticky tape and paper hexagons and pentagons is based on his own account in *The Sciences*.[52]

Giving a new discovery a name, be it phenomenon or theory, effect or molecule, is dear to the hearts of all scientists. It is not entirely surprising, therefore, that the naming of buckminsterfullerene is one of the major points of contention between Smalley and Kroto. The version I have given in the text is essentially Kroto's.[9,10,55] He believes the name buckminsterfullerene did not emerge until *after* Smalley presented his paper structure to the group on Tuesday, September 10, 1985. And he is convinced it was he who suggested the name. Smalley is equally convinced that the name was in general use within the group during the last few days of the previous week, together with the terms 'Bucky' (C_{60}) and 'Mrs Bucky' (C_{70}),[58,64–66] although Smalley's most recent description makes no mention of these last two names.[75] I have chosen to present Kroto's version (with Smalley's claims acknowledged in a footnote).

The paper announcing the discovery of buckminsterfullerene was published in *Nature* in November 1985.[76] The comments of one of the referees of the original manuscript were passed to me by Smalley.[77] The paper describing the results of experiments with lanthanum-impregnated graphite were published in the *Journal of the American Chemical Society*.[78]

The popular science press was quick to pick up the story. Malcolm Browne's article was published on December 3, 1985.[79] An article by Rudy Baum appeared in *Chemical and Engineering News* shortly afterwards.[80]

Chapter 6

Murphy's Second Law is an invention of my own. The article by David Jones (Daedalus) on hollow graphite balloons first appeared in *New Scientist* in 1966.[81] Poliakoff's role in drawing Kroto's attention to this article comes from Kroto's own personal account.[8] A discussion of Euler's formula can be found in *The inventions of Daedalus*[82] and a recent popular book by Devlin.[83]

D'Arcy Thompson's classic *On growth and form* has been republished in abridged form by Cambridge University Press.[84] Jones's reference to 'Daedalus's finest hour' is taken from the abstract of a lecture he delivered to the Royal Society in October 1992.[85] This lecture has now been published.[86]

My discussion of Fuller's energetic geometry, the geodesic domes and personal history are derived from a number of books about the Fuller phenomenon.[74,87,88] Many books by and about Fuller that are difficult to obtain through bookstores, together with maps, toys, posters and tapes, are available through the 'Fuller Supply Service'.[89] Fuller's contributions to architecture are set in perspective in a recent book by Jenks.[90]

Barth and Lawton's papers on corannulene were published in the *Journal of the American Chemical Society*.[91,92] Osawa's discovery of the soccer ball C_{60} structure is described in an abstract of a presentation he delivered to the Royal Society in October 1992.[93] Osawa's book (with Yoshida) was published (in Japanese) in 1971.[94] An English translation of a review of the work of Russian theoretician D. A. Bochvar and his colleagues (which includes a discussion of the Hückel calculations on soccer ball C_{60}) can be found in *Russian Chemical Reviews*.[95] Davidson's calculations were reported in *Theoretica Chimica Acta*.[96] Orville Chapman's comments on the state of organic chemistry in the early 1980s and his appeals to a higher authority are derived from the *Horizon* programme.[25] Chapman's discussion of the soccer ball concept with Diederich, and his subsequent efforts to synthesize C_{60}, were related to me in a letter from Diederich.[97] The structure of beta-rhombohedral boron by Hoard and his colleagues was reported in the *Journal of the American Chemical Society* in 1963.[98] I am grateful to M. J. Bunker for drawing this work to my attention.[99]

Chapter 7

The background to the agreement reached between Smalley and Kroto and their further pursuit of research on buckminsterfullerene is described in the article by Taubes.[70] The paper proposing a closed-cage structure for C_{70} and the icospiral nucleation mechanism for the formation of soot particles was published in the *Journal of Physical Chemistry*.[100] Reviews of this work were later published by Curl and Smalley[101] and by Kroto.[102]

Haymet's papers describing his rediscovery of the soccer ball structure and his Hückel calculations were published in *Chemical Physics Letters*[103] and the *Journal of the American Chemical Society*.[104] Schmalz and co-workers reported their calculations of the relative stabilities of five different possible structures for cage-like C_{60} in *Chemical Physics Letters*.[105]

The first attempt of the Exxon group to deflate the soccer ball proposal appeared in the *Journal of the American Chemical Society*.[106] The article by Brown and colleagues on negative and positive ion cluster distributions produced from the laser vaporization of graphite was published in *Chemical Physics Letters.*[107] Hahn, Whetten, and their colleagues made much the same points almost a year later.[108] The response of the Rice–Sussex group was published shortly afterwards.[109]

Chapter 8

The disintegration of Kroto and Smalley's relationship has been chronicled by Taubes,[70] and is apparent in many of the personal communications of both scientists with the author.[9,11,42,55,58,64] The theoretical predictions of Larsson, Volosov, and Rosén were published in *Chemical Physics Letters.*[110] The first reported experimental measurement of an absorption feature ascribed to C_{60} was published by Heath, Curl, and Smalley in the *Journal of Chemical Physics.*[111] The Rice group's work on the shrink-wrap mechanism and the effects of incorporating metal atoms was reviewed by Curl and Smalley.[101,112] I also received personal accounts from Smalley[58–60] and Curl.[40] Summaries of this work have appeared in the popular science press.[113,114]

Kroto attempted to obtain funding from four different sources in the UK to buy a mass spectrometer, but failed. The implications of this failure were examined in an article in *New Scientist.*[115] Kroto's efforts with the molecular models, his discovery of the isolated pentagon rule and general principles governing fullerene stability were described in his own personal account[8] and in notes prepared for the author.[9] The results were published in *Nature*[116] and subsequently reviewed in *Science.*[102] The development of simple relationships for identifying stable structures for the larger fullerenes from the many thousands of different possibilities was described in a series of papers by Fowler and his colleages.[117–121]

The war of words between Smalley, Kroto, and the community of soot specialists, represented by Frenklach and Ebert, has featured in a couple of popular science articles.[114,122] Ebert made his position clear in the journal *Science.*[123] The discovery of C_{60} in a sooting flame was reported by Homann and his colleagues in 1987.[124] Sumio Iijima's announcement that the '60-carbon cluster has been revealed!' was published in the *Journal of Physical Chemistry.*[125] There are several accounts of the cold fusion fiasco available in print: I can recommend the one by Frank Close.[126] Paquette and his colleagues reported the synthesis of dodecahedrane, $C_{20}H_{20}$, in 1983.[127]

Chapter 9

I have put together my description of the events in Heidelberg and Tucson from December 1985–September 1990 from a combination of personal interviews with Krätschmer,[26] Fostiropoulos,[128] and Huffman;[27] telephone conversations with Lamb;[129,130] letters from Krätschmer[38,131,132] and Fostiropoulos;[37] and published accounts.[28–31] This description is spread over Chapters 9–12.

Among the first reports of the calculated vibrational wavenumbers of soccer ball C_{60} were those by Wu, Jelski, and George,[133] Stanton and Newton,[134] and Cyvin, Brendsall, Cyvin, and Brunvoll.[135] The Krätschmer–Fostiropoulos–Huffman paper announcing their observation of the four infra-red lines in the soot produced by evaporating graphite was finally published in 1990.[136]

Kroto has described his reactions on receiving a copy of the abstract of this paper in his own personal account,[8] in notes prepared for the author[9] and in the *Horizon* programme.[25] Hare has described his role in the work carried out subsequently in Sussex in a personal account included in his Sussex University Ph.D. thesis,[137] and in letters to the author.[138,139] Hare's account contains copies of the letter he wrote to Krätschmer,[140] together with Krätschmer's reply.[141] Fostiropoulos[37] acknowledges the assistance he received from Dr Mayer-Komor at the University of Munich, and Dr Brey at the Max Planck Institute for Chemistry in Mainz.

Chapters 10–11

Kroto stressed the importance of the UCLA results in his own personal account[8] and in the many lectures on the discovery of buckminsterfullerene he has delivered since. That the UCLA scientists also recognized the importance of this work (and, indeed, were gearing up to try for a large scale synthesis of C_{60} through the coalescence reactions of C_{30}) was communicated to me in a letter from Diederich.[97] The paper announcing the synthesis and characterization of the cyclic polyyne C_{18} was published in *Science*.[142] The laser desorption studies of the large carbon oxides were subsequently published in the *Journal of the American Chemical Society*.[143]

Werner Schmidt's letter to Krätschmer[144] allowed the Heidelberg/Tucson group to stay ahead in the race to isolate buckminsterfullerene and was ultimately crucial to their later success. The compelling images of crystals of a totally new form of carbon forming right before the television camera were featured in the BBC's *Horizon* programme.[25]

For an overview of IBM's research strategy, see the article by Elizabeth Corcoran in *Scientific American*.[145] This undoubtedly needs modification in the light of IBM's recent record-breaking financial losses, and there is much gloomy speculation on the future of the company's basic research programmes. I have also made use of my personal views, formed through contacts in IBM Research and following visits to both the Almaden and Yorktown Heights Research Centers in 1987.

Bethune has provided a detailed personal account of the work of the IBM group on fullerenes which began in May 1990.[146] My description is derived largely from this account, Bethune's letters to me,[147–149] and a brief account and chronology put together by Mike Ross.[150] I also received copies of the overhead slides used for the in-house seminar on July 19 by Bethune (fully kitted out in the Dutch soccer team strip and wearing a wig of long black dreadlocks in honour of Dutch soccer star Ruud Gullit).[146,151] The first results obtained by Bethune and Meijer were published in December 1990 in the *Journal of Chemical Physics*.[152]

Comments on *Nature*'s editorial policy regarding referees, together with independent confirmation of the sequence of events surrounding submission of the Krätschmer–Lamb–Fostiropoulos–Huffman paper, were communicated to me in a letter from Philip Ball.[153] Curl's actions on receiving the paper as a referee were described to me in a telephone interview.[154] The IBM paper containing the first Raman spectrum of purified C_{60} and C_{70} was published in *Chemical Physics Letters* in November, 1990.[155]

Don Bethune never made it to Hawaii.[146]

Chapter 12

I have pieced together the comings and goings of the scientists involved in the fullerene drama in September 1990 from a number of sources, including a telephone interview with Smalley;[42] a sequence of faxes and letters from, and telephone conversations with Kroto;[10,55,156,157] a letter from Walton;[158] a letter from Fostiropoulos;[37] my interview with Krätschmer;[26] a telephone interview with Whetten;[159] a letter from Diederich;[97] and Bethune's personal account.[146] This chapter demonstrates that the real value of the international scientific conference circuit lies not so much in the opportunity it affords scientists to hear about the latest research efforts of their colleagues (or the opportunities it affords for travel). It is all about meeting colleagues with shared scientific interests, maintaining existing friendships, and building new ones. Modern information technology might have shrunk the globe, but when it comes to spreading news of a scientific breakthrough there is still nothing quite like a couple of conferences.

The Krätschmer–Lamb–Fostiropoulos–Huffman paper was published on September 27, 1990,[160] and acted as a trigger for the publication of a sequence of articles confirming and extending the results. There were the papers from the IBM group reporting the Raman spectra,[155] carbon-13 NMR spectra[161] and STM images;[162] the Sussex group's paper containing the one-line proof and the five-line NMR spectrum for C_{70};[163] the UCLA group's paper;[164] the Rice group's paper;[165] and the STM images obtained by the group in Missouri.[166] Detailed NMR studies by the IBM group were reported early in 1991.[167] The crystal structure of $C_{60}(OsO_4)$ (4-*tert*-butylpyridine)$_2$ was reported by Hawkins and his colleagues at Berkeley in April 1991.[168]

Chapter 13

As I was working on this book, I was fortunate to receive copies of Smalley's buckminsterfullerene bibliography, which I used in a (feeble) attempt to keep track of the flood of papers and to estimate the volumes of these publications in 1990 and 1991. My most recent version of this bibliography, now obtainable through the Arizona Fullerene Consortium, is dated January 1, 1993.[169] Readers interested to learn more of this bibliography should contact Frank A. Tinker, University of Arizona, Physics Building #81, Tucson, Arizona 85721, USA. Tinker can be contacted via the Internet electronic mail network on Tinker@physics.arizona.edu. A 'bucky' update service is available from John Fischer at Pennsylvania State University: contact bucky@sol1.lrsm.upenn.edu. I picked up the comment about correspondence with the US Patents Office from a book review by Thomas Ebbesen which appeared in January 1993.[170]

Details of Wudl's benchtop fullerene reactor can be found in the *Journal of Organic Chemistry*.[171] Commercial sources of fullerenes are listed in the buckminsterfullerene bibliography.[169]

A description of Michael Faraday's early experimental work on benzene can be found in the recent book by Thomas.[172] Controversy still surrounds the elucidation of benzene's cyclic structure. I have given the 'traditional' historial view which credits August Kekulé and his dreams. Others argue that Kekulé was, at best, influenced by the ideas of Austrian schoolteacher Josef Loschmidt and, at worst, plagiarized them. Still others credit 'forgotten genius' A. S. Couper. For a recent exchange of words on this subject, see the February and May 1993 issues of *Chemistry in Britain*.[173]

The paper by the group at AT&T Bell Laboratories describing their theoretical and experimental results on the magnetic susceptibility of C_{60} and C_{70} was published in March, 1991.[174] A useful review of this work and its relevance to the question of aromaticity in the fullerenes was provided by Patrick Fowler in the same issue.[175] A subsequent revision of the status of C_{60} as an aromatic molecule was published by the Bell scientists in September 1992[176] and March 1993.[177] I am grateful to Robert Haddon for drawing these papers to my attention.[178]

In reviewing the early work on the chemistry of C_{60}, I gratefully acknowledge the extremely useful summaries provided at regular intervals through 1991–93 by Rudy Baum for *Chemical and Engineering News*.[179–187] A similar summary by Leonard Lindoy was also useful.[188] In addition to these, I have drawn on the original literature and review articles relating to radical reactions of C_{60},[189,190] the potential for C_{60} as a lubricant,[191,192] its reactions with halogens,[193–198] 'inflation' reactions to give fulleroids,[199–201] C_{60}-containing polymers,[200,202] and the organometallic chemistry of C_{60} and C_{70}.[203–208] The chemistry of the fullerenes has been reviewed by Taylor and Walton.[209] I also acknowledge comments on a draft of this chapter made by Walton,[210] Wudl,[211] and Kroto.[212] It seems that the International Union of Pure and Applied Chemistry (IUPAC) has now ruled that fullerene derivatives formed by addition across a bond at the fusion of two six-membered rings will henceforth be called 'methanofullerenes'. Derivatives formed by addition across a bond at the fusion of a five- and six-membered ring will be called 'fulleroids'.[211]

My review of the production and characterization of metal-encapsulating fullerenes is based on press articles by Baum and Dagani,[181] and Baum,[182,185] and original papers.[213–221] Wudl[211] informed me of the successful isolation and characterization of La@C_{82} by Yohji Achiba and his colleagues at the Tokyo Metropolitan University. Bethune[148] informed me of the IBM group's success in isolating fullerenes containing scandium atoms.[222]

Chapter 14

I have derived much of the background history and science of superconductivity presented in this chapter from a series of books and recent review articles.[223–232] In May 1993, the new record transition temperature for a copper oxide ceramic superconductor was set at 133 kelvin by Schilling, Cantoni, Guo, and Ott at the Solid State Physics Laboratory, Swiss Federal Institute in Hönggerberg.[233]

The wager between Haddon and Raghavachari, and the Bell scientists' early work on doped fullerenes, were described to me in letters from Haddon.[178,234] The informal seminar that Haddon gave to the Materials Discussion Group was delivered on October 18, 1990.[235] The Bell group's report on the preparation and conductivity of alkali metal doped films of C_{60} and C_{70} was published in *Nature* in March 1991.[236] Philip Ball's reaction to the news that potassium-doped C_{60} films had been found to be superconducting was described in his letter.[153] The report on superconductivity in K_3C_{60} was published in *Nature* in April 1991,[237,238] and discussed at the American Chemical Society meeting in Atlanta.[239] Reports of superconductivity at 33 and 45 kelvin appeared later that year,[240,241] although the latter claim was subsequently retracted.[242] The work on conductivity and superconductivity in the fullerides was reviewed in 1992 by Haddon.[243]

Chapter 15

The idea of the 'paradigm', put to good use by Thomas Kuhn in his book *The structure of scientific revolutions*[244] became very trendy in the 1970s, so trendy that Kuhn himself regretted ever having used it. It has become a term to describe the way we are conditioned to think about something, in the sense that we may develop a 'mind-set' which makes it difficult for us to accept ideas that do not fit. For example, for the physicist in the 1920s conditioned to think of the behaviour of atoms only in terms of classical mechanics, the quantum revolution would have been profoundly disturbing.

Kuhn's philosophy of science, based on the distinction between 'normal' and 'revolutionary' scientific activity, belongs in a category known as *social constructivism*. In essence, scientists share a consensus (socially constructed) world view—the paradigm—which conditions both they way they approach scientific problems and they ways in which they interpret the results of experiments. Transcending the boundaries of the accepted paradigm by accident is a recognized way of achieving progress in science. This is so nicely depicted in the cartoon by Peter Schrank (drawn for an article of mine which was published in *New Scientist* in March, 1990[245]) that I purchased his original. This now hangs proudly in my living room.

I described Fowler and Manolopoulos's theoretical predictions[246–248] for the structures of C_{76}, C_{78}, and C_{84} in *New Scientist*.[249] Some background on the importance of the Hückel calculations and the application of Fowler's 'leapfrog' principle to these fullerene structures was provided in letters to me from Fowler and Manolopoulos.[250,251] Diederich and Whetten and their UCLA colleagues, together with Wudl and his colleagues in Santa Barbara, described the isolation and characterization of C_{76}, C_{84}, C_{90}, C_{94}, and $C_{70}O$ in the journal *Science* in April 1991.[252] They announced the structural determination of chiral C_{76} in the pages of *Nature* in September 1991.[253,254] Hawkins and Meyer announced that they had successfully separated the two mirror image forms of C_{76} in June 1993.[255,256] Diederich and Whetten have summarized their early work on the higher fullerenes in a review article published in 1992.[257] Kikuchi and Achiba and their colleagues at the Tokyo Metropolitan University described their results on C_{78}, C_{82}, and C_{84} in *Nature* in May 1992.[258]

Iijima's paper announcing the discovery of carbon nanotubes was published in *Nature* in November 1991,[259] and a method of bulk synthesis was announced in July 1992.[260,261] Dresselhaus[262] compared the method to Bacon's 1960 synthesis of graphite whiskers.[263] Iijima and Ichihashi[264] and scientists at IBM and the California Institute of Technology[265] described methods to produce large quantities of single-walled nanotubes in June 1993.[266] Iijima regards this as one of the most important developments in nanotube science.[267]

I first saw Ugarte's stunning electron micrographs during my visit with Krätschmer in Heidelberg in August 1992. They appeared in *Nature* the following October,[268] together with Kroto's commentary.[269] Baum reviewed the status of the science of nanotubes and hyperfullerenes in January 1993.[270] Through subsequent correspondence with Ugarte,[271,272] I received copies of several of his forthcoming publications.[273–277]

Following hard on the heels of these dramatic revelations, scientists at SRI International in California and Du Pont in Delaware announced they had found a way to encapsulate crystals of lanthanum carbide inside the hyperfullerenes.[278,279] Similar findings were reported by other scientists.[280,281] Weeks later, Ajayan and Iijima explained how they had filled the insides of carbon nanotubes with lead.[282,283] The results of experimental studies of the mechanisms by which the end caps of the tubes can be opened have subsequently been reported,[284,285] as have studies of the formation of the nanotubes[286] and its relationship to the formation of the fullerenes.[287] The possibility of negatively curved ('inside-out') fullerenes was raised by Leonosky, Gonze, Teter, and Elser in January 1992.[288]

Chapter 16

Smalley described his modified 'party line' mechanism at the Materials Research Society symposium in Boston in November 1990,[289] and in subsequent review articles, one of which was co-authored with Curl.[112,217] Heath described his mechanism at the American Chemical Society national meeting in Atlanta in 1991.[290] Curl has reviewed these, and other mechanisms.[291] Experimental evidence for fullerene formation through the collisonal heating and collapse of carbon rings was presented by von Helden, Gotts, and Bowers in May 1993.[292] Hunter, Fye, and Jarrold have also described the annealing of non-fullerene C_{60}^{+} ions into the closed-cage fullerene structure and what appears to be a large monocyclic ring.[293] Baum[294] has reviewed these results. Diederich and Whetten and their UCLA colleagues published evidence for the coalescence reactions of fullerenes in September 1992.[295] McElvany, Callahan, Ross, Lamb, and Huffman published their results on large odd-numbered carbon clusters in *Science* in June 1993.[296]

The MIT scientists led by Jack Howard reported their observation of C_{60} and C_{70} in flames in July 1991.[297] Smalley pondered on fate of all the buckminsterfullerene that had been produced in early cave fires during the symposium in Boston.[289] The Sussex group reported the results of their study of the photodegradation of C_{60} in *Nature* in May 1991.[298]

The first sighting of fullerenes in the geological environment was reported by Busek, Tsipursky, and Hettich in July 1992.[299,300] The announcement of the discovery of fullerenes in glassy rock formed by lightning was made nine months later.[301]

Huffman's intention to go right back to the original purpose of their 1982–83 experiments with the Heidelberg evaporator was described to me by Krätschmer.[26] Hare and Kroto's review of galactic carbon following the discovery of the fullerenes was published in March, 1992.[302] The results of a search for fullerenes in the Murchison and Allende meteorites were reported by the IBM/NASA Ames group in 1993.[303] Evidence for the presence of nanotubes and hyperfullerenes in the Allende meteorite was summarized by Becker, McDonald, and Bada in February, 1993.[304] de Heer and Ugarte have presented evidence linking the 217-nanometre interstellar band with hyperfullerenes containing between two to about eight graphitic shells in a paper to be published in *Chemical Physics Letters*.[305] The use of high pressures to squeeze solid C_{60} into polycrystalline diamond was described in *Chemical Engineering* in July 1992.[306]

The last word

The world of science does not stand and wait for the somewhat slow machinations of the publishing industry. Since submitting the final draft of the manuscript of this book to Oxford University Press in August 1993, there have been a number of notable developments in fullerene science. My publisher has graciously agreed to allow me the opportunity to tell you about these in this final 'stop press' section, added at the page proof stage in May 1994.

It is probably fair to say that the field has now gone relatively quiet, at least in terms measured by the appearance of breathless announcements or amazing stories in the pages of *Nature* or *Science*. This is to be expected. A period of quiet introspection is required to give the scientists time to draw breath, and to concentrate on the nitty gritty of their various specialisms. Beneath the placid exterior there is considerable activity in laboratories all around the world, manifested in a continually expanding fullerene literature in the specialist scientific press.

Apart from the few further advances described briefly below, a major concern of the new breed of fullerene scientists has been to find answers to the question: 'Yes, very nice. But what is it *good* for?' To a certain extent, the importance being attached to questions like these reveals something of a sense of insecurity on the part of the scientists and their financial backers. It is a sign of the times that a discovery of this magnitude is nowadays always accompanied by speculation about its commercial potential which seems to vary only in its level of outrage. This wave of hype, aided and abetted by science journalists and popularizers (such as myself) only serves to raise expectations that cannot later be fulfilled.

The simple truth is that, with relatively rare exceptions, translating a discovery of this kind into a commercial proposition can take many years. Even in the more pragmatic environment of commercial R&D, projects focused on clearly identified and thoroughly researched target markets can take up to five years or more. The fullerenes, superconducting fullerides, nanotubes, and hyperfullerenes provided no instantly marketable solutions to any problems that may have existed in our technology-hungry world. This does not mean that commercial applications cannot be found. It just takes a while.

Buckminster Fuller understood this process only too well. In 1927, Fuller predicted that it would take 25 years to translate the ideas behind his energetic geometry into marketable structures. This gestation period was necessary in part, he reasoned, because in order to turn his designs into buildings likely to be of interest in the commercial world he needed lightweight materials that were not then widely available. Fuller was spot on. Ford Motor Company became the first industrial organization to become licensed under his patents exactly 25 years after he had made his prediction.

Whilst we may not have to wait 25 years for something useful or practical from the fullerenes, anything that has so far come out of the discoveries described in this book are still matters for study and (more considered and less outrageous) speculation. Each of the scientists involved have their own opinions, and Rudy Baum captured these in an excellent survey published in November 1993.[307]

Despite the cautious nature of most industrial researchers, there continues to be a flurry of activity in patent offices around the world. Various patents have been either granted to or applied for by Exxon, AT&T, Xerox, Hughes, Hoechst, Sumitomo, Materials and Electrochemical Research, Rice University, and many others. The original Huffman–Krätschmer patent, filed in 1990 by Research Corporation Technologies on behalf of the University of Arizona and the Max Planck Society, is split into two parts. The first is a very broad application covering the production process and the second is a 'composition-of-matter' application on the fullerite itself. The patent has still not yet been granted,[308] a situation that some are claiming is holding up progress toward commercializing the fullerenes. It could be that the US Patent Office recognizes that these are landmark patents, with potentially vast implications for the development of new products. In the meantime, Fostiropoulos has gone public with a claim of co-inventorship.[309] Through a patent lawyer, he lodged an official complaint with the German Patent Office in August 1993.

Solar powered

Readers that may have become somewhat concerned about the environmental impacts of all those energy-intensive arc discharges now being employed to produce fullerenes in large quantities should take heart. Smalley and his colleagues at Rice University and a group at the National Renewable Energy Laboratory in Golden, Colorado, have both reported a new method of fullerene synthesis using a solar furnace.[310–312]

Dark grey, aromatic

The status of buckminsterfullerene as an aromatic molecule is now reasonably well confirmed. Studies of the ^{3}He NMR spectroscopy of $^{3}He@C_{60}$ and $^{3}He@C_{70}$ by a team of chemists and physicists at Yale University and UCLA revealed the existence of large ring currents characteristic of delocalised electrons.[313] These results were reconciled by Haddon[314] in terms of extensive theoretical and experimental work that points to a near cancellation of diagmagnetic ring currents in C_{60}'s hexagons by paramagnetic ring currents in its pentagons. C_{70} shows a much larger net ring current contribution to the magnetic susceptibility simply because it has more hexagons in its structure.

What the fullerenes have shown is that the more traditional view of aromaticity as a black-or-white phenomenon—a molecule is either aromatic or it's not—has to be modified to include shades of grey. Buckminsterfullerene is a dark grey aromatic molecule. Haddon[315] has argued that the planar analogues (such as pyracyclene) used by chemists to characterize C_{60} as an electron-deficient 'super-alkene' miss some very significant points. A much better approach, he claims, is to view the reactivity of buckminsterfullerene as the result of its somewhat ambiguous aromatic character combined with the drive to relieve the strain inherent in its spherical geometry.

The development of the chemistry of the fullerenes has continued pretty much along the lines described in Chapter 13. Both organic and organometallic chemists have learned enough of the basic rules of fullerene reactivity to allow them to play some increasingly sophisticated games. The chemists have also used their knowledge to sharpen the procedures used to purify C_{60} and C_{70}, and to invent some new ones.[316]

Buckyballs and AIDS

One way of telling when a young area of chemistry is beginning to approach puberty is to see how long it takes chemists to respond successfully when challenged to make something they have not tried to make before. When approached by chemists at the University of California at San Francisco (UCSF) for a water soluble C_{60} derivative, Fred Wudl and his colleagues at Santa Barbara sat down and worked out a synthesis for di(phenethylaminosuccinate)fulleroid.[317] This work led to one of the most astounding fullerene stories of 1993: buckyballs inhibit the AIDS virus.

Graduate student Simon Friedman at UCSF hit on the idea of using buckminsterfullerene to block the active site in a key enzyme in the human immunodeficiency virus known as HIV-1 protease. Computer models showed that C_{60} would fit snugly inside the roughly cylindrical, hydrophobic (water-repelling) active site of the protease. As an all-carbon molecule, C_{60} is by nature hydrophobic, suggesting an intriguing match with the active site in terms of shape, size, and physical characteristics. Wudl's derivative offered some solubility, allowing C_{60} and the protease to interact in solution. In further experiments at UCSF by Friedman, George Kenyon and Diane DeCamp, the derivative was indeed found to be active against the enzyme.[318] Although the derivative is itself not a candidate for an anti-AIDS drug, it does provide a good starting point for the possible development of such a drug based on fullerene chemistry.

The same derivative was also found by Craig Hill and Raymond Schinazi at Emory University in Atlanta to be active against another HIV enzyme known as reverse transcriptase. The Emory scientists found it to inhibit reproduction of the HIV virus in immune cells both acutely and chronically infected with HIV. The anti-AIDS drug AZT is effective in suppressing only acutely infected cells.[319,320]

Inside stories

The list of atoms that have been encapsulated in a fullerene cage continues to be extended to more members of the periodic table, and results indicating that small molecules such as carbon monoxide can be similarly encapsulated have also been reported. The field is still somewhat hampered by the lack of a suitable method to make pure forms of the atom- or molecule-encapsulating fullerenes in bulk quantities. Apart from La@C_{82}, the molecules Sc_2@C_{74}, Sc_2@C_{82}, and Sc_2@C_{84} have all been prepared and purified in milligram quantities using high-performance liquid chromatography. Bethune and his IBM colleagues reviewed the field in November 1993.[321] Martin Jarrold and his colleagues at the Department of Chemistry at Northwestern University have proposed a mechanism for the formation of metal-encapsulating fullerenes in the gas phase.[322]

Smalley and his colleagues at Rice University continue their studies of the metal-encapsulating fullerenes.[323] They have succeeded in isolating quantities of calcium- and yttrium-containing C_{60} (not C_{82}!). In these molecules, the carbon cage accepts electrons from the central metal atom which go into high-energy empty electron orbitals (see p. 211). Ca@C_{60} and Y@C_{60} are electron-rich and readily oxidized, and therefore extremely sensitive to air and moisture. Smalley believes that their earlier attempts to isolate metal-encapsulating C_{60} molecules failed simply because they did not know how to handle such reactive compounds.

In yet another rather fitting tribute to the imaginings of Daedalus, Robert Murry and Gustavo Scuseria have presented theoretical evidence for the formation of 'windows' in electronically excited fullerenes.[324] These windows may allow atoms present in the gas phase to enter the cages before they close up again as they lose their electronic excitation.

Superconductors heat up

Developments in high-temperature superconducting fullerides were rather overshadowed by the announcement, in December 1993, of superconductivity at an astounding 250 kelvin in a thin layer of bismuth–strontium–calcium–

copper oxide.[325,326] These results have yet to be independently confirmed but, if they can be reproduced, they reveal superconductivity taking place only 40 degrees or so below the holy grail of room temperature. Current trends in superconductivity were reviewed by Philip Yam in December 1993.[327]

Progress in raising in the transition temperature of fullerene derivatives was reported in September 1993. Yi-Han Kao and his colleagues at the State University of New York at Buffalo produced fullerenes doped with iodine monochloride which exhibited signs of superconductivity between 60–70 kelvin.[328,329] This is a *p*-type rather than an *n*-type superconductor, representing a whole new approach to the production of superconducting fullerene derivatives. The signs of superconductivity were tentative, but if they can be confirmed the field will have taken its own major step towards the all-important 77 kelvin target.

Nanotubes and more

A method of clearing out the carbonaceous dross produced in the manufacture of nanotubes, leaving only pure nanotubes behind, was reported in February 1994 by Ebbeson and his colleagues at NEC Corporation in Japan.[330] To these we must now add a couple more 'novel' forms of carbon: sea urchins and nanoworms.

Carbon sea urchins were discovered by a group of chemists from Du Pont's Experimental Station in Wilmington, Delaware and SRI International in Menlo Park, California. They are structures formed from the vaporization of graphite rods containing gadolinium or gadolinium oxide, and consist of central cores of either amorphous gadolinium carbide or single crystal GdC_2 with carbon nanotubes radiating outwards like . . . well, like a sea urchin.[331,332] Nanoworms are another discovery of the Du Pont group. Each structure consists of a 'head' made from a cubic palladium crystal inside a hyperfullerene and a segmented 'tail' of carbon tubes.[333]

Mesozoic fullerenes

Although nobody can say for sure how the fullerenes manage to assemble themselves, the scientists keep on coming up with ideas. Jarrold and his colleagues at Northwestern University added to the list of possible mechanisms in March 1994.[334,335] After studying the results of their experimental studies of large carbon clusters, the Northwestern scientists proposed that the large polycyclic rings rearrange at high temperatures to give polyyne chains spiralled into fullerene fragments. These fragments then 'zip up' to form closed, spheroidal fullerenes.

However they are formed, there is now some fairly conclusive evidence that the fullerenes are as ubiquitous as the original bucky pioneers thought they might be. Although Homann and Howard and their colleagues found evidence for the formation of C_{60} in sooting flames, these were generated under rather special laboratory conditions with little relevance to the production of soot from an average candle flame. Now Dieter Heymann at the Department of Geology and Geophysics at Rice University and Rick Smalley and their colleagues Felipe Chibante, Robert Brooks, and Wendy Wolbach have described an analytical method capable of detecting fullerenes in very low concentrations. They have confirmed that C_{60} is formed at concentrations between 1–10 ppm in soot from several free-burning flames, including flames from common decorative candles.[336] Something to think about next time you blow out the candles on your birthday cake.

But Heymann and Smalley and their colleagues were actually in search of older prey. About 65 million years ago, the extinction of the dinosaurs was triggered by a global cataclysmic event thought to be associated with the impact of a large meteor near what is now Chicxulub in the Yucatan Peninsula. The impact is thought to have caused massive wildfires which poured smoke and soot into the atmosphere, blocking the light from the sun. This event marks the end of the Mesozoic era. In geological terms, it is marked by a thin layer of soot at the Cretaceous/Tertiary (K/T) boundaries. The answer to your next question is yes. This layer contains some buckminsterfullerene.

Studies of geological samples from the K/T boundary layer taken from two sites in New Zealand revealed the presence of C_{60} at the level of 0.1–0.2 ppm.[336] No C_{60} could be detected in samples of Cretaceous limestone just below, or Tertiary shale just above the boundary layer. Fullerenes have also been found in 'abundant' quantities in samples from the K/T boundary layer taken from a site in Spain.[323] Although it is not impossible that this could be extraterrestrial C_{60} carried to earth by the meteor, the scientists believe it is more likely that the C_{60} was generated by the wildfires in the aftermath of the meteor impact. Smalley[323] likes to think that these 65 million-year-old fullerenes contain the last gasp of the dinosaurs.

Fullerenes in space

One last question: are there fullerenes in space? A definitive answer was provided in May 1994 by Filippo Radicati di Brozolo, Theodore E. Bunch, Ronald H. Fleming, and John Macklin.[337] They studied a tiny crater formed by the impact of a carbonaceous micrometeoroid on an aluminium panel of the Long Duration Exposure Facility (LDEF) spacecraft. The LDEF was retrieved in 1990 by astronauts of the space shuttle *Columbia* after six years in earth orbit, and sections were removed and studied in a 'clean room' environment. Analysis of the area of aluminium panel in and around the impact crater by laser ionization combined with time-of-flight mass spectrometry revealed the characteristic signatures of C_{60} and C_{70}. The scientists explored various explanations for the origin of the fullerenes, eliminating all the obvious potential experimental artefacts. Although they believe theirs is the first direct observation of fullerenes in space, they couldn't distinguish between two possible sources. Either the fullerenes were already present as components of the micrometeoroid or the fullerenes were formed during its impact.

Kroto[338] is inclined to believe that the impact of the micrometeroroid may recreate conditions obtained in more earth-bound experiments in which lasers have been used to drill holes in graphite targets. In these experiments, no C_{60} can be observed initially, but as the hole is bored deeper and deeper the atoms and ions formed in the plasma at the bottom have the opportunity to cluster together as they make their way to the top. Nevertheless, no matter how they are being formed in space, the fullerenes would at least seem to be *there*.

But what about the diffuse interstellar bands, the 'last great problem in astronomy' that led the scientists gathered at Rice in September 1985 to make their great discovery? The diffuse bands now number about 200, and B. H. Foing and P. Ehrenfreund added a further two near infra-red bands in May 1994.[339] These bands, detected in the light from seven stars at 9.577 and 9.632 microns, coincide with two bands observed in matrix-isolated C_{60}^{+} after correcting for the distorting effects of the inert gas matrix. Foing and Ehrenfreund estimate that about 0.3–0.9 per cent of interstellar carbon is present in the form of ionized C_{60}. Kroto has acclaimed these new results,[340] and hopes that he will soon be able to say with complete justification what he has always believed: that C_{60} is indeed a 'celestial sphere'.

Eyes on the Prize

The hallmark of the Great Victorian Novel is the epilogue, which describes the ultimate fates of the characters that have absorbed our intellectual energies for the last 500-odd pages. So what, we may ask, has become of our particular cast of characters, nearly nine years after the discovery of buckminsterfullerene?

Together with his colleagues at the University of Sussex, Harry Kroto is trying hard to stay at the forefront of research in fullerene chemistry, and is taking an increasing interest in the nanotubes. He feels that his work is going very well, but the competition is brutal and there is constant pressure to rush things into print. He keeps up with his interests in art, and was recently awarded a prize in a Science for Art competition.[338]

Apart from continuing with his interests in the metal-encapsulating fullerenes and searching for ancient C_{60}, Rick Smalley is busily establishing a major initiative on nanotechnology at Rice, with some $40 million of mostly private funds.[323] The aim of this initiative is to nurture the development of nanoscale science across all the scientific and engineering disciplines, using the new funding to reinvigorate some existing research programmes, set up some new ones and bring them all together. He sees the carbon nanotubes as central to the initiative, and claims to be close to producing single continuous fullerene fibres. These would find essentially instant applications as nanowires, nanopipettes, and 'proximate probes', nanometre-sized extensions of human fingertips which can 'touch' and 'feel' their way around structures such as living cells, transmitting information back along their lengths.

Smalley has been fascinated to watch the fullerene idea catch fire around the world. He believes that the C_{60} experience has made chemical physicists think about their field in a different way. No longer content with the why and how of internal molecular workings, more and more chemical physicists of Smalley's acquaintance have become interested in making or engineering things.

Bob Curl has drawn back from a big personal commitment to fullerene science, preferring to continue with the kind of science that interested him before the 1985 discovery and watching the fullerenes from the sidelines. He enjoyed being involved in the discovery and its aftermath, and kicks himself when he gets up every morning for not having pursued the potential for fullerene synthesis in an electrical discharge in the years before the Krätschmer–Lamb–Fostiropoulos–Huffman paper was published in *Nature*.[341]

Jim Heath, having spent some time at IBM's research centre in Yorktown Heights, is now back in California, at UCLA's Department of Chemistry and Biochemistry. He has also shied away from a big commitment to fullerene science, and instead concentrates his energies on problems associated with the controlled synthesis of very small semiconductor structures in solution. He looks back on the last nine years with a real sense of excitement at having been involved. But he also acknowledges that helping to establish a whole new field of science in his first year of graduate school has been a difficult act to follow.[342]

After completing his Ph.D. in 1988, Sean O'Brien worked with Jim Kinsey at Rice before moving to Texas Instruments' Semiconductor Process and Design Center in Dallas. He is currently developing new contamination removal techniques for CMOS technology.[343] He remains pretty impressed with what he and the other members of the 'soccer team' at Rice were able to achieve. Although he recognizes that commercialization of the fullerenes will be a long, drawn-out process, he is nonetheless disappointed that practical applications have not yet appeared.

Like Bob Curl, Wolfgang Krätschmer looks back over his involvement in the discovery of fullerite with a sense of having had some fun, despite all the setbacks. He is satisfied that he has helped to give the scientific community so much work to do, and is now content to sit back and watch. He continues his collaboration with Don Huffman, nagging away at the interstellar connection.[308]

Huffman himself is vigorously pursuing various collaborative projects in the chemistry, physics, and materials science of fullerenes, particularly the higher fullerenes and the nanotubes. He claims that he still finds it all great fun, providing a lot of pleasure and excitement, although he admits that he can no longer keep up with the inflated and increasingly specialized fullerene literature. Huffman is most outspoken over the issue of commercialization. He believes that the lack of a practical application is now seriously hurting, with peer reviewers and funding agencies and their advisers all stubbornly asking the same unhelpful questions.[344]

Konstaninos Fostiropoulos left Heidelberg in 1992 and moved to the Max Planck Institute for Radioastronomy in Jena. I had received reports that he was now working at the Fritz-Haber Institute in Berlin, but when I attempted to contact him there I was told that he was back in Heidelberg. I was unable to reach him in time for this stop press section.

Lamb has remained in Tucson as an assistant director of the Arizona Fullerene Consortium. Although he has been very happy in Tucson and feels that he has been treated very well, he recognizes the need to move on and is preparing to plunge himself into the academic job market. Echoing Heath, he admits that this is not a good time for science in the US and is wary of what he will find.[345]

Kroto, Smalley, Huffman, and Krätschmer came together for the very first time in Madrid in March, 1994, to receive the Hewlett-Packard Europhysics Prize, for outstanding achievements in condensed matter physics.[346] It was by and large a happy meeting.[308] Although the relationship between Kroto and Smalley functions satisfactorily on a purely scientific level, on a personal level the damage done over the last seven years would appear to be irreparable.[338] O'Brien would like to see a return to the happier days of late 1985.[343]

The next major event to impact the lives of these scientists will be the announcement of the Nobel Prize, which must come eventually. The Prize can be shared by only three scientists, and so the most difficult decision for the Swedish Academy and its advisers will be who to leave out.

References

1. Gribbin, John (1981). *Genesis*. Dell Publishing, New York.

2. Hawking, Stephen W. (1988). *A brief history of time*. Transworld Publishers, London.

3. Oparin, A. I. (1953). *Origin of life*, (trans. Serguis Morgulis). Dover Publications, New York.

4. Cheung, A. C., Rank, D. M., Townes, C. H., Thornton, D. D., and Welch, W. J. (1968). Detection of NH_3 molecules in the interstellar medium by their microwave emission. *Physical Review Letters*, **21**, 1701.

5. Turner, B. E. (1971). Detection of interstellar cyanoacetylene. *Astrophysical Journal*, **163**, L35.

6. Hoyle, Fred and Wickramasinghe, Chandra (1978). *Lifecloud*. J.M. Dent and Sons, London.

7. Hoyle, Fred and Wickramasinghe, Chandra (1993). *Our place in the cosmos: the unfinished revolution*. J.M. Dent, London. (See the review by Tony Jones in *New Scientist*, July 10, 1993.)

8. Kroto, Harold W. (1992). C_{60}: buckminsterfullerene, the celestial sphere that fell to earth. *Angewante Chemie (international edition)*, **31**, 111.

9. Kroto, Harry. Notes prepared for Jim Baggott by Harry Kroto. March 12, 1991.

10. Kroto, Harry. Letter to the author. January 4, 1993.

11. Kroto, Harry. Interview with the author. November 16, 1991.

12. Alexander, A. J., Kroto, H. W., and Walton, D. R. M. (1976). The microwave spectrum, substitution structure and dipole moment of cyanobutadiyne, H—C≡C—C≡C—C≡N. *Journal of Molecular Spectroscopy*, **62**, 175.

13. Kirby, C, Kroto, H. W., and Walton, D. R. M. (1980). The microwave spectrum of cyanohexatriyne, H—C≡C—C≡C—C≡C—C≡N. *Journal of Molecular Spectroscopy*, **83**, 261.

14. Eastmond, R., Johnson, T. R., and Walton, D. R. M. (1972). Silylation as a protective method for terminal alkynes in oxidative couplings. A general

synthesis of the parent polyynes H(C≡C)$_n$H ($n = 4$–10, 12). *Tetrahedron*, **28**, 4601.

15. Herbst, Eric and Klemperer, William (1973). The formation and depletion of molecules in dense interstellar clouds. *Astrophysical Journal*, **185**, 505.

16. Dalgarno, A. and Black, J. H. (1976). Molecule formation in the interstellar gas. *Reports on Progress in Physics*, **39**, 573.

17. Kroto, Harold. Chemistry between the stars. *New Scientist*, August 10, 1978.

18. Kroto, H. W. (1982). Tilden Lecture: Semistable molecules in the laboratory and in space. *Chemical Society Reviews*, **11**, 435.

19. Bell, M. B., Feldman, P. A., Kwok, Sun, and Matthews, H. E. (1982). Detection of $HC_{11}N$ in IRC+10°216. *Nature*, **295**, 389.

20. Hintenberger, H, Franzen, J., and Schuy, K. D. (1963). Die periodizitaten in den häufigkeitsverteilungen der positiv und negativ geladenen vielatomigen kohlenstoffmolekülionen C_n^+ und C_n^- im hochfrequenzfunken zwischen graphitelektroden. *Zeitschrift für Naturforschung A*, **18**, 1236.

21. Huffman, D. R. (1977). Interstellar grains: The interaction of light with a small particle system. *Advances in Physics*, **26**, 129.

22. Wynn-Williams, Gareth (1992). *The fullness of space*. Cambridge University Press.

23. Henbest, Nigel. Dust in space, *Inside Science No. 45. New Scientist*, May 18, 1991.

24. Douglas, A. E. (1977). Origin of diffuse interstellar bands. *Nature*, **269**, 130.

25. Molecules with sunglasses (produced by John Lynch), first shown on *Horizon*, BBC2, January 20, 1992.

26. Krätschmer, Wolfgang. Interview with the author. August 28, 1992.

27. Huffman, Donald. Telephone interview with the author. November 4, 1992.

28. Krätschmer, W. (1991). How we came to produce C_{60}-fullerite. *Zeitschrift für Physik D*, **19**, 405.

29. Krätschmer, Wolfgang and Fostiropoulos, Konstantinos. Fullerite - neue modifikationen des kohlenstoffs. *Physik in Unserer Zeit*. January 23, 1992.

30. Huffman, Donald R. Solid C_{60}. *Physics Today*. November 1991.

31. Krätschmer, Wolfgang and Huffman, Donald R. Fullerites: new forms of crystalline carbon. Preprint. August 1992.

32. Stecher, Theodore P. (1969). Interstellar extinction in the ultra-violet. II. *Astrophysical Journal*, **157**, L125.

33. Heidelberg. City Guide in Colour. (1991). Edm. von König-Verlag, Heidelberg.

34. Krätschmer, W., Sorg, N., and Huffman, Donald R. (1985). Spectroscopy of

matrix-isolated carbon cluster molecules between 200 and 850 nm wavelength. *Surface Science*, **156**, 814.

35. El Goresy, A. and Donnay, G. (1968). A new allotropic form of carbon from the Ries Crater. *Science*. **161**, 363.

36. Whittaker, A. Greenville (1978). Carbon: a new view of its high temperature behaviour. *Science*. **200**, 763.

37. Fostiropoulos, Kosta. Letter to the author. January 1993.

38. Krätschmer, Wolfgang. Letter to the author. January 18, 1993.

39. Curl, Bob. Letter to the author. February 18, 1992.

40. Curl, Bob. Interview with the author. September 30, 1992.

41. The Smithsonian guide to historic America: Texas and the Arkansas River Valley. (1990). Stewart, Tabori, and Chang, New York.

42. Smalley, Rick. Telephone interview with the author. January 12, 1993.

43. Dietz, T. G., Duncan, M. A., Powers, D. E., and Smalley, R. E. (1981). Laser production of supersonic metal cluster beams. *Journal of Chemical Physics*, **74**, 6511.

44. Powers, D. E., Hansen, S. G., Geusic, M. E. Pulu, A. C., Hopkins, J. B., Dietz, T. G. Duncan, M. A., and Smalley, R. E. (1982). Supersonic metal cluster beams: laser photoionization of Cu_2. *Journal of Physical Chemistry*, **86**, 2556.

45. Michaelopoulos, D. L., Geusic, M. E., Hansen, S. G., Powers, D. E., and Smalley, R. E. (1982). The bond length of Cr_2. *Journal of Physical Chemistry*, **86**, 3914.

46. Hopkins, J. B., Langridge-Smith, P. R. R., Morse, M. D., and Smalley, R. E. (1982). Supersonic metal cluster beams of refractory metals: spectral investigations of ultracold Mo_2. *Journal of Chemical Physics*, **78**, 1627.

47. Michaelopoulos, D. L., Geusic, M. E., Langridge-Smith, P. R. R., and Smalley, R. E. (1983). Visible spectroscopy of jet-cooled SiC_2: geometry and electronic structure. *Journal of Chemical Physics*, **80**, 3556.

48. Heath, J. R., Liu, Yuan, O'Brien, S. C., Zhang, Qing-Ling, Curl, R. F., Tittel, F. K., and Smalley, R. E. (1985). Semiconductor cluster beams: one and two colour ionization studies of Si_x and Ge_x. *Journal of Chemical Physics*, **83**, 5520.

49. O'Brien, S. C., Liu, Y., Zhang, Q., Heath, J. R., Tittel, F. K., Curl, R. F., and Smalley, R. E. (1986). Supersonic cluster beams of III-V semiconductors: Ga_xAs_y. *Journal of Chemical Physics*, **84**, 4074.

50. Rohlfing, E. A., Cox, D. M., and Kaldor, A. (1984). Production and characterization of supersonic carbon cluster beams. *Journal of Chemical Physics*, **81**, 3322.

51. Rohlfing, Eric A. Letter to the author. October 14, 1992.

52. Smalley, Richard E. (1991). Great balls of carbon: the story of buckminsterfullerene. *The Sciences*, **31**, 22.

53. Baggott, Jim. Great balls of carbon. *New Scientist*, July 6, 1991.

54. Kroto, H. W. (1986). Chemistry between the stars. *Proceedings of the Royal Institution*, **58**, 45.

55. Kroto, Harold. Letter to the author. April 21, 1991.

56. Kroto, Harold. Letter to Paul Hoffman (*Discover* magazine). September 13, 1990.

57. Smalley, Rick. Letter to the author. February 13, 1991.

58. Smalley, Rick. Letter to the author. April 11, 1991.

59. Smalley, R. E. Letter to Paul Hoffman (*Discover* magazine). August 28, 1990.

60. Smalley, Rick. Letter to Gary Taubes (*Discover* magazine). September 5, 1990.

61. Kroto, Harold. Letter to the author. July 8, 1993.

62. Curl, Robert. Letter to the author. April 12, 1991.

63. Kurosawa, Akira and Richie, Donald (1987). *Rashomon*. Rutgers University Press.

64. Smalley, Rick. Letter to the author. June 13, 1991.

65. Smalley, Rick. Letters to Kroto, Curl, Heath, and O'Brien. May 31 and June 6, 1991.

66. Smalley, Rick, Heath, Jim, Curl, Bob and O'Brien, Sean. AP2 chronology 1985. May 1992.

67. Heath, Jim. Letter to the author. October 2, 1992.

68. O'Brien, Sean. Letter to the author. July 1, 1993.

69. Alford, Mike. Interview with the author. September 18, 1992.

70. Taubes, Gary (1992). The disputed birth of buckyballs. *Science*, **253**, 1476.

71. Wolfe, Tom (1980). *The right stuff*. Black Swan, London.

72. Heath, J. R., Zhang, Q., O'Brien, S. C., Curl, R. F., Kroto, H. W., and Smalley, R. E. (1987). The formation of long carbon chain molecules during laser vaporization of graphite. *Journal of the American Chemical Society*, **109**, 359.

73. Kroto, H. W., Heath, J. R., O'Brien, S. C., Curl, R. F., and Smalley, R. E. (1987). Long carbon chain molecules in circumstellar shells. *Astrophysical Journal*, **314**, 352.

74. Marks, Robert and Fuller, R. Buckminster (1973). *The Dymaxion world of Buckminster Fuller*. Anchor Press/Doubleday, New York.

75. Smalley, R. E. in Billups, W. Edward and Ciufolini Marco, A. (eds.) (1993). *Buckminsterfullerenes*, VCH Publishers, Inc.

76. Kroto, H. W., Heath, J. R., O'Brien, S. C., Curl, R. F., and Smalley, R. E. (1985). C_{60}: Buckminsterfullerene. *Nature*, **318**, 162.

77. Smalley, R. E. Letter to the author. January 12, 1993.

78. Heath, J. R., O'Brien, S. C., Zhang, Q., Liu, Y., Curl, R. F., Kroto, H. W., Tittel, F. K., and Smalley, R. E. (1985). Lanthanum complexes of spheroidal carbon shells. *Journal of the American Chemical Society*, **107**, 7779.

79. Browne, Malcolm. Molecule is shaped like a soccer ball. *New York Times*. December 3, 1985.

80. Baum, Rudy M. Laser vaporization of graphite gives stable 60-carbon molecules. *Chemical and Engineering News*, December 23, 1985.

81. Jones, David (writing as Daedalus). Ariadne. *New Scientist*, November 3, 1966.

82. Jones, David E. H. (1982). *The inventions of Daedalus: A compendium of plausible schemes*. W.H. Freeman, New York.

83. Devlin, Keith (1988). *Mathematics: The new Golden Age*. Penguin, London.

84. Thompson, D'Arcy (1961). *On growth and form* (abridged edn) (Bonner, J. T., ed.). Cambridge University Press.

85. Jones, D. E. H. Dreams in a charcoal fire. Abstract for the Discussion Meeting, *A postbuckminsterfullerene view of the chemistry, physics and astrophysics of carbon*. Royal Society, October 1-2, 1992.

86. Jones, David E. H. (1993). Dreams in a charcoal fire: predictions about giant fullerenes and graphite nanotubes. *Philosophical Transactions of the Royal Society of London, A*, **343**, 9.

87. Meller, James (ed.) (1972). *The Buckminster Fuller reader*. Penguin, London.

88. Fuller, R. Buckminster (1969). *50 years of the design science revolution and the world game*. World Resources Inventory, Philadelphia.

89. A Fuller Supply Service. Roger Golten, 15 Water Lane, Kings Langley, Herts. WD4 8HP, England.

90. Jencks, Charles (1985). *Modern movements in architecture* (2nd edn). Penguin, London.

91. Barth, Wayne E. and Lawton, Richard G. (1966). Dibenzo[*ghi,mno*]fluoranthene. *Journal of the American Chemical Society*, **88**, 380.

92. Barth, Wayne E. and Lawton, Richard G. (1971). The synthesis of corannulene. *Journal of the American Chemical Society*, **93**, 1730.

93. Osawa, E. The evolution of the football structure for the C_{60} molecule: an historical perspective. Abstract for the Discussion Meeting, *A postbuckminsterfullerene view of the chemistry, physics and astrophysics of carbon*. Royal Society, October 1-2, 1992.

94. Yoshida, Z. and Osawa, E. (1971). *Aromaticity*. Kagakudojin, Kyoto.

95. Stankevich, I. V., Nikerov, M. V., and Bochvar, D. A. (1984). The structural chemistry of crystalline carbon: geometry, stability and electronic spectrum. *Russian Chemical Reviews*, **53**, 640.

96. Davidson, Robert A. (1981). Spectral analysis of graphs by cyclic automorphism subgroups. *Theoretica Chimica Acta*, **58**, 193.

97. Diederich, François. Letter to the author. November 6, 1992.

98. Hughes, R. E., Kennard, C. H. L., Sullenger, D. B., Weakliem, H. A., Sands, D. E., and Hoard, J. L. (1963). The structure of β-rhombohedral boron. *Journal of the American Chemical Society*, **85**, 361.

99. Bunker, M. J. Letter to the author. February 24, 1992.

100. Zhang, Q. L., O'Brien, S. C., Heath, J. R., Liu, Y., Curl, R. F., Kroto, H. W., and Smalley, R. E. (1986). Reactivity of large carbon clusters: spheroidal carbon shells and their possible relevance to the formation and morphology of soot. *Journal of Physical Chemistry*, **90**, 525.

101. Curl, Robert F. and Smalley, Richard E. (1988). Probing C_{60}. *Science*, **242**, 1017.

102. Kroto, Harold (1988). Space, stars, C_{60} and soot. *Science*, **242**, 1139.

103. Haymet, A. D. J. (1985). C_{120} and C_{60}: Archimedean solids constructed from sp^2 hydridized carbon atoms. *Chemical Physics Letters*, **122**, 421.

104. Haymet, A. D. J. (1986). Footballene: A theoretical prediction for the stable, truncated icosahedral molecule C_{60}. *Journal of the American Chemical Society*, **108**, 319.

105. Schmalz, T. G., Seitz, W. A., Klein, D. J., and Hite, G. E. (1986). C_{60} carbon cages. *Chemical Physics Letters*, **130**, 203.

106. Cox, D. M., Trevor, D. J., Reichmann, K. C., and Kaldor, A. (1986). C_{60}La: A deflated soccer ball? *Journal of the American Chemical Society*, **108**, 2457.

107. Bloomfield, L. A., Geusic, M. E., Freeman, R. R., and Brown, W. L. (1985). Negative and positive cluster ions of carbon and silicon. *Chemical Physics Letters*, **121**, 33.

108. Hahn, M. Y., Honea, E. C., Paguia, A. J., Schriver K. E., Camarena, A. M. and Whetten, R. L. (1986). Magic numbers in C_N^+ and C_N^- abundance distributions. *Chemical Physics Letters*, **130**, 12.

109. O'Brien, S. C., Heath, J. R., Kroto, H. W., Curl, R. F., and Smalley, R. E. (1986). A reply to 'Magic numbers in C_N^+ and C_N^- abundance distributions' based on experimental observations. *Chemical Physics Letters*, **132**, 99.

110. Larsson, Sven, Volosov, Andrey, and Rosén, Arne (1987). Optical spectrum of the icosahedral C_{60}—'Follene-60'. *Chemical Physics Letters*, **137**, 501.

111. Heath, J. R., Curl, R. F., and Smalley, R. E. (1987). The UV absorption

spectrum of C_{60} (buckminsterfullerene): A narrow band at 3860 Å. *Journal of Chemical Physics*, **87**, 4236.

112. Curl, Robert F. and Smalley, Richard E. Fullerenes. *Scientific American*, October 1991.

113. Baum, Rudy M. Studies support spherical structure, aromaticity of C_{60} carbon clusters. *Chemical and Engineering News*, August 29, 1988.

114. Taubes, Gary. Great balls of carbon. *Discover*, September 1990.

115. Baggott, Jim. How the molecular football was lost. *New Scientist*, July 6, 1991.

116. Kroto, H. W. (1987). The stability of the fullerenes C_n, with $n = 24$, 28, 32, 36, 50, 60, and 70. *Nature*, **329**, 529.

117. Fowler, P. W. and Woolrich, J. (1986). π-systems in three dimensions. *Chemical Physics Letters*, **127**, 78.

118. Fowler, P. W. (1986). How unusual is C_{60}? Magic numbers for carbon clusters. *Chemical Physics Letters*, **131**, 444.

119. Fowler, P. W. and Steer, J. I. (1987). The leapfrog principle: A rule for electron counts of carbon clusters. *Chemical Communications*, 1403.

120. Fowler, P. W., Cremona, J. E., and Steer, J. I. (1988). Systematics of bonding in non-icosahedral carbon clusters. *Theoretica Chimica Acta*, **73**, 1.

121. Fowler, P. W. (1990). Carbon cylinders: A class of closed-shell clusters. *Journal of the Chemical Society, Faraday Transactions*, **86**, 2073.

122. Baum, Rudy M. Ideas on soot formation spark controversy. *Chemical and Engineering News*, February 5, 1990.

123. Ebert, Lawrence B. (1990). Is soot composed predominantly of carbon clusters? *Science*, **247**, 1468.

124. Gerhardt, P., Löffler, S., and Homman, K. H. (1987). Polyhedral carbon ions in hydrocarbon flames. *Chemical Physics Letters*, **137**, 306.

125. Iijima, Sumio (1987). The 60-carbon cluster has been revealed! *Journal of Physical Chemistry*, **91**, 3466.

126. Close, Frank (1992). *Too hot to handle: the race for cold fusion*. Penguin, London.

127. Paquette, Leo A., Teransky, Robert J., Balogh, Douglas W., and Kentgen, Gary (1983). Total synthesis of dodecahedrane. *Journal of the American Chemical Society*, **105**, 5446.

128. Fostiropoulos, Kosta. Interview with the author. August 28, 1992.

129. Lamb, Lowell. Telephone interview with the author. November 2, 1992.

130. Lamb, Lowell. Telephone interview with the author. January 11, 1993.

131. Krätschmer, Wolfgang. Letter to the author. February 12, 1991.

132. Krätschmer, Wolfgang. Letter to the author. April 11, 1991.

133. Wu, Z. C., Jelski, Daniel A., and George, Thomas F. (1987). Vibrational motions of buckminsterfullerene, *Chemical Physics Letters*, **137**, 291.

134. Stanton, Richard E. and Newton, Marshall D. (1988). Normal vibrational modes of buckminsterfullerene. *Journal of Physical Chemistry*, **92**, 2141.

135. Cyvin, S. J., Brendsall, E., Cyvin, B. N., and Brunvoll, J. (1988). Molecular vibrations of footballene. *Chemical Physics Letters*, **143**, 377.

136. Krätschmer, W., Fostiropoulos, K., and Huffman, D. R. (1990). Search for the UV and IR spectra of C_{60} in laboratory-produced carbon dust. *Dusty objects in the universe*. (Bussoletti, E. and Vittone, A.A., eds). Kluwer Academic Publishers, Amsterdam.

137. Hare, Jonathan. (1993). A tale of two fullerenes. Ph.D. Thesis, University of Sussex (provided to the author as a preprint, October 5, 1992).

138. Hare, Jonathan. Letter to the author. December 18, 1992.

139. Hare, Jonathan. Letter to the author. July 7, 1993.

140. Hare, Jonathan. Letter to Krätschmer, W. February 28, 1990.

141. Krätschmer, Wolfgang. Letter to Hare, J. March 19, 1990.

142. Diederich, François, Rubin, Yves, Knobler, Carolyn B., Whetten, Robert L., Schriver, Kenneth E, Houk, Kendall N., and Li, Yi (1989). All-carbon molecules: evidence for the generation of cyclo[18]carbon from a stable organic precursor. *Science*, **245**, 1088.

143. Rubin, Yves, Kahr, Michael, Knobler, Carolyn B., Diederich, François, and Wilkins, Charles L. (1991). The higher oxides of carbon $C_{8n}O_{2n}$ ($n=3$–5): synthesis, characterization, and X-ray crystal structure. Formation of cyclo[n]-carbon ions C_n^+ ($n=18$, 24), C_n^- ($n=18$, 24, 30), and higher carbon ions including C_{60}^+ in laser desorption Fourier transform mass spectrometric experiments. *Journal of the American Chemical Society*, **113**, 495.

144. Schmidt, W. Letter to Krätschmer, Wolfgang. April 24, 1990.

145. Corcoran, Elizabeth. Redesigning research. *Scientific American*, June 1992.

146. Bethune, Donald S. The origins and early course of fullerene research at IBM Almaden Research Center. Final revision. December 22, 1992.

147. Bethune, Donald S. Letter to the author. December 18, 1992.

148. Bethune, Donald S. Letter to the author. July 19, 1993.

149. Bethune, Don. Electronic mail message to the author, via Oxford University Press. July 21, 1993.

150. Ross, Michael. Fullerene research at IBM Almaden Research Center. Faxed to the author on April 16, 1991.

151. Bethune, Don and Meijer, Gerard. World Cup madness or a new approach to the study of carbon clusters? July 19, 1990.

152. Meijer, Gerard and Bethune, Donald S. (1990). Laser deposition of carbon clusters on surfaces: a new approach to the study of fullerenes. *Journal of Chemical Physics*, **93**, 7800.

153. Ball, Philip. Letter to the author. February 25, 1993.

154. Curl, Bob. Telephone interview with the author. December 1992.

155. Bethune, Donald S., Meijer, Gerard, Tang, Wade C., and Rosen, Hal, J. (1990). The vibrational Raman spectra of purified solid films of C_{60} and C_{70}. *Chemical Physics Letters*, **174**, 219.

156. Kroto, Harold. C_{60} buckminsterfullerene: the third form of carbon. Faxed to the author (from Zagreb), September 19, 1990.

157. Kroto, Harold. Telephone interview with the author. October 5, 1990.

158. Walton, David. Letter to the author. January 2, 1993.

159. Whetten, Robert. Telephone interview with the author. January 14, 1993.

160. Krätschmer, W., Lamb, Lowell D., Fostiropoulos, K., and Huffman, Donald R. (1990). Solid C_{60}: a new form of carbon. *Nature*, **347**, 354.

161. Johnson, Robert D., Meijer, Gerard, and Bethune, Donald S. (1990). C_{60} has icosahedral symmetry. *Journal of the American Chemical Society*, **112**, 8983.

162. Wilson, R. J., Meijer, G., Bethune, D. S., Johnson, R. D., Chambliss, D. D., de Vries, M. S., Hunziker, H. E., and Wendt, H. R. (1990). Imaging C_{60} clusters on a surface using a scanning tunneling microscope. *Nature*, **348**, 621.

163. Taylor, Roger, Hare, Jonathan P., Abdul-Sada, Ala'a K., and Kroto, Harold W. (1990). Isolation, separation and characterization of the fullerenes C_{60} and C_{70}: the third form of carbon. *Chemical Communications*, 1423.

164. Ajie, Henry, Alvarez, Marcos M., Anz, Samir J., Beck, Rainer D., Diederich, François, Fostiropoulos, K., Huffman, Donald R., Krätschmer, Wolfgang, Rubin, Yves, Schriver, Kenneth E., Sensharma, Dilip, and Whetten, Robert L. (1990). Characterization of the soluble all-carbon molecules C_{60} and C_{70}. *Journal of Physical Chemistry*, **94**, 8630.

165. Haufler, R. E., Conceicao, J., Chibante, L. P. F., Chai, Y., Byrne, N. E., Flanagan, S., Haley, M. M., O'Brien, S. C., Pan, C. Xiao, Z., Billups, W. E., Ciufolini, M. A., Hauge, R. H., Margrave, J. L., Wilson, L. J., Curl, R. F., and Smalley, R. E. (1990). Efficient production of C_{60} (buckminsterfullerene), $C_{60}H_{36}$ and the solvated buckide ion. *Journal of Physical Chemistry*, **94**, 8634.

166. Wragg, J. L., Chamberlain, J. E., White, H. W., Krätschmer, W., and Huffman, Donald R. (1990). Scanning tunneling microscopy of solid C_{60}/C_{70}. *Nature*, **348**, 623.

167. Yannoni, C. S., Johnson, R. D., Meijer, G., Bethune, D. S., and Salem, J. R. (1991). ^{13}C NMR study of the C_{60} cluster in the solid state: molecular motion and carbon chemical shift anisotropy. *Journal of Physical Chemistry*, **95**, 9.

168. Hawkins, Joel M., Meyer, Axel, Lewis, Timothy A., Loren, Stefan, and

Hollander, Frederick J. (1991). Crystal structure of osmylated C_{60}: confirmation of the soccer ball framework. *Science*, **252**, 312.

169. The almost (but never quite) complete buckminsterfullerene bibliography. Arizona Fullerene Consortium, January 1, 1993.

170. Ebbesen, Thomas W. (1993). Making bucks (book review). *Nature*, **361**, 218.

171. Koch, A. S., Khemani, K. C., and Wudl, F. (1991). Preparation of fullerenes with a simple benchtop reactor. *Journal of Organic Chemistry*, **56**, 4543.

172. Thomas, John Meurig (1991). *Michael Faraday and the Royal Institution*. Adam Hilger, Bristol.

173. Noe, Christian R., and Bader, Alfred (1993). Facts are better than dreams. *Chemistry in Britain*, **29**, 126; Rocke, Alan J. (1993). Waking up to the facts? *ibid*, **29**, 401; see also letters from Lee, Donald and Hutchison, William C. *ibid*, **29**, 396.

174. Haddon, R. C., Schneemeyer, L. F., Waszczak, J. V., Glarum, S. H., Tycko, R., Dabbagh, G., Kortan, A. R., Muller, A. J., Mujsce, A. M., Rosseinsky, M. J., Zahurak, S. M., Makhija, A. V., Thiel, F. A., Raghavachari, K., Cockayne, E., and Elser, V. (1991). Experimental and theoretical determination of the magnetic susceptibility of C_{60} and C_{70}. *Nature*, **350**, 46.

175. Fowler, Patrick (1991). Aromaticity revisited. *Nature*, **350**, 20.

176. Pasquarello, Alfredo, Schlüter, Michael, and Haddon, R. C. (1992). Ring currents in icosahedral C_{60}. *Science*, **257**, 1660.

177. Pasquarello, Alfredo, Schlüter, Michael, and Haddon, R. C. (1993). Ring currents in topologically complex molecules: application to C_{60}, C_{70}, and their hexa-anions. *Physical Review A*, **47**, 1783.

178. Haddon, Robert C. Letter to the author. July 3, 1993.

179. Baum, Rudy. Buckminsterfullerene: structure of C_{60} adduct determined. *Chemical and Engineering News*, April 15, 1991.

180. Baum, Rudy M. Research on buckminsterfullerene continues to proliferate. *Chemical and Engineering News*, June 10, 1991.

181. Baum, Rudy and Dagani, Ron. Metals in fullerenes: laser method yields bulk samples. *Chemical and Engineering News*, September 2, 1991.

182. Baum, Rudy M. Systematic chemistry of C_{60} beginning to emerge. *Chemical and Engineering News*, December 16, 1991.

183. Baum, Rudy. Metal interactions with fullerenes probed. *Chemical and Engineering News*, April 27, 1992.

184. Baum, Rudy. First C_{60}-containing copolymer synthesized. *Chemical and Engineering News*, May 18, 1992.

185. Baum, Rudy M. Flood of fullerene discoveries continues unabated. *Chemical and Engineering News*, June 1, 1992.

186. Baum, Rudy M. More buckminsterfullerene derivatives prepared. *Chemical and Engineering News*, June 22, 1992.

187. Baum, Rudy. Sugar derivatives, new production methods among fullerene advances. *Chemical and Engineering News*, November 9, 1992.

188. Lindoy, Leonard F. (1992). C_{60} chemistry expands. *Nature*, **357**, 443.

189. Krusic, P. J., Wasserman, E., Parkinson, B. A., Malone, B., and Holler Jr., E. A. (1991). Electron spin resonance study of the radical reactivity of C_{60}. *Journal of the American Chemical Society*, **113**, 6274.

190. Krusic, P. J., Wasserman, E., Keizer, P. N., Morton, J. R., and Preston, K. F. (1991). Radical reactions of C_{60}. *Science*, **254**, 1183.

191. Kovski, Alan (1990). Bucky balls: are they a new super lubricant material? *The Oil Daily*, December 6.

192. Taylor, Roger, Avent, Anthony G., Dennis, T. John, Hare, Jonathan P., Kroto, Harold W., and Walton, David R. M. (1992). No lubricants from fluorinated C_{60}. *Nature*, **355**, 27.

193. Selig, H., Lifshitz, C., Peres, T., Fischer, J. E., McGhie, A. R., Romanow, W. J., McCauley Jr, J. P., amd Smith, A. B. (1991). Fluorinated fullerenes. *Journal of the American Chemical Society*, **113**, 5475.

194. Holloway, John H., Hope, Eric G., Taylor, Roger, Langley, G. John, Avent, Anthony G., Dennis, T. John, Hare, Jonathan P., Kroto, Harold W., and Walton, David R. M. (1991). Fluorination of buckminsterfullerene. *Chemical Communications*, 966.

195. Olah, George A., Bucsi, Imre, Lambert, Christian, Aniszfled, Robert, Trivedi, Nirupam J., Sensharma, Dilip K., and Prakash, G. K. Surya (1991). Chlorination and bromination of fullerenes. Nucleophilic methoxylation of polychlorofullerenes and their aluminium trichloride catalysed Friedel–Crafts reaction with aromatics to polyarylfullerenes. *Journal of the American Chemical Society*, **113**, 9385.

196. Taylor, Roger, Holloway, John H., Hope Eric G., Avent, Anthony G., Langley, G. John., Dennis, T. John, Hare, Jonathan P., Kroto, Harold W., and Walton, David R. M. (1992). Nucleophilic substitution of fluorinated C_{60}. *Chemical Communications*, 665.

197. Taylor, Roger, Langley, G. John, Meidine, Mohamed F., Parsons, Jonathan P., Abdul-Sada, Ala'a K., Dennis, T. John, Hare, Jonathan P., Kroto, Harold W., and Walton, David R. M. (1992). Formation of $C_{60}Ph_{12}$ by electrophilic aromatic substitution. *Chemical Communications*, 667.

198. Birkett, Paul R., Hitchcock, Peter B., Kroto, Harold W., Taylor, Roger and

Walton, David R. M. (1992). Preparation and characterization of $C_{60}Br_6$ and $C_{60}Br_8$. *Nature*, **357**, 479.

199. Suzuki, T., Li, Q., Khemani, K. C., Wudl, F., and Almarsson, Ö. (1991). Systematic inflation of buckminsterfullerene C_{60}: synthesis of diphenyl fulleroids C_{61} to C_{66}. *Science*, **254**, 1186.

200. Wudl, F. (1992). The chemical properties of buckminsterfullerene (C_{60}) and the birth and infancy of fulleroids. *Accounts of Chemical Research*, **25**, 157.

201. Wudl, F., Hirsch, A., Khemani, K. C., Suzuki, T., Allemand, P.-M., Kock, A., Eckert, H., Srdanov, G., and Webb, H. M. (1992). Survey of chemical reactivity of C_{60}, electrophile and dieno-polarophile par excellence. *Fullerenes*, Hammond, George S. and Kuck, Valerie J. (eds.). ACS Symposium Series 481, American Chemical Society.

202. Nagashima, Hideo, Nakaoka, Akihito, Saito, Yahachi, Kato, Masanao, Kawanishi, Teruhiko, and Itoh, Kenji (1992). $C_{60}Pd_n$: The first organometallic polymer of buckminsterfullerene. *Chemical Communications*, 377.

203. Fagan, Paul, J., Calabrese, Joseph, C., and Malone, Brian (1991). The chemical nature of buckminsterfullerene (C_{60}) and the characterization of a platinum derivative. *Science*, **252**, 1160.

204. Balch, Alan L., Catalano, Vincent J., and Lee, Joong W. (1991). Accumulating evidence for the selective reactivity of the 6-6 ring fusion of C_{60}. Preparation and structure of $(\eta^2\text{-}C_{60})Ir(CO)Cl(PPh_3) . 5C_6H_6$. *Inorganic Chemistry*, **30**, 3980.

205. Balch, Alan J., Catalano, Vincent J., Lee, Joong W., Olmstead, Marilyn M., and Parkin, Sean R. (1991). $(\eta^2\text{-}C_{70})Ir(CO)Cl(PPh_3)_2$: The synthesis and structure of an organometallic derivative of a higher fullerene. *Journal of the American Chemical Society*, **113**, 8953.

206. Koefod, Robert S., Hudgens, Mark F., and Shapley, John R. (1991). Organometallic chemistry with buckminsterfullerene. Preparation and properties of an indenyliridium(I) complex. *Journal of the American Chemical Society*, **113**, 8957.

207. Fagan, Paul J., Calabrese, Joseph C., and Malone, Brian (1992). Metal complexes of buckminsterfullerene (C_{60}). *Accounts of Chemical Research*, **25**, 134.

208. Fagan, Paul J., Calabrese, Joseph C., and Malone, Brian (1992). The chemical nature of C_{60} as revealed by the synthesis of metal complexes. *Fullerenes*, Hammond, George S. and Kuck, Valerie J. (eds.). ACS Symposium Series 481, American Chemical Society.

209. Taylor, Roger and Walton, David R. M. (1993). The chemistry of the fullerenes. *Nature*, **363**, 685.

210. Walton, D. R. M. Letter to the author. July 12, 1993.

211. Wudl, Fred. Letter to the author. July 8, 1993.

212. Kroto, Harold. Telephone interview with the author. July 11, 1993.

213. Guo, Ting, Jin, Changming, and Smalley, R. E. (1991). Doping bucky: formation and properties of boron-doped buckminsterfullerene. *Journal of Physical Chemistry*, **95**, 4948.

214. Chai, Yan, Guo, Ting, Jin, Changming, Haufler, Robert E., Chibante, L. P. Felipe, Fure, Jan, Wang, Lihong, Alford, J. Michael, and Smalley, Richard E. (1991). Fullerenes with metals inside. *Journal of Physical Chemistry*, **95**, 7564.

215. Johnson, Robert D., de Vries, Mattanjah S., Salem, Jesse, Bethune, Donald S., and Yannoni, Constantino S. (1992). Electron paramagnetic resonance studies of lanthanum-containing C_{82}. *Nature*, **355**, 239.

216. Yannoni, Constantino S., Hoinkis, Mark, de Vries, Mattanjah S., Bethune, Donald S., Salem, Jesse R., Crowder, Mark S., and Johnson, Robert D. (1992). Scandium clusters in fullerene cages. *Science*, **256**, 1191.

217. Smalley, R. E. (1992). Self-assembly of the fullerenes. *Accounts of Chemical Research*, **25**, 98.

218. Smalley, R. E. (1992). Doping the fullerenes. *Fullerenes*, Hammond, George S. and Kuck, Valerie J. (eds.). ACS Symposium Series 481, American Chemical Society.

219. Guo, Ting, Diener, M. D., Chai, Yan, Alford, M. J., Haufler, R. E., McClure, S. M., Ohno, T., Weaver, J. H., Scuseria, G. E., and Smalley, R. E. (1992). Uranium stabilization of C_{28}: a tetravalent fullerene. *Science*, **257**, 1661.

220. Kroto, H. W. (1992). $(T_d)C_{28}$: Fullerene-28 and tetravalent analogues. Preprint.

221. Saunders, Martin, Jiménez-Vázquez, Hugo A., Cross, R. James, and Poreda, Robert J. (1993). Stable compounds of helium and neon: $He@C_{60}$ and $Ne@C_{60}$. *Science*, **259**, 1428.

222. Bethune, D. S., de Vries, M. S., Johnson, R. D., Salem, J. R., and Yannoni, C. S. (1993). Carbon-caged atoms: progress in endohedral fullerene research. Preprint.

223. Asimov, Isaac (1978). *Asimov's guide to science 1: The physical sciences*, Penguin, London.

224. Hey, Tony and Walters, Patrick (1987). *The quantum universe*. Cambridge University Press.

225. Kittel, Charles (1976). *Introduction to solid state physics* (5th edn). John Wiley & Sons, New York.

226. Feynman, Richard P., Leighton, Ralph B., and Sands, Matthew (1965). *The Feynman lectures on physics*. Vol. III. Addison-Wesley, Reading, Massachusetts.

227. Wolsky, Alan M., Giese, Robert F., and Daniels, Edward J. The new superconductors: prospects for applications. *Scientific American*, February, 1989.

228. Cava, Robert J. Superconductors beyond 1-2-3. *Scientific American*, August, 1990.

229. Carlson, Douglas and Williams, Jack. Superconductors go organic. *New Scientist*, November 14, 1992.

230. Adrian, Frank J. and Cowan, Dwaine O. The new superconductors. *Chemical and Engineering News*, December 21, 1992.

231. Amato, Ivan (1993). New superconductors: a slow dawn. *Science*, **259**, 306.

232. Bishop, David J., Gammel, Peter L., and Huse, David A. Resistance in high-temperature superconductors. *Scientific American*, February, 1993.

233. Schilling, A., Cantoni, M., Guo, J. D., and Ott, H. R. (1993). Superconductivity above 130 K in the Hg–Ba–Ca–Cu–O system. *Nature*, **363**, 56

234. Haddon, Robert C. Letter to the author. February 9, 1993.

235. Haddon, Robert C. LiC_{60}: The next organic superconductor and the first 3-D organic metal. Poster for a presentation to the Materials Discussion Group, AT&T Bell Laboratories, October 18, 1990.

236. Haddon, R. C., Hebard, A. F., Rosseinsky, M. J., Murphy, D. W., Duclos, S. J., Lyons, K. B., Miller, B., Rosamilia, J. M., Fleming, R. M. Kortan, A. R., Glarum, S. H., Makhija, A. V., Muller, A. J., Eick, R. H., Zahurak, S. M., Tycko, R., Dabbagh, G., and Thiel, F. A. (1991). Conducting flims of C_{60} and C_{70} by alkali-metal doping. *Nature*, **350**, 320.

237. Hebard, A. F., Rosseinsky, M. J., Haddon, R. C., Murphy, D. W., Glarum, S. H., Palstra, T. T. M., Ramirez, A. P., and Kortan, A. R. (1991). Superconductivity at 18 K in potassium-doped C_{60}. *Nature*, **250**, 600.

238. Sleight, Arthur W. (1991). Sooty superconductors. *Nature*, **350**, 557.

239. Baum, Rudy. ACS Fullerene Symposium: C_{60} superconductivity highlighted. *Chemical and Engineering News*, April 22, 1991.

240. Tanigaki, K., Ebbesen, T. W., Saito, S., Mizuki, J., Tsai, J. S., Kubo, Y., and Kuroshima, S. (1991). Superconductivity at 33 K in $Cs_xRb_yC_{60}$. *Nature*, **352**, 222.

241. Iqbal, Zafar, Baughman, Ray H., Ramakrishna, B. L., Khare, Sandeep, Murthy, N. Sanjeeva, Bornemann, Hans, J., and Morris, Donald E. (1991). Superconductivity at 45 K in Rb/Tl codoped C_{60} and C_{60}/C_{70} mixtures. *Science*, **254**, 826.

242. Iqbal, Z., Baughman, R. H., Khare, S., Murthy, N. S., Ramakrishna, B. L., Bornemann, H. J., and Morris, D. E. (1992). Superconducting transition temperature of doped C_{60}: retraction. *Science*, **256**, 950.

243. Haddon, R. C. (1992). Electronic structure, conductivity and superconductivity of alkali metal doped C_{60}. *Accounts of Chemical Research*, **25**, 127.

244. Kuhn, Thomas S. (1970). *The structure of scientific revolutions.* (2nd edn). The University of Chicago Press, Chicago, Illinois.

245. Baggott, Jim. Serendipity and scientific progress. *New Scientist,* March 3, 1990

246. Fowler, Patrick W. (1991). Three candidates for the structure of C_{84}. *Journal of the Chemical Society, Faraday Transactions,* **87**, 1945.

247. Manolopoulos, David E. (1991). Proposal of a chiral structure for the fullerene C_{76}. *Journal of the Chemical Society, Faraday Transactions,* **87**, 2861.

248. Fowler, Patrick W., Batten, Robin C., and Manolopoulos, David E. (1991). The higher fullerenes: a candidate for the structure of C_{78}. *Journal of the Chemical Society, Faraday Transactions,* **87**, 3103.

249. Baggott, Jim. Chemists predict structures of 'natural' fullerenes. *New Scientist,* October 26, 1991.

250. Fowler, Patrick. Letter to the author. September 20, 1991.

251. Manolopoulos, D. E. Note faxed to the author. September 30, 1991.

252. Diederich, François, Ettl, Roland, Rubin, Yves, Whetten, Robert L., Beck, Rainer, Alvarez, Marcos, Anz, Samir, Sensharma, Dilip, Wudl, Fred, Khemani, Kishan C., and Kock, Andrew (1991). The higher fullerenes: isolation and characterization of C_{76}, C_{84}, C_{90}, C_{94}, and $C_{70}O$, an oxide of D_{5h}-C_{70}. *Science,* **252**, 548.

253. Ettl, Roland, Chao, Ito, Diederich, François and Whetten, Robert L. (1991). Isolation of C_{76}, a chiral (D_2) allotrope of carbon. *Nature,* **353**, 149.

254. Baum, Rudy. C_{76} fullerene molecule determined to be chiral. *Chemical and Engineering News,* September 16, 1991.

255. Hawkins, Joel M. and Meyer, Axel (1993). Optically active carbon: kinetic resolution of C_{76} by asymmetric osmylation. *Science,* **260**, 1918.

256. Baum, Rudy. C_{76} enantiomers resolved by osmylation technique. *Chemical and Engineering News,* June 28, 1993.

257. Diederich, François and Whetten, Robert L. (1992). Beyond C_{60}: The higher fullerenes. *Accounts of Chemical Research,* **25**, 119.

258. Kikuchi, Koichi, Nakahara, Nobuo, Wakabayashi, Tomonari, Suzuki, Shinzo, Shiromaru, Haruo, Miyake, Yoko, Saito, Kazuya, Ikemoto, Isao, Kainosho, Masatsune, and Achiba, Yohji (1992). NMR characterization of isomers of C_{78}, C_{82}, and C_{84}. *Nature,* **357**, 142.

259. Iijima, Sumio (1991). Helical microtubules of graphitic carbon. *Nature,* **354**, 56.

260. Ebbesen, T. W. and Ajayan, P. M. (1992). Large-scale synthesis of carbon nanotubes. *Nature,* **358**, 220.

261. Dagani, Ron. Graphitic microtubules: bulk synthesis opens up research field. *Chemical and Engineering News*, July 20, 1992.

262. Dresselhaus, M. S. (1992). Down the straight and narrow. *Nature*, **358**, 195.

263. Bacon, Roger (1960). Growth, structure and properties of graphite whiskers. *Journal of Applied Physics*, **31**, 283.

264. Iijima, Sumio and Ichihashi, Toshinari (1993). Single-shell carbon nanotubes of 1-nm diameter. *Nature*, **363**, 603.

265. Bethune, D. S., Kiang, C. H., de Vries, M. S., Gorman, G., Savoy, R., Vazquez, J., and Beyers, R. (1993). Cobalt-catalyzed growth of carbon nanotubes with single-atomic-layer walls. *Nature*, **363**, 605.

266. Dagani, Ron. Carbon nanotubes: recipes found for simplest variety. *Chemical and Engineering News*, June 21, 1993.

267. Iijima, Sumio. Letter to the author. July 12, 1993.

268. Ugarte, Daniel (1992). Curling and closure of graphitic networks under electron beam irradiation. *Nature*, **359**, 707.

269. Kroto, H. W. (1992). Carbon onions introduce new flavour to fullerene studies. *Nature*, **359**, 670.

270. Baum, Rudy. Fullerenes broaden scientists' view of molecular structure. *Chemical and Engineering News*, January 4, 1993.

271. Ugarte, Daniel. Letters to the author. March 4 and May 4, 1993.

272. Ugarte, Daniel. Letter to the author. June 30, 1993.

273. Ugarte, Daniel (1993). Structure of carbon particles formed by curved graphene sheets (fullerenes, nanotubes): an electron microscopy study. Proceedings of ISSPIC 6, to be published in *Zeitschrift für Physik, D.*

274. Ugarte, Daniel (1993). Morphology and structure of graphitic soot particles generated in arc-discharge C_{60} production. Preprint.

275. Ugarte, Daniel (1993). Canonical structure of large carbon clusters: $C_n > 100$. Preprint.

276. Ugarte, Daniel (1993). Formation mechanism of quasi-spherical carbon particles induced by electron bombardment. Preprint.

277. Ugarte, Daniel (1993). How to fill or empty a graphitic onion. Preprint.

278. Ruoff, Rodney S., Lorents, Donald C., Chan, Bryan, Malhotra, Ripudaman, and Subramoney, Shekhar (1993). Single crystal metals encapsulated in carbon nanoparticles. *Science*, **259**, 346.

279. Baum, Rudy. Metal encapsulated in carbon particles. *Chemical and Engineering News*, January 18, 1993.

280. Ross, Philip E. Faux fullerenes. *Scientific American*, February, 1993.

281. Ball, Philip (1993). New horizons in inner space. *Nature*, **361**, 297.

282. Ajayan, P. M. and Iijima, Sumio (1993). Capillarity-induced filling of carbon nanotubes. *Nature*, **361**, 333.

283. Baum, Rudy. Liquid lead fills carbon nanotubes. *Chemical and Engineering News*, February 8, 1993.

284. Tsang, S. C., Harris, P. J. F., and Green, M. L. H. (1993). Thinning and opening of carbon nanotubes by oxidation using carbon dioxide. *Nature*, **362**, 520.

285. Ajayan, P. M., Ebbesen, T. W., Ichihashi, T., Iijima, S., Tanigaki, K., and Hiura, H. (1993). Opening carbon nanotubes with oxygen and implications for filling. *Nature*, **362**, 522.

286. Ge, Maohui and Sattler, Klaus (1993). Vapor-condensation generation and STM analysis of fullerene tubes. *Science*, **260**, 515.

287. Dravid, V. P., Lin, X., Wang, Y., Wang, X. K., Yee, A., Ketterson, J. B., and Chang, R. P. H. (1993). Buckytubes and derivatives: Their growth and implications for buckyball formation. *Science*, **259**, 1601.

288. Leonosky, Thomas, Gonze, Xavier, Teter, Michael, and Elser, Veit (1992). Energetics of negatively curved graphitic carbon. *Nature*, **355**, 333.

289. Haufler, R. E., Chai, Y., Chibante, L. P. F., Conceicao, J., Jin, Changming, Wang, Lai-Sheng, Maruyama, Shigao, and Smalley, R. E. Carbon arc generation of C_{60}. Materials Research Society Symposium, Boston, November 29, 1990.

290. Heath, James R. (1992). Synthesis of C_{60} from small carbon clusters. *Fullerenes*, Hammond, George S. and Kuck, Valerie J. (eds.). ACS Symposium Series 481, American Chemical Society.

291. Curl, Robert F. (1993). Collapse and growth. *Nature*, **363**, 14.

292. von Helden, Gert, Gotts, Nigel G., and Bowers, Michael T. (1993). Experimental evidence for the formation of fullerenes by collisional heating of carbon rings in the gas phase. *Nature*, **363**, 60.

293. Hunter, Joanna, Fye, James, and Jarrold, Martin F. (1993). Annealing C_{60}^{+}: Synthesis of fullerenes and large carbon rings. *Science*, **260**, 784.

294. Baum, Rudy M. Research suggests alternative route to fullerenes. *Chemical and Engineering News*, May 17, 1993.

295. Yeretzian, Chahan, Hansen, Klavs, Diederich, François and Whetten, Robert L. (1992). Coalescence reactions of fullerenes. *Nature*, **359**, 44.

296. McElvany, Stephen W., Callahan, John H., Ross, Mark M., Lamb, Lowell D., and Huffman, Donald R. (1993). Large odd-numbered carbon clusters from fullerene–ozone reactions. *Science*, **260**, 1632.

297. Howard, Jack B., McKinnon, Thomas, Makarovsky, Yakov, Lafleur, Arthur L., and Johnson, M. Elaine (1991). Fullerenes C_{60} and C_{70} in flames. *Nature*, **352**, 139.

298. Taylor, Roger, Parsons, Jonathan P., Avent, Anthony G., Rannard, Steven

P., Dennis, T. John, Hare, Jonathan P., Kroto, Harold W., and Walton, David R. M. (1991). Degradation of C_{60} by light. *Nature,* **351**, 277.

299. Busek, Peter R., Tsipursky, Semeon J., and Hettich, Robert (1992). Fullerenes from the geological environment. *Science,* **257**, 215.

300. Amato, Ivan (1992). A first sighting of buckyballs in the wild. *Science,* **257**, 167.

301. Daly, Terry K., Bušek, Peter R., Williams, Peter, and Lewis, Charles F. (1993). Fullerenes from a fulgurite. *Science,* **259**, 1599.

302. Hare, J. P. and Kroto, H. W. (1992). A postbuckminsterfullerene view of carbon in the galaxy. *Accounts of Chemical Research,* **25**, 106.

303. de Vries, M. S., Wendt, H. R., Hunziker, H. E., Peterson, E. and Chang, S. (1993). High molecular weight polycyclic aromatic hydrocarbons, carbonaceous meteorites and fullerenes in interstellar space. *Geochimica and Cosmochimica Acta,* **57**, 933.

304. Becker, L., McDonald, G. D., and Bada, J. L. (1993). Carbon onions in meteorites. *Nature,* **361**, 595.

305. de Heer, Walt A. and Ugarte, Daniel. (1993). Carbon onions produced by heat treatment of carbon soot and their relation to the 217.5 nm interstellar absorption feature. Preprint.

306. Moore, Stephen, Ondrey, Gerald and Samdani, Gulam. The shape of things to come? *Chemical Engineering,* July 1992.

307. Baum, Rudy M. Commercial uses of fullerenes slow to develop. *Chemical and Engineering News,* November 22, 1993.

308. Krätschmer, Wolfgang. Telephone interview with the author. May 18, 1994.

309. Clery, Daniel (1993). Patent dispute goes public. *Science,* **261**, 978.

310. Chibante, L. P. F., Thess, Andreas, Alford, J. M., Diener, M. D., and Smalley, R. E. (1993). Solar generation of the fullerenes. *Journal of Physical Chemistry,* **97**, 8696.

311. Fields, C. L., Pitts, J. R., Hale, M. J., Bingham, C., Lewandowski, A., and King, D. E. (1993). Formation of fullerenes in highly concentrated solar flux. *Journal of Physical Chemistry,* **97**, 8701.

312. Baum, Rudy. Fullerenes produced by harnessing sunlight. *Chemical and Engineering News,* August 30, 1993.

313. Saunders, Martin, Jiménez-Vázquez, Hugo A., Cross, R. James, Mroczkowski, Stanley, Freedberg, Darón I., and Anet, Frank A. L. (1994). Probing the interior of fullerenes by ^{3}He NMR spectroscopy of endohedral $^{3}He@C_{60}$ and $^{3}He@C_{70}$. *Nature,* **367**, 256.

314. Haddon, Robert C. (1994). From the outside in. *Nature,* **367**, 214. See also

Baum, Rudy. ^{3}He NMR study looks at fullerene aromaticity. *Chemical and Engineering News*, February 28, 1994.

315. Haddon, R. C. (1993). Chemistry of the fullerenes: the manifestation of strain in a class of continuous aromatic molecules. *Science*, **261**, 1545.

316. Baum, Rudy M. Chemists increasingly adept at modifying, manipulating the fullerenes. *Chemical and Engineering News*, September 20, 1993.

317. Wudl, F., Sijbesma, Rint, Srdanov, Gordana, Wilkins, Charles L., and Castoro, J.A. (1993). *Journal of the American Chemical Society*, **115**, 6510.

318. Friedman, Simon H., Kenyon, George L., and DeCamp, Diane L. (1993). *Journal of the American Chemical Society*, **115**, 6506.

319. Hill, Craig L. and Schinazi, Raymond F. (1993). *Antimicrobial Agents and Chemotherapy*, **37**, 1707.

320. Baum, Rudy. Fullerene bioactivity: C_{60} derivative inhibits AIDS viruses. *Chemical and Engineering News*, August 2, 1993.

321. Bethune, D. S., Johnson, R. D., Salem, J. R., deVries, M. S., and Yannoni, C. S. (1993). Atoms in carbon cages: the structure and properties of endohedral fullerenes. *Nature*, **366**, 123.

322. Clemmer, David E., Shelimov, Konstantin B., and Jarrold, Martin F. (1994). Gas-phase self-assembly of endohedral metallofullerenes. *Nature*, **367**, 718.

323. Smalley, Rick. Telephone interview with the author. May 23, 1994.

324. Murry, Robert L. and Scuseria, Gustavo E. (1994). Theoretical evidence for a C_{60} "window" mechanism. *Science*, **263**, 791.

325. Laguës, Michel, Xie, Xiao Ming, Tebbji, Hassan, Xu, Xiang Zhen, Mairet, Vincent, Hatterer, Christophe, Beuran, Cristian, F., Deville-Cavellin, Catherine. (1993). Evidence suggesting superconductivity at 250 K in a sequentially deposited cuprate film. *Science*, **262**, 1850.

326. Dagani, Ron. Cuprate superconductivity: French team finds telltale signs at 250 K. *Chemical and Engineering News*, December 20, 1993. See also Of Meissner men, *The Economist*, January 22, 1994.

327. Yam, Philip. Current events. *Scientific American*, December, 1993.

328. Kao, Yi-Han, *et al.* (1993). *Solid State Communications*, **87**, 387.

329. Travis, John. (1993). Fullerene superconductors heat up. *Science*, **261**, 1392.

330. Ebbesen, T. W., Ajayan, P. M., Hiura, H., and Tanigaki, K. (1994). Purification of nanotubes. *Nature*, **367**, 519.

331. Subramoney, Shekar, Ruoff, Rodney S., Lorents, Donald C., and Malhotra, Ripudman. (1993). Radial single-layer nanotubes. *Nature*, **366**, 637.

332. Baum, Rudy. "Sea-urchins"—another novel form of carbon. *Chemical and Engineering News*, January 3, 1994.

333. Wang, Ying. (1994). Encapsulation of palladium crystallites in carbon and the formation of wormlike nanostructures. *Journal of the American Chemical Society*, **116**, 397.

334. Hunter, Joanna M., Fye, James L., Roskamp, Eric J., and Jarrold, Martin F. (1994). Annealing carbon cluster ions: a mechanism for fullerene synthesis. *Journal of Physical Chemistry*, **98**, 1810.

335. Haggin, Joseph. Mechanism for fullerene formation proposed. *Chemical and Engineering News*, March 7, 1994. See also Bradley, David. How to make a buckyball—it's a wind up. *New Scientist*, April 23, 1994.

336. Heymann, Dieter, Chibante, L. P. Felipe, Brooks, Robert R., Wolbach, Wendy S., and Smalley, Richard E. Fullerenes in the K/T boundary layer. Manuscript submitted to *Science*, March 24, 1994.

337. di Brozolo, Filippo Radicati, Bunch, Theodore E., Fleming, Ronald H., and Macklin, John. (1994). Fullerenes in an impact crater on the LDEF spacecraft. *Nature*, **369**, 37.

338. Kroto, Harry. Telephone interview with the author. May 19, 1994.

339. Foing, B. H. and Ehrenfreund, P. (1994). Detection of two interstellar absorption bands coincident with spectral features of C_{60}^{+}. *Nature*, **369**, 296.

340. Kroto, Harold. (1994). Fullerene's faint fingerprint? *Nature*, **369**, 274.

341. Curl, Bob. Telephone interview with the author. May 6, 1994.

342. Heath, Jim. Telephone interview with the author. May 19, 1994.

343. O'Brien, Sean. Telephone interview with the author. May 19, 1994.

344. Huffman, Don. Telephone interview with the author. May 19, 1994.

345. Lamb, Lowell. Telephone interview with the author. May 19, 1994.

346. See *Physics World*, May 1994, p. 61.

Name index

Subject index